AF389514

Advances in Synthesis of Metallic, Oxidic and Composite Powders

Advances in Synthesis of Metallic, Oxidic and Composite Powders

Editors

Srecko Stopic
Bernd Friedrich

MDPI • Basel • Beijing • Wuhan • Barcelona • Belgrade • Manchester • Tokyo • Cluj • Tianjin

Editors
Srecko Stopic
RWTH Aachen University
Germany

Bernd Friedrich
RWTH Aachen University
Germany

Editorial Office
MDPI
St. Alban-Anlage 66
4052 Basel, Switzerland

This is a reprint of articles from the Special Issue published online in the open access journal *Metals* (ISSN 2075-4701) (available at: https://www.mdpi.com/journal/metals/special_issues/adv_synth_met_oxid_compos_powders).

For citation purposes, cite each article independently as indicated on the article page online and as indicated below:

LastName, A.A.; LastName, B.B.; LastName, C.C. Article Title. *Journal Name* **Year**, *Volume Number*, Page Range.

ISBN 978-3-03943-929-4 (Hbk)
ISBN 978-3-03943-930-0 (PDF)

Contents

About the Editors

Srecko Stopic, Privat Dozent Dr.-Ing. habil

Born 03.04.1965, in Uzice/Serbia (former Yugoslavia)

Education:

1984—High school graduation in Uzice/Serbia

1986–1991—Study of non-ferrous metallurgy at the Faculty of Technology and Metallurgy, University of Belgrade, Serbia

Degree: Diploma in Engineering

1991–1994—Magister and doctoral study at the Faculty of Technology and Metallurgy, University of Belgrade, Serbia (former Yugoslavia)

1997—PhD in Engineering: Examination, dissertation on "mechanism and kinetics of reduction of nickel chloride"

Professional career:

1991–1999—Scientific assistant at Department of Nonferrous Metallurgy of the Faculty of Technology and Metallurgy of the University of Belgrade

1999–2001—Assistant Professor of Non-Ferrous Metallurgy at the Faculty of Technology and Metallurgy of the University in Belgrade

2002–2003—Research Fellowship with Alexander von Humboldt Foundation at the IME Process Metallurgy and Metal Recycling, RWTH Aachen University

Since April 2003—Scientific IME Process Metallurgy and Metal Recycling, Chair of Institute of RWTH Aachen University

30.04.2014: Professional thesis on Synthesis of Metallic Nanosized Particles by Ultrasonic Spray Pyrolysis (Privat Dozent at the RWTH Aachen University)

June 2020—Visiting Professor, Technical Faculty, Cacak, University in Kragujevac, Serbia

Bernd Friedrich, Prof. Dr. Ing. Dr. h. c.

Born 01.02.1958, in Dillenburg/Hessen, Germany

Education:

1977—High school graduation in Aachen/Germany

1978–1983—Study of non-ferrous metallurgy at RWTH Aachen University

Degree: Diploma in Engineering

1984–1987—Scientist at IME Process Metallurgy and Metal-Recycling RWTH Aachen University

1988—PhD in Engineering: Examination, dissertation on "electrolytic refining of recycling-tin using three dimensional-electrodes"

Professional career:

1988–1992—Head of R&D Institute at GfE, Nuremberg/Germany (refractory metals, ferro-alloys, advanced materials, metal recycling and residue utilisation)

1992–1995—Head of profit centre "hydride-technology and advanced materials" at GfE, Nuremberg/Germany

1995–1999—Plant manager NiCad/NiMH at Varta Batterie AG in Hagen/Germany and Ceska Lipa/Czech. Republic, Head of R&D Center "innovative rechargeable battery systems" R&D coordinator at 3C-Alliance (Varta-Toshiba-Duracel)

Since July 1999—Director of IME Process Metallurgy and Metal Recycling, Chair of Institute of the
RWTH Aachen University

Preface to "Advances in Synthesis of Metallic, Oxidic and Composite Powders"

The high demand for new materials, such as metals, oxides, and composites, raises the need for an advanced synthesis of different materials, which are crucial for technological applications. Different process synthesis routes, such as atomization, reduction in aqueous phase, crystallization, chemical precipitation, high pressure reaction in autoclave, and electrolysis, can be used to create controlled powder characteristics with specific properties for a particular application or industry. Advances in synthesis explore a range of materials and techniques used for powder metallurgy and the use of this technology across a variety of application areas, such as medicine, catalysis and automotive industry. This Special Issue, "Advances in the Synthesis of Metallic, Oxidic and Composite Powders", is dedicated to the latest scientific achievements in the efficient preparation of metals, oxides and composite materials. In this issue, we are focused on description of the synthesis of metal, oxide and composite particles from the water, metalorganic and colloid solutions using different synthesis methods. The main challenge of this issue is the controlled synthesis via process parameters (conditions and modes atomization, the concentration of solution, residence time of aerosol in a reactor, presence of additives, flow rate, decomposition and reduction temperature, different precursors with reducing agents, and surrounding atmosphere), in order to guide the process to obtain powders with such a morphology that satisfies more and more complex requirements for the properties of advanced engineering materials. The synthesis of powders has two different strategies: "Top-Down" and "Bottom-Up". The meaning of "Top-Down" is based on the mechanical grinding of initial materials to small dimensions. It is necessary to decrease the powder size in order to perform Hall-Petch strengthening and apply a severe plastic deformation to powder particles to perform work hardening. High energy milling has a potential for realizing the new ideas of materials designers. The meaning of "Bottom-Up" is related to the physico-chemical preparation methods in gas phase (ultrasonic spray pyrolysis, flame pyrolysis and chemical vapor deposition) and in liquid phase (sol gel, hydrothermal processes, precipitation, electrolytic synthesis, high pressure reactions in an autoclave and crystallization). The precipitation methods are usually used for the purification of spent solution. In this regard, new approaches in material and synthesis design, structural engineering and morphological characteristics are provided. The preparation of metal particles by spray pyrolysis of metal salts is especially challenging. Using aerosol synthesis, a single-step and multistep preparation process of different core-shall particles is possible, thus avoiding several steps like drying, shrinkage, solute precipitation, thermolysis, and sintering to form uniform spherical particles in a nanosized and submicron range. Technical limitations of this technique, as well as a comparison with other synthesis methods (difficulty in controlling morphology-porous or hollow particles, relatively low production rate and process of large volume of gas), will be partly considered in order to prevent or solve these problems. Especially, the newest results in the synthesis of nanosized core-shell particles by ultrasonic spray pyrolysis method will be published. The HDH process consists of the following sequence: surface conditioning of the turnings, hydrogenation, ball milling (for powder production), and dehydrogenation. This Special Issue contains 17 papers from Europe, Asia, Australia, South Africa and Balkan countries, which confirms that there is a high interest for this research subject worldwide. The advances in the synthesis of metallic, oxidic and composite powders were presented via the following methods: ultrasound-assisted leaching process, ultrasonic spray pyrolysis, hydrogenation, dehydrogenation, ball milling, molten salt electrolysis, galvanostatic electrolysis, hydrogen reduction, thermochemical decomposition, inductively coupled thermal plasma, precipitation and high pressure carbonation in an autoclave. The synthesis methods are focused on metals: Co, Cu; Re; oxides: ZnO, MgO, SiO$_2$; V$_2$O$_5$; sulfides: MoS$_2$, core shell material: Cu-Al$_2$O$_3$, Pt/TiO$_2$; Ca$_{0.75}$Ce$_{0.25}$ZrTi$_2$O$_7$, and compounds: Mo$_5$Si$_3$, Ti$_6$Al$_4$V.

The environment friendly strategy was presented at the carbonation of olivine, nuclear waste immobilization via the stability of zirconolite and treatment of acid mine drainage water. The application of the flotation tailings as an alternative material for an acid mine drainage remediation was successfully applied for an extremely acidic Lake Robule in Serbia.

Ultrasonic spray pyrolysis mentioned in three papers was applied for the synthesis of ZnO [2], core shell particles Pt/TiO$_2$ [11] and MoS$_2$ [7]. In addition, the ultrasonic spray pyrolysis of ammonium meta-tungstate hydrate (AMT) was used for the production of WO$_3$ particles at 650 °C in air. The synthesis of tungsten disulfide (WS$_2$) powder was performed by the sulfurization of tungsten trioxide (WO$_3$) particles in the presence of additive potassium carbonate (K$_2$CO$_3$) in nitrogen (N$_2$) atmosphere, first at a lower temperature (200 °C) and followed by reduction at higher temperature (900 °C). Nanostructured zinc oxide (ZnO) particles were synthesized by the one-step ultrasonic spray pyrolysis (USP) process from nitrate salt solution (Zn(NO$_3$)$_2$·6H$_2$O). A flexible USP formation model was proposed, ending up in various ZnO morphologies rather than only ideal spheres, which is highly promising to target a wide application area. USP-synthesized Pt/TiO$_2$ composites were generated in the form of a solid mixture, morphologically organized in nesting huge hollow and small solid spheres, or TiO$_2$ core/Pt shell regular spheroids by in situ or ex situ methods, respectively. This paper exclusively reports on characteristic mechanisms of the formation of novel two-component solid composites, which are intrinsic from the USP approach, and controlled precursor composition, as shown in Figure 1.

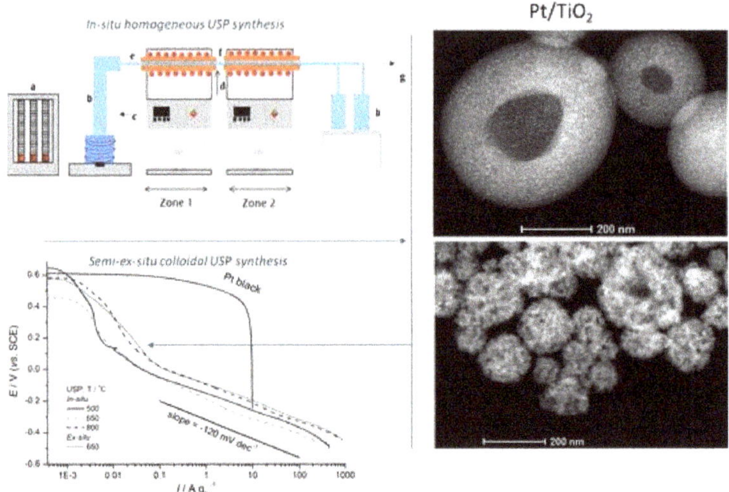

Figure 1: Experimental setup for ultrasonic spray pyrolysis method with SEM analysis of produced particles and electrochemical measurements [11].

The carbonation process under high pressure conditions in autoclave is mentioned in two papers and applied for the synthesis of magnesium oxide (2–5 μm) and nanosized silica [4,8]. The subject of both studies is the carbonation of an olivine (Mg_2SiO_4) and synthetic magnesia sample (>97 wt% MgO) under high pressure and temperature in an autoclave. Early experiments have studied the influence of some additives, such as sodium bicarbonate, oxalic acid and ascorbic acid, solid/liquid ratio, and particle size on the carbonation efficiency. The obtained results for carbonation of olivine have confirmed the formation of magnesium carbonate in the presence of additives and complete carbonation of the MgO sample in the absence of additives. Differently to the traditional methods of the synthesis of nanosilica such as sol gel, ultrasonic spray pyrolysis method and hydrothermal synthesis using some acids and alkaline solutions; this synthesis method takes place in water solution at 175 °C and above 100 bar. The obtained particles of magnesium carbonate and nanosilica were shown in Figure 2:

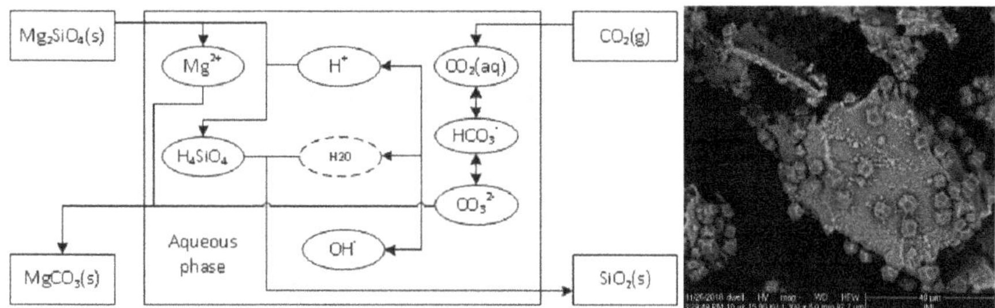

Figure 2: Reaction path of direct forsterite carbonation in aqueous solution and SEM analysis of the obtained $MgCO_3$ and spherical nanosilica [4, 8].

Electrochemical synthesis was mentioned in two papers using two methods: galvanostatic electrolysis and molten salt electrolysis. Al-Ti alloys were electrodeposited from equimolar chloroaluminate molten salts containing up to 0.1 M of titanium ions, which were added to the electrolyte by the potentiostatic dissolution of metallic Ti. Titanium dissolution and titanium and aluminium deposition were investigated by linear sweep voltammetry and chronoamperometry at 200 and 300 °C [13]. The obtained deposits were characterized by SEM, energy-dispersive spectrometry and XRD. In the deposits on the glassy carbon electrode, the analysis identified an Al and $AlTi_3$ alloy formed at 200 °C and an Al_2Ti and Al_3Ti alloy obtained at 300 °C. Three different forms of copper powder particles obtained by either galvanostatic electrolysis or a non-electrolytic method were analyzed by a scanning electron microscope (SEM), X-ray diffraction (XRD) and particle size distribution (PSD). Electrolytic procedures were performed under different hydrogen evolution conditions, leading to the formation of either 3D branched dendrites or disperse cauliflower-like particles. The third type of particles were compact agglomerates of the Cu grains, whose structural characteristics indicated that they were formed by a non-electrolytic method [6].

Ultrasound-assisted leaching process and hydrogen reduction were mentioned in three papers [14, 16, 17] describing the synthesis of metallic powders such as rhenium and cobalt. The preparation of rhenium powder by a hydrogen reduction of ammonium perrhenate is the only industrial production method. However, due to the uneven particle size distribution and large particle size of rhenium powder, it is difficult to prepare high-density rhenium ingot. Moreover, the existing process requires a secondary high-temperature reduction and the deoxidization process is complex and requires a high-temperature resistance of the equipment.

The leaching of industrial polycrystalline diamond (PCD) blanks in aqua regia at atmospheric pressure between 60 °C and 80 °C was performed using an ultrasound to improve the rate of cobalt removal, in order to be able to reuse very expensive polycrystalline diamond [16,17]. A transition from a reaction-controlled to a diffusion-controlled shrinking core model was observed for PCD with a thickness greater than 2.8–3.4 mm. Intermittent ultrasound doubles the reaction rate constant, and the full use of ultrasound provides a 1.5-fold further increase. The obtained maximum activation energy between 60 °C and 80 °C is 20 kJ/mol, for a leaching of diamond blank with grain size of 5 µm. Some results are shown in Figure 3.

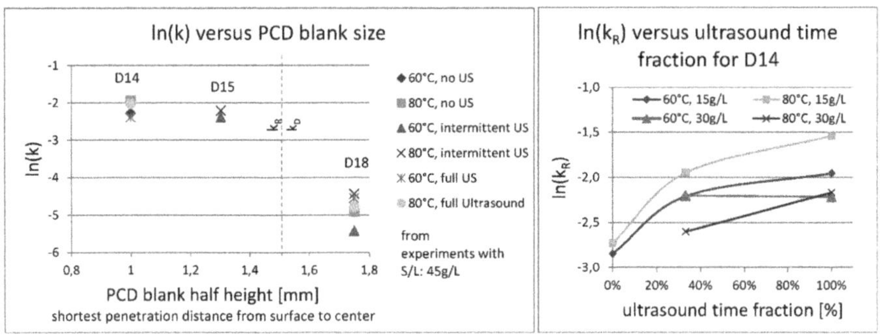

Figure 3: Plots of ln(k) over PCD blank size and ultrasound time fraction (D14) [17].

Leaching and precipitation were mentioned in four papers describing the synthesis of poly-alumino-ferric sulphate (AMD-PAFS) [10], vanadium oxide [5] and hydroxide based on Al, Mn and Co [12]. Tests conducted in Erlenmeyer flasks showed that after neutralization of the lake water in Serbia to pH 7, over 99% of aluminum (Al), iron (Fe), and copper (Cu) precipitated, as well as 92% of Zn and 98% of Pb. In order to remove residual Mn and Ag, the water was further treated with NaOH. Flotation tailings rich in carbonate minerals from the tailings deposit of the copper mine Majdanpek (Serbia) were applied for neutralization of the water taken from the extremely acidic Lake Robule (Bor, Serbia). The co-precipitation of iron and aluminium from acid mine drainage water (AMD) from South Africa is conducted at pH values of 5.0, 6.0 and 7.0, using sodium hydroxide in order to evaluate the recovery of iron and aluminium as hydroxide precipitates, while minimizing the co-precipitation of the other heavy metals. The precipitation at pH 5.0 yields iron and aluminium recovery of 99.9 and 94.7%, respectively. An increase in the pH from 5.0 to 7.0 increases the recovery of aluminium to 99.1%, while the recovery of iron remains the same. The production of the coagulant is carried out by dissolving the precipitate in 5.0% (w/w) sulphuric acid. Subsequently, the treatment of the brewery wastewater shows that the AMD-PAFS coagulant is as efficient as the conventional poly ferric sulphate (PFS) coagulant. In contrast, to use ammonium solution for precipitation, an eco-friendly technology was investigated to prepare vanadium oxides from a typical vanadium (IV) strip liquor, obtained after the hydrometallurgical treatment of a vanadium-bearing shale. Thermodynamic analysis demonstrated that VO(OH)$_2$ could be prepared as a precursor over a suitable solution pH range. Experimental results showed that by adjusting the pH to around 5.6, at room temperature, 98.6% of the vanadium in the strip liquor was formed into hydroxide, in 5 min. After obtaining the VO(OH)$_2$, it was washed with dilute acid to minimize the level of impurities. VO$_2$ and V$_2$O$_5$ were then produced by reacting the VO(OH)$_2$

with air or argon, in a tube furnace. Consequently, this process could promote the sustainable development of the vanadium chemical industry. Synthetic zirconolite samples with a target composition $Ca_{0.75}Ce_{0.25}ZrTi_2O_7$, prepared using two different methods, were used to study the stability of zirconolite for nuclear waste immobilization [9]. Particular focus was on plutonium, with cerium used as a substitute. The testing of destabilisation was conducted under conditions previously applied to other highly refractory uranium minerals that have been considered for safe storage of nuclear waste, brannerite and betafite. Acid (HCl, H_2SO_4) leaching for up to 5 h and alkaline ($NaHCO_3$, Na_2CO_3) leaching for up to 24 h was done to enable comparison with brannerite leached under the same conditions. Ferric ion was added as an oxidant. Given the demonstrated durability of zirconolite, long term criticality risks in the disposal environment seem a remote possibility, which supports its selection, above brannerite or betafite, as the optimal waste form for the disposition of nuclear waste, including of surplus plutonium.

Milling and thermal decomposition and hydrogenation process were mentioned in three papers and used for the synthesis of Mo_5Si_3 [3], $Cu-Al_2O_3$ [15] and Ti6Al4V [1] powders. A method was developed to fabricate spherical Mo_5Si_3 powder by milling and spheroidizing using inductively coupled thermal plasma. A Mo_5Si_3 alloy ingot was fabricated by vacuum arc melting, after which it was easily pulverized into powder by milling due to its brittle nature. The milled powders had an irregular shape, but after being spheroidized by the thermal plasma treatment, they had a spherical shape. Sphericity was increased with increasing plasma power. After plasma treatment, the percentage of the Mo_3Si phase had increased due to Si evaporation. The hydrogenation–dehydrogenation (HDH) process for synthesis of Ti6Al4V consists of the following sequence: surface conditioning of the turnings, hydrogenation, ball milling (for powder production), and dehydrogenation. Promising results were obtained regarding the potential of the recycled powders in additive manufacturing after making minor adjustments in the HDH process. Thermochemical synthesis of copper/alumina nanocomposites in a $Cu-Al_2O_3$ system with 1–2.5 wt.% of alumina and their characterization, which included: transmission electron microscopy: focused ion beam (FIB), analytical electron microscopy (AEM) and high resolution transmission electron microscopy (HRTEM), confirming high potential for using this process in nanotechnology. Thermodynamic analysis was used to study the formation mechanism of desirable products during drying, thermal decomposition and reduction processes. Upon the synthesis of powders, samples were cold pressed (2 GPa) in tools dimension 8 × 32 × 2 mm and sintered at temperatures within the range 800–1000 °C for 15 to 120 min in a hydrogen atmosphere.

Additionally, the HSC Chemistry® software package 9.0 and FactSage were used for the analysis of chemistry and thermodynamic parameters of the processes for powder synthesis [15].

Scanning electron microscopy (SEM), high resolution transmission electron microscopy (HRTEM), selected area electron diffraction (SAED), focused ion beam (FIB), analytical electron microscopy (AEM), inductively coupled plasma optical emission spectroscopy (ICP OES), thermal gravimetric analysis (TGA), X-ray analysis, differential thermal analysis (DTA) and differential scanning calorimetry (DSC) were used for the characterization of morphology, structure and chemical phase and composition.

We hope that this Special Issue will offer new information and shed light on advances in the synthesis of metallic, oxidic, and composite powders.

Srecko Stopic, Bernd Friedrich

Guest Editors

References

1. Gökelma, M., Celik, D., Tazegul, O., Cimenoglu, H., Friedrich, B. Characteristics of Ti6Al4V Powders Recycled fromTurnings via the HDH Technique. *Metals* **2018**, *8*, 336.

2. Emil, E.; Alkan, G.; Gurmen, S.; Rudolf, R.; Jenko, D.; Friedrich, B. Tuning the Morphology of ZnO Nanostructures with the Ultrasonic Spray Pyrolysis Process. *Metals* **2018**, *8*, 569.

3. Kang, J.-W.; Park, J.M.; Choe, B.H.; Lee, S.; Park, J.H.; Park, K.B.; Kim, H.K.; Na, T.-W.; Seo, B.; Park, H.-K. Preparation of Spherical Mo_5Si_3 Powder by Inductively Coupled Thermal Plasma Treatment. *Metals* **2018**, *8*, 604.

4. Stopic, S.; Dertmann, C.; Modolo, G.; Kegler, P.; Neumeier, S.; Kremer, D.; Wotruba, H.; Etzold, S.; Telle, R.; Rosani, D.; Knops, P.; Friedrich, B. Synthesis of Magnesium Carbonate via Carbonation under High Pressure in an Autoclave. *Metals* **2018**, *8*, 993.

5. Ma, Y.; Wang, X.; Stopic, S.; Wang, M.; Kremer, D.; Wotruba, H.; Friedrich, B. Preparation of Vanadium Oxides from a Vanadium (IV) Strip Liquor Extracted from Vanadium-Bearing Shale Using an Eco-Friendly Method. *Metals* **2018**, *8*, 994.

6. Avramović, L.; Maksimović, V.M.; Baščarević, Z.; Ignjatović, N.; Bugarin, M.; Marković, R.; Nikolić, N.D. Influence of the Shape of Copper Powder Particles on the Crystal Structure and Some Decisive Characteristics of the Metal Powders. *Metals* **2019**, *9*, 56.

7. Gajić, N.; Kamberović, Ž.; Anđić, Z.; Trpčevská, J.; Plešingerova, B.; Korać, M. Synthesis of Tribological WS_2 Powder from WO_3 Prepared by Ultrasonic Spray Pyrolysis (USP). *Metals* **2019**, *9*, 277.

8. Stopic, S.; Dertmann, C.; Koiwa, I.; Kremer, D.; Wotruba, H.; Etzold, S.; Telle, R.; Knops, P.; Friedrich, B. Synthesis of Nanosilica via Olivine Mineral Carbonation under High Pressure in an Autoclave. *Metals* **2019**, *9*, 708.

9. Nikoloski, A.N.; Gilligan, R.; Squire, J.; Maddrell, E.R. Chemical Stability of Zirconolite for Proliferation Resistance under Conditions Typically Required for the Leaching of Highly Refractory Uranium Minerals. *Metals* **2019**, *9*, 1070.

10. Mwewa, B.; Stopić, S.; Ndlovu, S.; Simate, G.S.; Xakalashe, B.; Friedrich, B. Synthesis of Poly-Alumino-Ferric Sulphate Coagulant from Acid Mine Drainage by Precipitation. *Metals* **2019**, *9*, 1166.

11. Košević, M.G.; Zarić, M.M.; Stopić, S.R.; Stevanović, J.S.; Weirich, T.E.; Friedrich, B.G.; Panić, V.V. Structural and Electrochemical Properties of Nesting and Core/Shell Pt/TiO_2 Spherical Particles Synthesized by Ultrasonic Spray Pyrolysis. *Metals* **2020**, *10*, 11.

12. Petronijević, N.; Stanković, S.; Radovanović, D.; Sokić, M.; Marković, B.; Stopić, S.R.; Kamberović, Ž. Application of the Flotation Tailings as an Alternative Material for an Acid Mine Drainage Remediation: A Case Study of the Extremely Acidic Lake Robule (Serbia). *Metals* **2020**, *10*, 16.

13. Cvetković, V.S.; Vukićević, N.M.; Milićević-Neumann, K.; Stopić, S.; Friedrich, B.; Jovićević, J.N. Electrochemical Deposition of Al-Ti Alloys from Equimolar $AlCl_3$ + NaCl Containing Electrochemically Dissolved Titanium. *Metals* **2020**, *10*, 88.

14. Tang, J.; Sun, Y.; Zhang, C.; Wang, L.; Zhou, Y.; Fang, D.; Liu, Y. Reaction Mechanism and Process Control of Hydrogen Reduction of Ammonium Perrhenate. *Metals* **2020**, *10*, 640.

15. Korać, M.; Kamberović, Ž.; Anđić, Z.; Stopić, S. Advances in Thermochemical Synthesis and Characterization of the Prepared Copper/Alumina Nanocomposites. *Metals* **2020**, *10*, 719.

16. Kießling, F.; Stopic, S.; Gürmen, S.; Friedrich, B. Recovery of Diamond and Cobalt Powder from Polycrystalline Drawing Die Blanks via Ultrasound-Assisted Leaching Process—Part 1: Process Design and Efficiencies. *Metals* **2020**, *10*, 731.

17. Kießling, F.; Stopic, S.; Gürmen, S.; Friedrich, B. Recovery of Diamond and Cobalt Powders from Polycrystalline Drawing Die Blanks via Ultrasound Assisted Leaching Process—Part 2: Kinetics and Mechanisms. *Metals* **2020**, *10*, 741.

Article

Recovery of Diamond and Cobalt Powders from Polycrystalline Drawing Die Blanks via Ultrasound Assisted Leaching Process—Part 2: Kinetics and Mechanisms

Ferdinand Kießling [1], Srecko Stopic [2,*], Sebahattin Gürmen [3] and Bernd Friedrich [2]

[1] Redies Deutschland GmbH & Co.KG, Metzgerstr. 1, 52070 Aachen, Germany; fer87den@gmail.com
[2] IME Process Metallurgy and Metal Recycling, RWTH Aachen University, Intzestrasse 3, 52056 Aachen, Germany; bfriedrich@ime-aachen.de
[3] Metallurgical and Materials Engineering Department, Istanbul Technical University, Ayazaga Campus, Istanbul 34469, Turkey; gurmen@itu.edu.tr
*   Correspondence: sstopic@metallurgie.rwth-aachen.de; Tel.: +49-241-809-5860

Received: 30 April 2020; Accepted: 26 May 2020; Published: 3 June 2020

Abstract: The leaching of industrial polycrystalline diamond (PCD) blanks in aqua regia at atmospheric pressure between 60 °C and 80 °C was performed using an ultrasound to improve the rate of cobalt removal in order to be able to reuse very expensive polycrystalline diamond. Because cobalt (20 wt.%) is used as a solvent catalyst in the production of PCD, its recovery is very important. The cleaned PCD are returned to the production process. Kinetic models were used in the study of cobalt dissolution from polycrystalline diamond blanks by measuring the declining ferromagnetic properties over time. For a better understanding of this leaching process, thermochemical aspects are included in this work. The lowest free Gibbs energy value was obtained with a low solid/liquid ratio and the full use of an ultrasound. A transition from a reaction-controlled to a diffusion-controlled shrinking core model was observed for PCD with a thickness greater than 2.8–3.4 mm. Intermittent ultrasound doubles the reaction rate constant, and the full use of ultrasound provides a 1.5-fold further increase. The obtained maximum activation energy between 60 °C and 80 °C is 20 kJ/mol, for a leaching of diamond blank with grain size of 5 μm.

Keywords: cobalt; aqua regia; polycrystalline diamond blanks; kinetics; thermochemistry

---

Part 1: Experimental Design and Efficiencies

This study attempts to achieve optimal recovery of diamond and cobalt from polycrystalline diamond (PCD) blanks. In nine experimental runs of 5 days' duration, cobalt-containing PCD was leached in aqua regia at atmospheric pressure between 60 °C and 80 °C. Using two reactors in parallel, the temperature, ultrasound irradiation time, solid-to-liquid ratio, and PCD size were varied to find out which parameters are beneficial and could possibly accelerate the process. PCD weights and cobalt content in solution were monitored as well. It was found that aqua regia accumulated more dissolved cobalt at 60 °C than at 80 °C, probably due to volatile reagents being less available over time. With added ultrasound and at low S/L ratios, i.e., close to 15 g/L, the leaching time for D14 to reach a 90% leach mark was reduced to three days, a significant shortening. PCD type D18 with a thickness of 3.5 mm were not leached to completion after five days. Leaching temperature had more impact on the results than ultrasound. These findings were reinforced by the mass balance in which a small discrepancy was found. The PCD lost a fraction of weight that could not be explained by the weight of dissolved cobalt. From EDS (Energy Dispersive Spectroscopy) data and the nature of PCD, this fraction probably consisted of, oxygen from oxides in the PCD, iron or single diamond grains that were broken off by the impact of the ultrasound.

1. Introduction

The term polycrystalline diamond (PCD) describes a variety of amorphous compounds mostly or wholly consisting of microscopically small diamond grains. A single crystal of natural diamond is anisotropic in terms of its mechanical and thermodynamic properties, including tensile strength and thermal conductivity, for instance. Most PCD will have a random arrangement of individual grains, resulting in a quasi-isotropic compound. However, there are forms of PCD that are made in a different way and that have different properties. Binderless PCD (Sumidia), CVD crystals [1], and monocrystalline dies in general will not be discussed herein. The conditions needed for diamond powder to form a framework are extreme. Only in the region of 50 kbar and at a temperature of 2000 °C will the desired reaction happen on reasonable time scales [2–4]. Cobalt is used as a solvent catalyst in the production of PCD; without it, the reaction would require even more pressure and a higher temperature.

The leaching solution in this case has the colloquial name "aqua regia" because it was found to dissolve noble metals such as gold or platinum; early records of its use date back centuries [5]. Aqua regia ensures an oxidation environment. More specifically, the aqua regia was mixed from 3 parts Merck KGaA fuming hydrochloric acid 37%, Emsure ACS/ISO quality and one part PanReac ApplicChem nitric acid 65% ISO analysis quality.

The solution is a mixture of hydrochloric acid, HCl, and nitric acid, HNO$_3$. Both are strong acids, and at a ratio of 3:1, reactions (1) to (3) occur [6].

$$3HCl_{(aq)} + HNO_{3(aq)} \leftrightarrow NOCl_{(aq)} + Cl_{2(g)} + 2H_2O_{(l)} \tag{1}$$

$$NOCl_{(aq)} + H_2O_{(l)} \leftrightarrow HNO_{2(aq)} + HCl_{(aq)} \tag{2}$$

$$2HNO_{2(aq)} \leftrightarrow NO_{(aq)} + NO_{2(aq)} + H_2O_{(l)} \tag{3}$$

The formation and transport of molecular chlorine gas and NOCl has been found to occur within minutes to hours [6]. Baghalha et al. [7] concluded that the 3:1 mixing ratio maximizes the production of chlorine per unit mass of reactants, and is to be favored when chlorine is the desired oxidizing agent. However, if the desired reaction requires only low pH or different oxidizing agents, this ratio or aqua regia itself may not be suitable. This study intends to extract cobalt from PCD as a chloride, CoCl$_{2(aq)}$, and therefore, uses aqua regia, or NOCl to be precise. The desired reaction in this case is the oxidation and dissolution of cobalt into aqua regia, which is achieved in the following redox-reactions:

$$Co_{metal} + 2NO_{aq}^+ + 2Cl_{aq}^- \leftrightarrow Co_{aq}^{2+} + 2Cl_{aq}^- + 2NO_g \tag{4}$$

$$CoO_s + 2NO_{aq}^+ + 2Cl_{aq}^- \leftrightarrow Co_{aq}^{2+} + 2Cl_{aq}^- + NO_g + NO_{2g} \tag{5}$$

The Pourbaix diagrams by Huang et al. [8] show that the equilibrium for this reaction should be on the right side of the balance, since the divalent cobalt cation is not only stable at pH << 1, but also at pH > 1. In many cases, the most cost- and energy-efficient way to extract metal from gangue or scraps is to oxidize and dissolve it in a leaching solution. There are many examples; the well-established Caron process is one of them [9]. It is used to treat lateritic nickel ores by reduction roasting and subsequent leaching for the purpose of obtaining a nickel-bearing solution while separating nickel from iron [10,11].

The polycrystalline diamond in aqua regia can be seen as a solid compound particle where its metallic components react with the solution. The reaction front moves inwards and leaves a layer of inert diamond grains behind. That is why the model of the shrinking unreacted core (SCM) is applied [12,13]. According to the model, the leaching rate may depend on the reaction or diffusion of educts and products from the reaction site; in reality, it is often a mixture of both effects. If a linear relationship is found in the plots of Equations (6) and (7), over time, the model is confirmed and the apparent rate constant can be extracted from the slope. Equation (6) is the Ginstling-Brounshtein (D4) model. The D4 model is another type of diffusion three-dimensional model, in contrast to widely used Jander model, as reported by Khawan and Flanagan [14]. If a solid particle has a spherical or cubical shape, a contracting sphere/cube model can be applied, as shown in Equation (7).

$$1 - \frac{2}{3}X - (1-X)^{\frac{2}{3}} = k_D \times t \tag{6}$$

$$1 - (1-X)^{\frac{1}{3}} = k_R \times t \tag{7}$$

X is a dimensionless variable that represents the relative change in the amount of substance or concentration, which is why in leaching processes, the yield is taken for X, for example. In this study, k will be indexed with "D" or "R", depending on whether the diffusion- or chemical reaction-controlled model was applied.

From this apparent rate constant, the activation energy for the reaction can be obtained by plotting the natural logarithm of k over the reciprocal temperature in an Arrhenius plot, or by using the following equation [15]:

$$E_A = \partial \ln(k) / \frac{\partial}{T} \tag{8}$$

$$\Delta G^0 = -RT \ln(k) \tag{9}$$

The slope of the Arrhenius plot delivers the activation energy. Furthermore, the k-rate coefficient can be used to calculate the Gibbs energy accompanying this reaction using Equation (9)

The main aim of this work was to study the kinetics of cobalt removal from polycrystalline diamond blanks using an ultrasound-assisted leaching process; no reports of such a process exist in the literature. Two mathematical models will be tested in order to determine the activation energy and rate coefficient. An additional thermochemical analysis was included to provide a better explanation of the behavior of cobalt in a water solution at different pH-Eh values using an Eh-pH diagram.

2. Thermochemistry of Cobalt Leaching

A thermochemical analysis of Eh-pH diagram was performed using the HSC-software (Outotec, Espoo, Finland), as shown in Figure 1, where an increase of temperature from 25 °C to 80 °C did not affect the presence of cobalt ions at different pHs.

As shown in Figure 1, the Pourbaix diagram (potential Eh-pH) of cobalt in a water solution at 25 °C and 80°C confirms the presence of cobalt in the form of Co^{2+} and Co^{3+} at pHs below 0. At an increased potential between 2.0 and 3.0 V, cobalt is available only as Co^{3+}.

Regarding the leaching of cobalt, Han and Meng [16] found that the dissolution of cobalt is dependent on diffusion, while the dissolution of divalent oxides is reaction controlled. They reported that the leaching rate of cobalt is generally faster than that of the respective oxides. Huang et al. [8] conducted experiments on the precipitation of cobalt and molybdenum from effluents and used HSC software to compute the potential pH diagrams for a $Co-H_2O$ system at temperatures of 20 °C, 40 °C, 60 °C and 80 °C [8]. Within the parameters of this study, namely, without external potential, Eh = 0, and at pH values close to zero, the stable form of cobalt is a divalent cation within this temperature range, as was observed in our work.

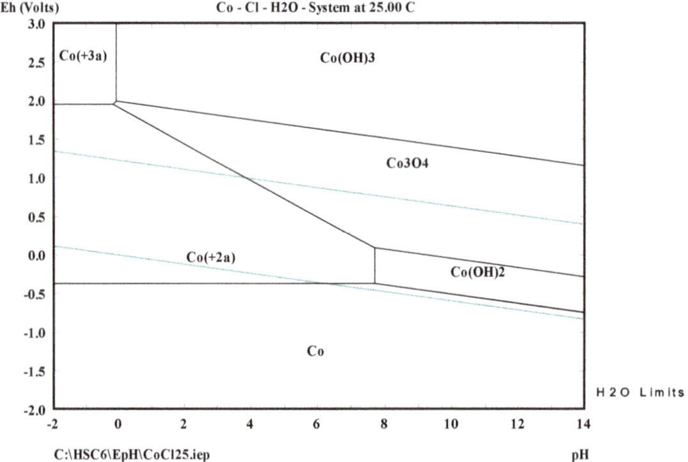

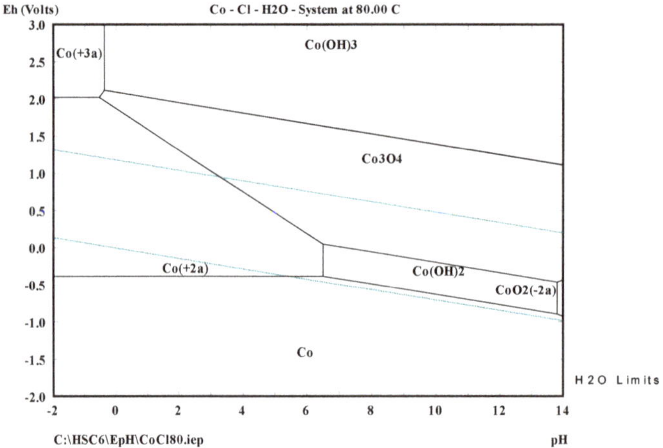

Figure 1. eH-pH diagram for Co-Cl-H_2O at 25 °C and 80 °C.

3. Experiment

This study is focused on polycrystalline diamond (PCD) blanks made by Redies Deutschland GmbH & Co. KG (Aachen, Germany), a manufacturer of wire drawing dies. The material characteristics and procedure are shown in a paper by Kiessling et al. [17]. A SEM (Scanning Electron Microscopy) analysis was performed using a JEOL JSM 7000F Field Emission Scanning Electron Microscope, (2003, JEOL Inc, Peabody, MA, USA) with Energy Dispersive Spectroscopy (EDS), Wavelength Dispersive Spectroscopy (WDS) and Electron back-scattered diffraction (EBSD, JEOL JSM 7000F SEM). The SEM and EDS analysis after polishing the PCD surface is shown in Figure 2 and Table 1.

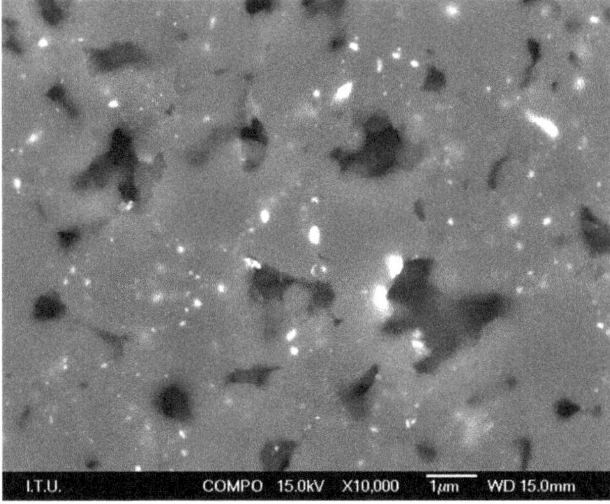

Figure 2. SEM image of ground and polished PCD surface, 5 μm class (dark grey areas are the bridged diamond grains with cavities where traces of cobalt also show up as lighter shades).

The maximal value of cobalt in the analyzed sample was 67 wt. %, as shown in Table 1.

Table 1. Energy dispersive X-ray spectroscopy image of ground and polished PCD surface, also 5 μm class.

Content (Weight %)	C	O	Fe	Co
Spectrum 1	86.81	12.94	-	0.25
Spectrum 2	88.31	10.42	0.21	1.05
Spectrum 3	89.35	10.32	-	0.34
Spectrum 4	87.41	10.78	0.14	1.67
Max.	89.35	12.94	0.21	1.67
Min.	86.81	10.32	0.00	0.25

Different types of samples were used in our work, as shown in Table 2. The columns "volume", "surface area" and "surface-to-volume ratio" contain calculated, and not measured, values.

Table 2. Dimensions of PCD samples with grain size of 5 μm.

Blank Type	Symbol	Diameter [mm]	Height [mm]	Weight [g]	Volume [mm^3]	Surface Area [mm^2]	Surface/Volume [mm^{-1}]
Mant®MSD-14-005	D14	4.05 ± 0.09	2.00 ± 0.04	0.099	25.71 ± 1.40	51.14 ± 1.97	1.99 (+0.20|−0.18)
Mant®MSD-15-005	D15	5.2	2.5	0.241	53.093	83.315	1.569
Mant®MSD-18-005	D18	5.22 ± 0.02	3.50 ± 0.02	0.299	74.90 ± 0.77	100.22 ± 0.68	1.338 (±0.023)

3.1. Changes in Magnetic Properties of the PCD

Samples were weighed and measured with a teslametric probe beforehand. During the experiment, liquid samples were taken and a sample of 5 PCD of each type was measured with a teslameter built by Projekt Elektronik GmbH, Berlin, Germany as shown in Figure 3. These ultrasound baths have a nominal frequency of 35 kHz and put out 60 W effectively, while output peaks can occur at up to 240 W. The output level is fixed and the ultrasound irradiation was altered by using it intermittently.

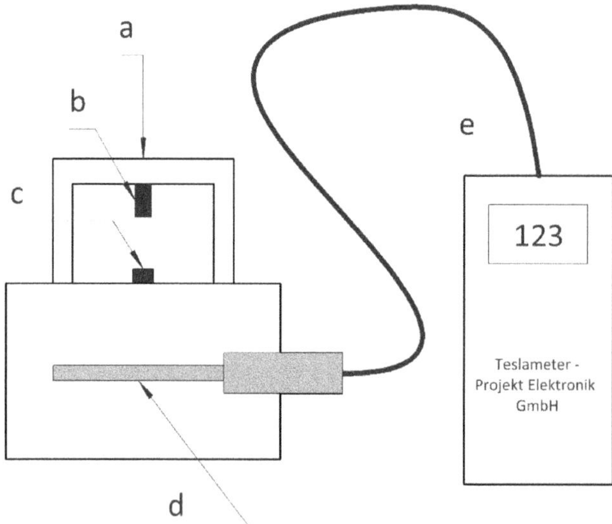

Figure 3. Measuring magnetic properties of PCD a: magnet support, b: ∅ 5 mm by 8 mm Nd-Alloy magnet, c: PCD blank, d: probe, e: handheld teslameter.

To investigate the influence of different parameters on the leaching of PCD with aqua regia, the experiments were performed in the following way; see Table 3.

The abbreviations in the first column are successive week number and R1 and R2, representing reactors 1 and 2, respectively. These codes also served as stems for sample identification. The column header S/L is short for solid-to-liquid ratio in units of grams per liter. Constant parameters were stirring speed and batch time. The duration of each batch was between 90 and 100 h. Sample names D14, D15 and D18 are the abbreviated product names of MSD-14-005, MSD-15-005, and MSD-18-005, respectively, all of which are self-supported PCD blanks with diamond grain sizes of 5 μm.

Table 3. Parameters for the leaching of cobalt from polycrystalline diamond blank.

ID	T_{Bath} [K]	PCD Type	Leaching Time [h/d] in the Presence of Ultrasound	S/L [g/L]
W1R1	333	D14	0	15
W1R2	353	D14	0	15
W2R1	333	D14 + D18	0	30
W2R2	353	D14 + D18	0	30
W3R1	333	D14 + D18	0	45
W3R2	353	D14 + D18	0	45
W4R1	333	D14 + D18	8 h/d	15
W4R2	353	D14 + D18	8 h/d	15
W5R1	333	D14 + D18	8 h/d	30
W5R2	353	D14 + D18	8 h/d	30
W6R1	333	D15 + D18	8 h/d	45
W6R2	353	D15 + D18	8h/d	45
W7R1	333	D14 + D18	24 h/d	15
W7R2	353	D14 + D18	24 h/d	15
W8R1	333	D14 + D18	24 h/d	30
W8R2	353	D14 + D18	24 h/d	30
W9R1	333	D14 + D18	24 h/d	45
W9R2	353	D14 + D18	24 h/d	45

3.2. Preparation of Samples

The PCD blanks were weighed and measured with the teslameter beforehand to obtain a '100%' reference value for evaluation. Then, 240 mL of hydrochloric acid and 80 mL nitric acid were prepared in covered beakers for each reactor. Meanwhile, the ultrasound baths had been filled with tap water and their heaters were set to 60 °C and 80 °C to ensure that the bath temperature was nominal at the beginning of the experiment. The gas tightness of the apparatus was checked daily.

As stated, the presence of cobalt in the PCD can be determined electromagnetically. The teslameter consists of a handheld unit and a probe which is sensitive to magnetic field changes of ± 2 mT. To offset any noise or ambient disruptions, a magnetic field from a ⌀ 5 mm by 8 mm rare earth magnet was introduced. The magnet was held at a constant distance to the probe by enclosing the probe in plastic housing and attaching the magnet to a mild steel support structure made of 10 mm × 10 mm square bars.

Data from daily samples was taken and plotted over time. For a better comparison, the measurements were normalized to an initial value of 1, and a reference value without a PCD blank of zero. Then, the daily measurements were plotted as a percentage relative to the initial measurement. For example, in Figure 4, there is the original data plot on the left and the normalized plot on the right.

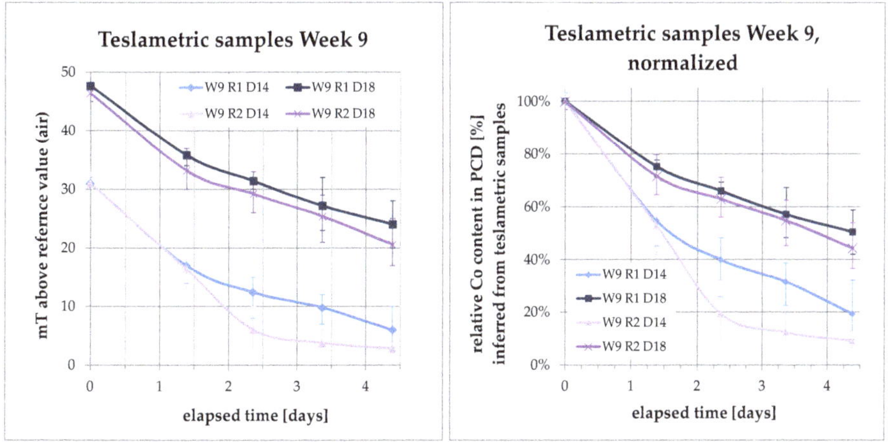

Figure 4. Data plot of teslametric samples, (**left**): as measured, (**right**): normalized to initial value.

What's more, this plot can be understood to display the content of ferromagnetic cobalt inside the PCD relative to its initial content. The error bars indicate the extreme values of the sample, while the data points and curve mark the average of the five PCDs sampled per reactor and type.

All data sets have positive curvature, indicating a slowing rate of change over time. The difference in bath temperatures is also visible in this plot, as is the fact that the PCDs in the warmer reactor number 2 were leached to lower values than those in reactor 1.

The effects of ultrasound-assisted leaching on the teslametric measurements were compiled and plotted in order to make a comparison under different conditions and with respect to PCD size, as shown in Figure 5.

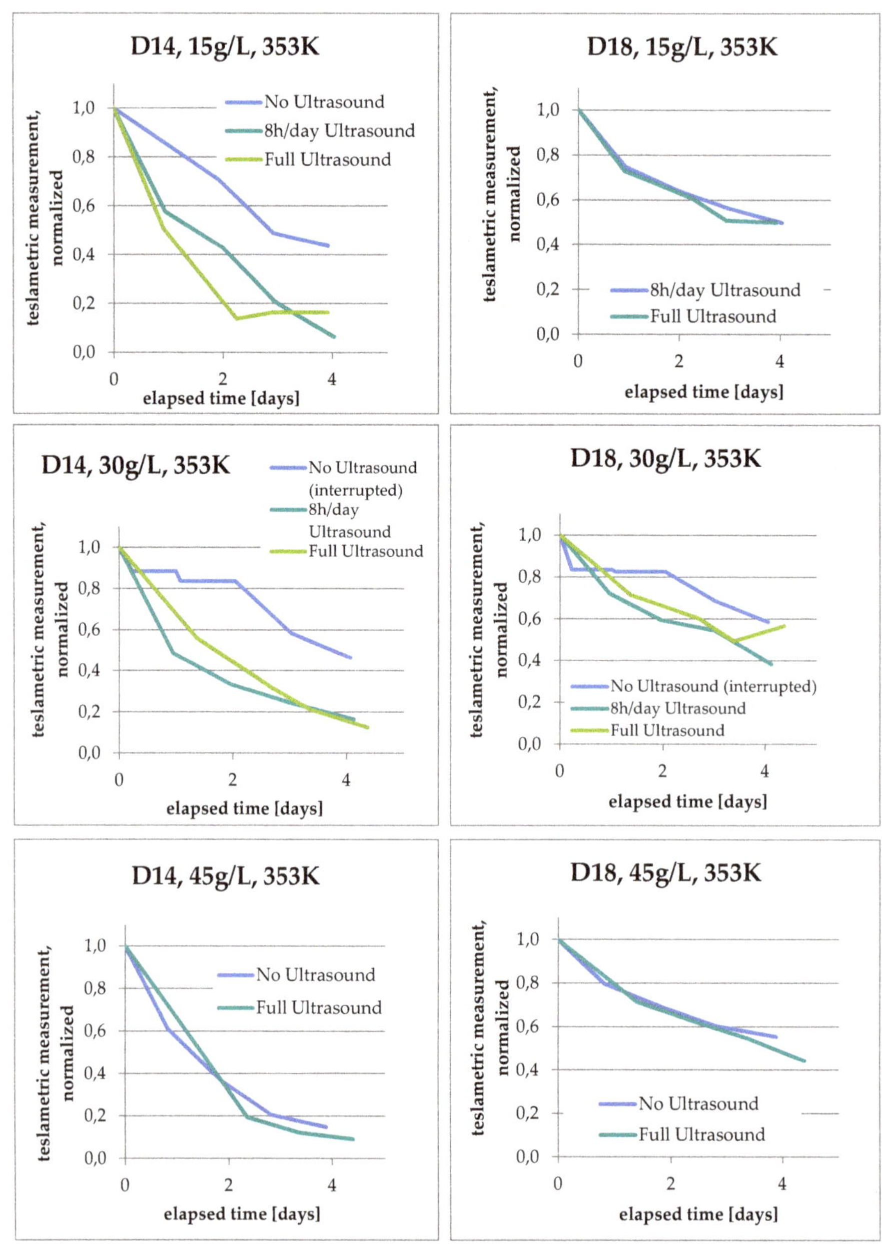

Figure 5. Teslametric measurements—comparison of process parameters. "Full Ultrasound" refers to the full-time use of ultrasound.

An additional comparison was made to study the effects of bath temperature with respect to PCD sizes D14 and D18, as shown in Figure 6. At a higher temperature and with full-time ultrasound, the initial drop in the plot is steeper than at lower temperatures and with intermittent ultrasound. When looking at the final values, intermittent ultrasound accomplishes the PCD to achieve lower readings on the teslameter. The difference between leaching D14 and D18 becomes clear in both images. The decrease of magnetic effects in D18 happens significantly more slowly than in D14, where they seem to reach desaturation within the timeframe. Desaturation in this case means that the remaining magnetic effect is less than 10% of the initial effect. It can be inferred that 90% of the ferromagnetic contents of the PCD have been leached in those cases.

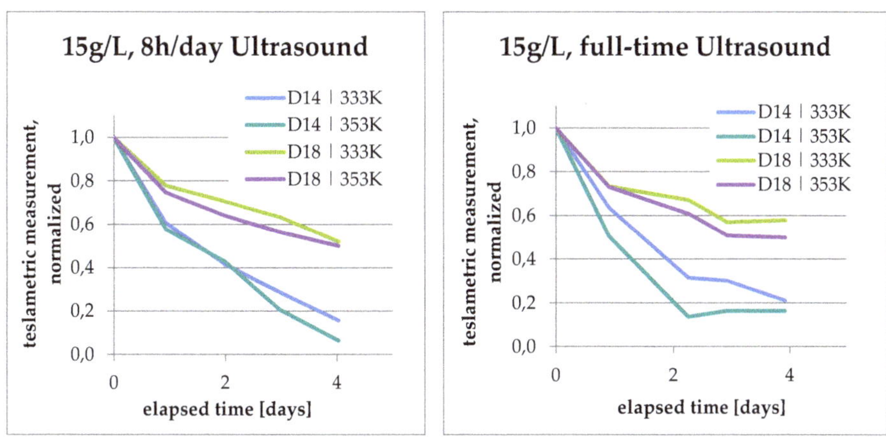

Figure 6. Teslametric measurements—comparison with respect to bath temperature

3.3. Increase of Co Content in Solution

Considering the analyses of the leaching solutions throughout this study, some trends can be observed. For one, there is the fact that the solution under full ultrasound and at 80 °C accumulates less cobalt than the experiments at 60 °C. This may be due to evaporation losses in the form of hydrochloric acid or water vapor. As stated, the remaining liquid was measured only in weeks 5, 7, 8, and 9. In the extreme case of week nine, in reactor 2, approximately 194 mL remained, which is only 60% of the initial volume. So, the seemingly lower solubility for cobalt at higher temperatures might be due to the reduced available liquid volume and fewer available chlorine anions. Moreover, the formation and escape of nitrous compounds into the off-gas stream is likely to be faster at higher temperatures. The redox potential was measured using a pH meter 7310, InoLAb, WTW, Weilheim, Germany. The redox potential amounted to 460.5 mV for a dissolution of 45 g/L of PCD using full ultrasound. As stated in the literature, aqua regia ensures an oxidation environment. For pH = 0 and Eh = 460.5 mV, cobalt is present as Co^{2+}, according to the Eh-pH diagram.

3.4. Analysis of Kinetics and Thermochemistry of Cobalt Dissolution

Assuming the PCDs can be viewed as solid particles that lose substance during leaching, according to the unreacted shrinking core model with constant particle size, calculations can be made using two different approaches. First, there are the declining ferromagnetic properties over time, and second, there is the rising cobalt content in solution. Taking equations from the value for X can be calculated in the following way:

$$X_{tesla} = 1 - \left(\frac{\mu_t}{\mu_{t=0}} \right) \tag{10}$$

$$X_{liquid} = 1 - \left(\frac{c_{Co}}{c_{Co,\,final}} \right) \tag{11}$$

where: μ is the value from the teslametric measurements. X_{liquid} is calculated from the final cobalt concentration in the leaching solution, since a reference is needed and there were no initial analyses of the cobalt content in the PCD. Furthermore, the expected concentration was surpassed in some weeks, leaving *X* outside the interval {0;1} which would not return realistic results.

Figure 7 shows the respective plots of the diffusion- and reaction-controlled SCM for D14 and D18; these data are exemplary for the weeks denoted in the diagram captions.

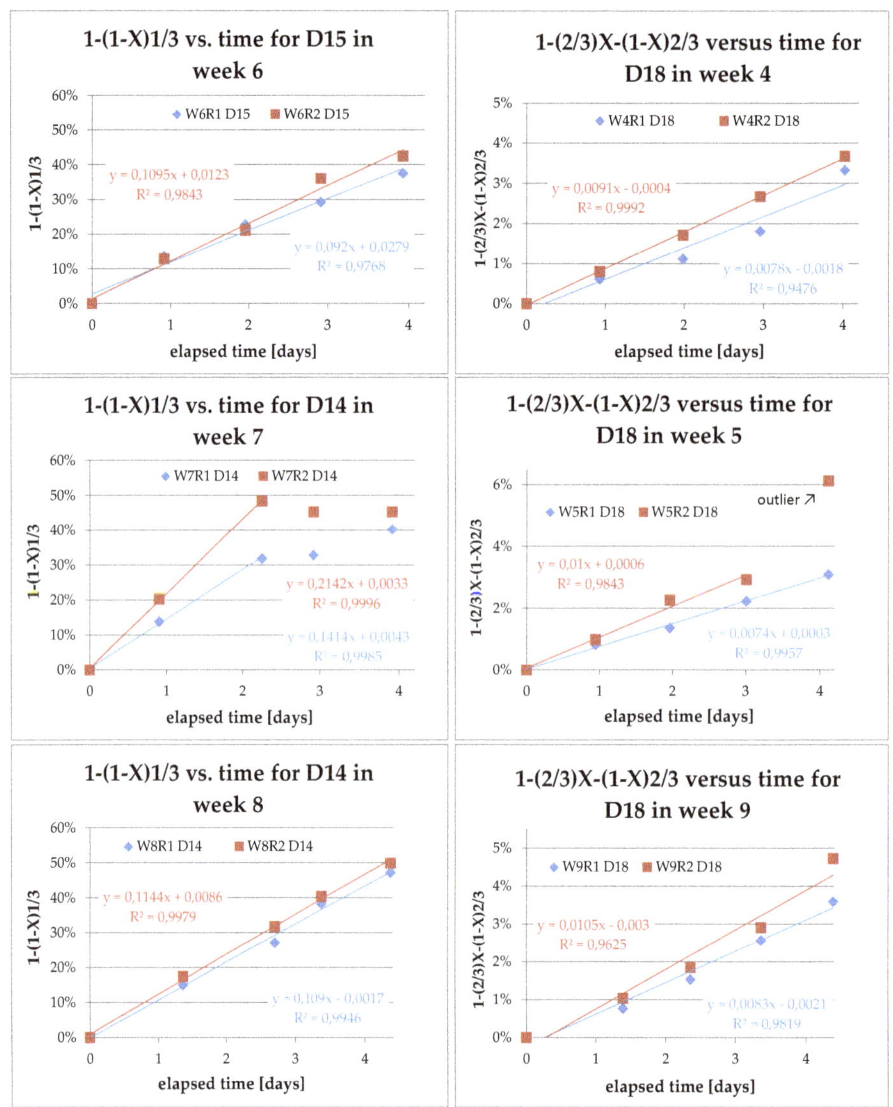

Figure 7. SCM plots for D14 and D18, red: 80 °C, blue: 60 °C.

When applying the two models to the teslametric data, it became clear that for D14, the reaction-controlled model was satisfactory, in contrast to D18, which seemed to follow the diffusion-controlled model described by the Ginstling-Brounshtein equation. In all cases, the goodness of the linear regression, R^2, was above 0.9, indicating that the fit was satisfactory. In the data from D14 in week 7, there was a very clear change in the reaction rate toward the end of the experiment run. This is another indicator of either approaching desaturation of ferromagnetic compounds in the PCD or of the saturation with cobalt of the leaching solution. The latter is not very likely, since this observation was made while leaching at a low solid-to-liquid ratio, and the teslametric values for D18 in the same batch were declining at the same time. When the models were applied to the data from the ICP-OES analyses, no linear relationship could be found (as shown in Figure 8). This might indicate that there was a chemical balance prior to the oxidation and dissolution of cobalt, such as the formation of NOCl, which is very important for the subsequent leaching process.

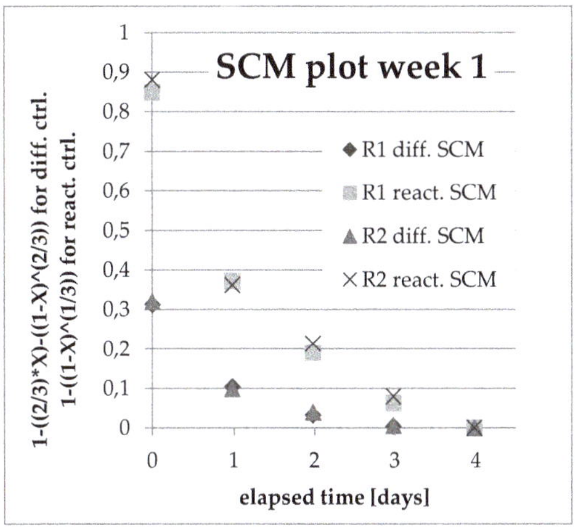

Figure 8. SCM plot from liquid samples week 1.

For both D14 and D18, Arrhenius plots were made to see if there were trends concerning the activation energy calculated using Equations (6)–(8). In Figure 9, the results are shown for a select group of experiments for clarity, and to check the extreme cases with regards to the solid-to-liquid ratio and ultrasound irradiation. Leaching D14 at low solid/liquid (S/L) ratios seems to benefit from the use of ultrasound. Temperature, on the other hand, has a bigger effect than ultrasound when leaching at high S/L ratios. For D14, the reaction rates were, in general, higher at 80 °C, as would be expected from a reaction-dependent kinetic model. Unfortunately, the results for D18 were inconclusive. Considering that the experiments were used for orientation and optimization, rather than for a detailed thermochemical analysis, this is an aspect wherein more research is needed.

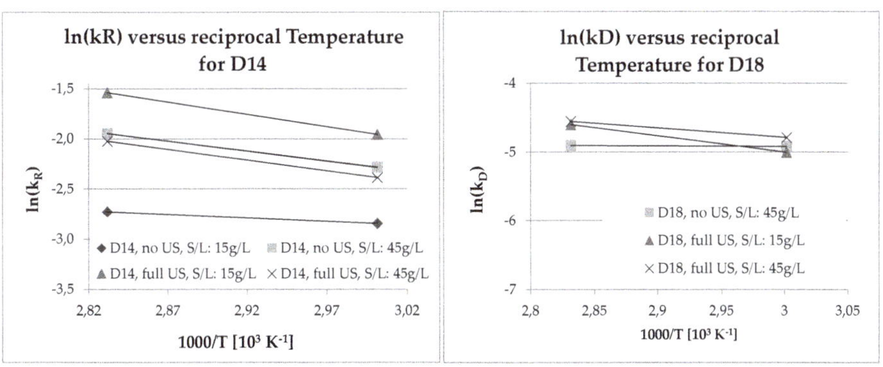

Figure 9. Arrhenius plots for D14 and D18.

Regarding the apparent rate constant, other interesting results emerged from the data, as shown in Figure 10. For one thing, there is the question of which particle size, or better, at which depth of penetration into the PCD structure the change from reaction controlled to diffusion controlled SCM might happen. Secondly, looking at the mass balance, the D14 seemed to benefit from intermittent ultrasound just as much as full ultrasound, which should also be recognizable in the rate constant.

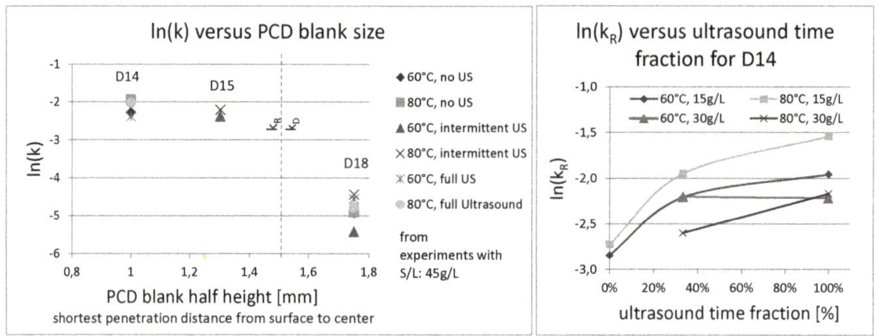

Figure 10. Plots of ln(k) over PCD blank size and ultrasound time fraction (D14).

Somewhere in the region of 1.4 to 1.7 mm penetration depth, the leaching of PCD seems to change from the reaction-controlled shrinking core model to diffusion-controlled mechanisms, as described by Equation (6). Moreover, the beneficial effects of ultrasound are reflected in the plot on the right side. For D14, intermittent ultrasound doubles the reaction rate constant and full ultrasound provides a further increase by a factor of 1.5 (see Table 4 below). The maximal obtained activation energy between 60 °C and 80 °C was 20 kJ/mol; below this value, a diffusion controlled-process occurred.

Table 4. Activation energy E_A (J/mol) derived from the apparent rate constants.

Week No.	PCD Type	$\delta\ln(k) = \ln(k_{353K}) - \ln(k_{333K})$	$\delta/T = (1/353K) - (1/333K)$	R_{Ideal} [$J{*}K^{-1}{*}mol^{-1}$]	$E_A = -R_{Ideal}{*}(\delta\ln(k)/(\delta/T))$
1	D14	0.1183566	−0.000170142	8.314	5783.8
3	D14	0.3347918	−0.000170142	8.314	16360.5
7	D14	0.4153174	−0.000170142	8.314	20295.6
9	D14	0.3621971	−0.000170142	8.314	17699.8
3	D18	0.0136057	−0.000170142	8.314	664.9
7	D18	0.4004776	−0.000170142	8.314	19570.4
9	D18	−0.2351197	−0.000170142	8.314	11489.8

When plotting the calculated Gibbs Energy for the dissolution of cobalt, again, the results for D18 are inconclusive, since there is too little data, and only two different temperatures were investigated. The only statement that could be made is that 80 °C bath temperatures have the tendency to increase ΔG^0 compared to 60 °C. This may again be explained by the more rapid evaporation of the reactants in the warmer reactor. For D14, the same conclusions can be made as with the Arrhenius plot. The lowest ΔG^0 corresponds to a low S/L ratio and the full use of ultrasound in the process, as shown in Figure 11. At higher temperature, the reaction requires less additional energy input.

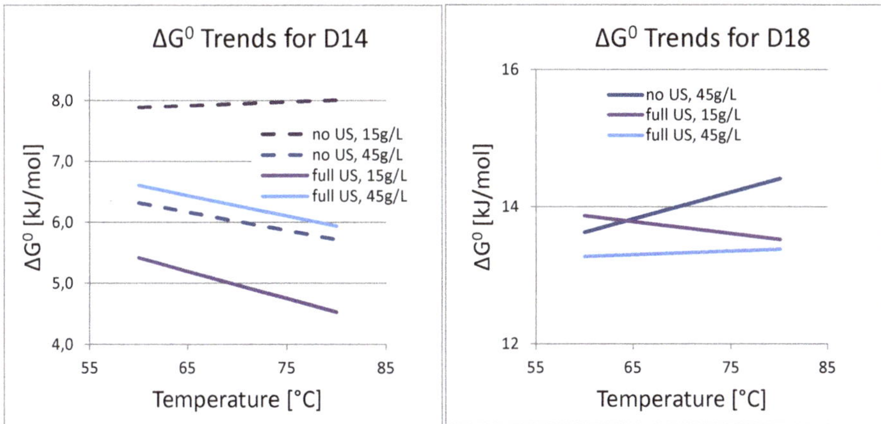

Figure 11. ΔG^0 trends for D14 and D18.

For PCD with thicknesses smaller than 2.8 mm, the diffusion-controlled, unreacted shrinking core model according to Equation (6) was confirmed with teslametric data. Similarly, Grénnman et al. [18] used the Ginstling-Brounshtein equation to represent mass transfer across a nonporous product layer. An increase in temperature to above 80 °C does not influence the kinetics of cobalt leaching. Diffusion might be affected by the mixing rate using a mixer and argon gas. Han and Lawson [19] reported a leaching study on metallic cobalt in an acidic medium. This was carried out using a rotating disc geometry to confirm the surface reaction involving hydrogen discharge and the diffusion of hydrogen ions through the boundary layer; together, these phenomena were found to be responsible for the rate of dissolution. The apparent activation energy for cobalt dissolution under the conditions of the experiments was found to be 16.7 kJ/mol, which represents a diffusion-controlled mechanism. Our maximum calculated activation energy amounted 20 kJ for the dissolution of cobalt from PCD blanks, which is in accordance with this.

4. Conclusions

This study sought to accelerate the leaching of cobalt from polycrystalline diamond using an ultrasound-assisted leaching process. It seems that varying the ultrasound frequency might be a new strategy to decrease the leaching time for maximal cobalt removal from PCD blanks.

In nine experimental runs of 5 days' duration, cobalt-containing PCD were leached in aqua regia at atmospheric pressure at between 60 °C and 80°. Using two reactors in parallel, temperature, ultrasound irradiation time, solid-to-liquid ratio, and PCD size were varied to determine the optimal parameters and possibly accelerate this process. The redox potential amounted to 460.5 mV for a dissolution of 45 g/L of PCD using full ultrasound, thereby confirming the presence of Co^{2+}-ions in solution with a pH value close to zero. Two kinetics models were tested in this work. In both cases, the goodness of the linear regression, R^2, was above 0.9, indicating that the fit was satisfactory. Intermittent ultrasound doubles the reaction rate constant, and full ultrasound further increases the rate by a factor of 1.5. The obtained activation energy between 60 °C and 80 °C was 20 kJ/mol, which corresponds to a diffusion-controlled process for dissolution of cobalt. The lowest ΔG^0 corresponds to a low S/L ratio and the full use of ultrasound in the process.

Author Contributions: F.K. and S.S. conceptualized and managed the research. S.S. co-wrote the paper. S.G. contributed the SEM and EDS analysis of the PCD surface. B.F. supervised personnel, coordinated resources, and co-wrote the paper. F.K. performed the experiments and wrote the paper. All authors have read and agreed to the published version of the manuscript.

Funding: This research was funded by Projektträger (PtJ) Jülich, Grant Number 005-1902-0147.

Acknowledgments: We would like to thank to Redies Deutschland GmbH & Co. KG for providing PCD samples as well as additional equipment.

Conflicts of Interest: The authors declare no conflict of interest.

References

1. Dekempeneer, E.; Jacobs, R.; Smeets, J.; Meneve, J.; Eersels, L.; Blanpain, B.; Roos, J.; Oostra, D.J. Rf plasma-assisted chemical vapour deposition of diamond-like carbon: Physical and mechanical properties. *Thin Solid Films* **1992**, *217*, 56–61. [CrossRef]
2. General Electric Research Laboratory. Man-made diamonds. *Nature* **1955**, *176*, 50–56.
3. Bovenkerk, H.P.; Bundy, F.P.; Hall, H.T.; Strong, H.M.; Wentorf, R.H., Jr. Preparation of Diamond. *Nature* **1959**, *184*, 1094–1098. [CrossRef]
4. Strong, H.M.; Chrenko, R.M. Diamond growth rates and physical properties of laboratory-made diamond. *J. Phys. Chem.* **1971**, *75*, 838–1843. [CrossRef]
5. Lavoisier, A.L. Elements of Chemistry, in a New Systematic Order. Containing All the Modern Discoveries. In *Dover Histories and Classics of Science*; Dover Publications: New York, NY, USA, 1965.
6. Massucci, M.; Clegg, S.L.; Brimblecombe, P. Equilibrium Partial Pressures, Thermodynamic Properties of Aqueous and Solid Phases, and Cl_2 Production from Aqueous HCl and HNO3 and Their Mixtures. *J. Phys. Chem. A* **1999**, *21*, 4209–4226. [CrossRef]
7. Baghalha, M.; Khosravian, G.H.; Mortaheb, H.R. Kinetics of platinum extraction from spent reforming catalysts in aqua-regia solutions. *Hydrometallurgy* **2009**, *95*, 247–253. [CrossRef]
8. Huang, J.H.; Kargl-Simard, C.; Oliazadeh, M.; Alfantazi, A.M. pH-Controlled precipitation of cobalt and molybdenum from industrial waste effluents of a cobalt electrodeposition process. *Hydrometallurgy* **2004**, *75*, 77–90. [CrossRef]
9. Moskalyk, R.R.; Alfantazi, A.M. Nickel laterite processing and electrowinning practice. *Miner. Eng.* **2002**, *15*, 593–605.
10. Olanipekun, E.O. Kinetics of leaching laterite. *Int. J. Miner. Process.* **2000**, *60*, 9–14. [CrossRef]
11. De Graaf, J.E. The treatment of lateritic nickel ores—A further study of the caron process and other possible improvements. Part, I. Effect of reduction conditions. *Hydrometallurgy* **1979**, *5*, 47–65. [CrossRef]
12. Chang, J.; Zhang, E.-D.; Zhang, L.-B.; Peng, J.-H.; Zhou, J.-W.; Srinivasakannan, C.; Yang, C.-J. A comparison of ultrasound-augmented and conventional leaching of silver from sintering dust using acidic thiourea. *Ultrason. Sonochem.* **2017**, *34*, 222–231. [CrossRef] [PubMed]
13. Wen, C.Y. Noncatalytic heterogeneous solid-fluid reaction models. *Ind. Eng. Chem.* **1968**, *9*, 34–54. [CrossRef]
14. Khawam, A.; Flanagan, D.R. Solid state kinetic models- Basics and Mathematical Fundamentals. *J. Phys. Chem. B* **2006**, *110*, 17315–17328. [CrossRef] [PubMed]
15. Wedler, G. *Lehrbuch der Physikalischen Chemie*; Wiley-VCH: Weinheim, Germany, 2010.
16. Han, K.N.; Meng, X. The Leaching Behavior of Nickel and Cobalt from Metals and Ores—A review. *Extr. Metall. Copp. Nickel Cobalt* **1993**, *1*, 709–733.
17. Kiessling, F.; Gürmen, S.; Stopic, S.; Friedrich, B. Advances in synthesis of metallic powders using ultrasound assisted leaching process from polycrystalline diamond blanks—Process design, (First part). *Metals* **2020**, in press.
18. Grénnman, H.; Salmi, T.Y.; Murzin, D. Solid-liquid reaction kinetics—Experimental aspects and model development. *Rev. Chem. Eng.* **2011**, *27*, 53–77. [CrossRef]
19. Han, K.N.; Lawson, F. Leaching behavior of cobalt in acid solutions. *J. Less Common. Metals* **1974**, *38*, 19–29. [CrossRef]

Article

Recovery of Diamond and Cobalt Powder from Polycrystalline Drawing Die Blanks via Ultrasound-Assisted Leaching Process—Part 1: Process Design and Efficiencies

Ferdinand Kießling [1], Srecko Stopic [2,*], Sebahattin Gürmen [3] and Bernd Friedrich [2]

[1] Redies Deutschland GmbH & Co.KG, Metzgerstr. 1, 52070 Aachen, Germany; fer87den@gmail.com
[2] IME Process Metallurgy and Metal Recycling, RWTH Aachen University, Intzestrasse 3, 52056 Aachen, Germany; bfriedrich@ime-aachen.de
[3] Metallurgical and Materials Engineering Department, Istanbul Technical University, Ayazaga Campus, Istanbul 34469, Turkey; gurmen@itu.edu.tr
*   Correspondence: sstopic@metallurgie.rwth-aachen.de; Tel.: +49-24-1809-5860

Received: 30 April 2020; Accepted: 27 May 2020; Published: 1 June 2020

Abstract: The treatment of industrial polycrystalline diamond (PCD) blanks in aqua regia at atmospheric pressure between 333 K and 353 K was performed via the ultrasound-assisted leaching process to investigate whether the influence of ultrasound is beneficial. Cobalt content in the solution and in the blanks was monitored as well as the effects of leaching temperature, solid-to-liquid ratio, and PCD blank size. The use of intermittent and permanent ultrasound helped reduce the leaching time and thus energy consumption by up to 50%. In all trials with ultrasound, higher temperature only has a slight effect. Solid-to-liquid ratio does not have a positive or negative impact. A new process design was tested using an innovative experimental setup for ultrasound-assisted leaching aiming at maximum cobalt and diamond recovery from PCD and final reuse of fine PCD for cutting and polishing other hard materials in different important industrial applications.

Keywords: polycrystalline diamond; leaching; cobalt; ultrasound

---

1. Introduction

Cobalt is a ferromagnetic transition metal which is located between iron and nickel in the periodic table of elements and mostly available in lateritic ores [1–3]. Cobalt is used as a solvent catalyst in the production of polycrystalline diamond (PCD) that would otherwise take even more pressure and a higher temperature to achieve. [4]. The wide use of cobalt relative to other metals is associated with the high solubility of carbon in its melt during thermobaric treatment [5]. Unfortunately, at temperatures above 800 °C, which are developed during operation of a polycrystalline diamond tool, the cobalt promotes the formation of microcracks and results in significant heat-resistance reduction and subsequent reduction in the abrasion resistance of the polycrystalline diamond [6]. Thus, the reduction of the cobalt content in the sintered polycrystalline diamond (PCD) greatly improves the performance of a superhard composite. Finally, the removal of cobalt from the used PCD is the most important request in the industry of wire-drawing tools made from diamond materials and tungsten carbide.

Therefore, recycling is a chosen strategy for cobalt recovery in contrast to traditional primary metallurgy [7]. Diamond, to this date, is the hardest material that is put to use on commercial and industrial scales. It appears as the working edges of cutting tools or the grains in the hardest abrasives, among other uses. Due to advancements in the field of high strength steels and super alloys, the demand for hard cutting and forming tools will be increased in the future [8]. Industrially used diamonds can be found in the form of a naturally grown and mined crystals or as a man-made products with mono- or poly-crystalline microstructures. Die blanks are usually made at high temperatures and high pressure processes which require cobalt (Co) as the solvent catalyst [9]. Cobalt is incorporated in the final product resulting in a multiphase compound (PCD). This multiphase characteristic renders the PCD vulnerable to thermal stress since cobalt and diamond have different thermal expansion coefficients. The only way to make these PCDs stable for industrial applications at high temperatures, such as hot forming of metals, is to remove the inclusions from the cavities in the framework of diamond grains. Because of its strong oxidizing properties, aqua regia, or more precisely the forming nitrosyl chloride (NOCl), was used for the leaching of cobalt. This research also acknowledges the importance of using cobalt responsibly, possibly in a closed-loop recycling, not only because of price and toxicity but rather socioeconomic problems and difficulties linked to the conflict mineral columbite–tantalite, often referred to as 'coltan' [10,11].

Since the middle of the twentieth century, there have also been investigations into whether ultrasound can increase the chemical turnover in leaching processes. Many researchers have found that the reaction rate as well as the overall leaching efficiency can be increased significantly [12–14]. This positive effect is attributed to cavitation and specifically the Kelvin impulse [15–17]. Ultrasound is capable of creating such dynamic pressure changes in liquids that they will pass the phase boundary to their gaseous phase. When located near a solid surface, these cavities collapse and create an energetic jet pointed at the surface, because there is less instreaming liquid from the direction of the surface and overall impulse has to be conserved. These microjets have the ability to pierce through diffusion layers, effectively constantly renewing them. The positive influence of ultrasound was confirmed using ultrasound for the synthesis of nanosized particles by ultrasonic spray pyrolysis (USP). Due to its easy feasibility, flexibility and cost-efficiency, the USP method is an important alternative to the chemical vapor deposition (CVD) and other synthesis methods. Cobalt nanoparticles were successfully prepared from cobalt nitrate solution formed after an acidic treatment of the cemented tungsten carbide using the ultrasonic spray pyrolysis method [18,19]. An increase of ultrasound from 0.8 to 2.5 MHz decreases an aerosol diameter of cobalt nitrate to 2.2 μm that leads to the formation of submicron cobalt particles after drying and precipitation above 500 °C in a furnace using a hydrogen reduction atmosphere.

Shortening the process time for the cobalt removal from PCD in the presence of ultrasound is the main motivation for this study. In solid–liquid reactions, there is always an obstacle in the form of a diffusion layer where the mass transfer is inhibited because adsorption reaction and desorption take additional time and energy; even more so if a phase transition is involved. What's more, the component's concentrations can be quite different from the bulk solution, hindering further reaction. One way to mitigate this effect is to mechanically agitate the solution and decrease the thickness of this layer. For this reason, a stirrer was used to create sufficient turbulence in the reaction vessel.

An increase in temperature has an effect on most chemical reactions as it measures a substance's inner energy that is potentially available for reactions, changes the substance's activity or simply aids the mass transfer processes. In the leaching process, it has been found that increased temperature can accelerate the reaction speed [20–22]. The dissolution process with an acid needs a high activation energy to begin this solid–liquid reaction. Since this study is carried out at ambient pressure with the leaching solution described, care has to be taken because increasing the temperature also increases vapor pressures of the liquids involved [23].

Finally, the main aim of this work is optimizing leaching of cobalt from polycrystalline diamond blanks with grain size 5 μm in the presence of ultrasound. The influence of temperature and ultrasound on the leaching of cobalt will be studied in one ultrasound leaching assisted process. This study will take a look at the effects of ultrasound on the leaching efficiency but also at the penetration depth into the PCD, proposing a new experimental setup. Being capable of making any statements about the metallic constituents of PCD without breaking the blanks is an advantage and the reason why this method has been implemented. A challenge of this work is to maximize efficiency of cobalt leaching and diamond recovery in a shorter time than traditional hydrometallurgical methods.

2. Experimental

2.1. Material

As shown in Figure 1, this study is centered on the polycrystalline diamond (PCD) blanks made by Redies GmbH & Co. KG, Aachen, Germany, a manufacturer of wire drawing dies.

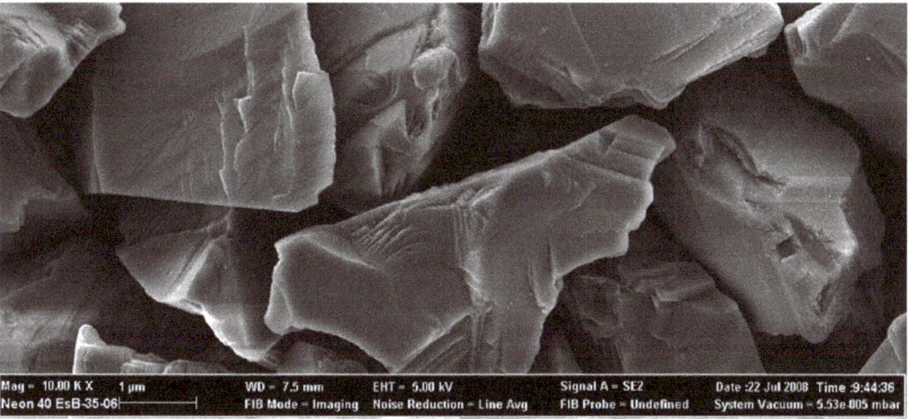

Figure 1. Scanning electron microscopy (SEM) image of raw diamond powder 5 μm class before the high temperature and high pressure (HTHP) process.

The SEM (Scanning electron microscopy) and EDS (Energive Dispersive Spectroscopy) analysis of the PCD surface after being polished is shown in Figure 2 and Table 1.

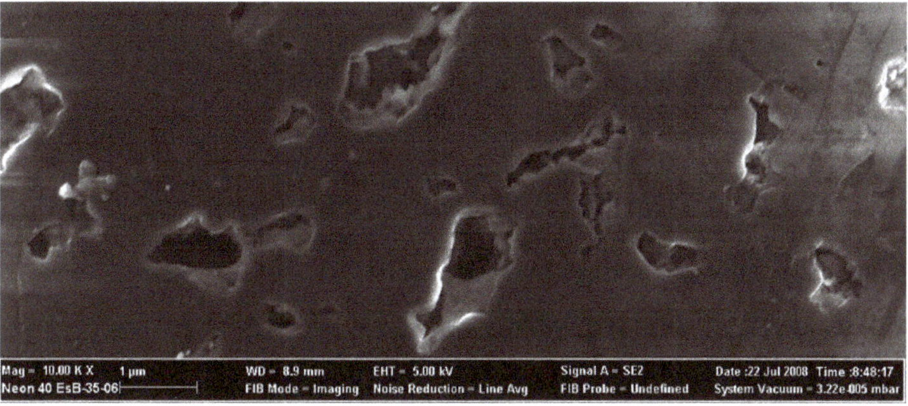

Figure 2. SEM image of ground and polished PCD surface, 5 μm class (dark grey areas are the bridged diamond grains with cavities where also traces of cobalt show up in lighter shades).

In contrast to the cobalt content of 0.1 wt.% in nickel lateritic ore, its average content in PCD is about 5–20 wt.%, which makes it a very promising material for the recycling of cobalt. The maximal value of cobalt in analyzed sample amounts is 1.67 wt.%, as shown in Table 1.

Table 1. Energy dispersive X-ray spectroscopy image of ground and polished PCD surface, also 5 µm class.

Content (Weight %)	C	O	Fe	Co
Spectrum 1	86.81	12.94	-	0.25
Spectrum 2	88.31	10.42	0.21	1.05
Spectrum 3	89.35	10.32	-	0.34
Spectrum 4	87.41	10.78	0.14	1.67
Max.	89.35	12.94	0.21	1.67
Min.	86.81	10.32	0.00	0.25

Different types of samples were used in our work as shown in Table 2. The columns "volume", "surface area", and "surface-to-volume ratio" in Table 2 do not contain measured but calculated values. The values for volume should be seen as the apparent outer volume of a porous body, not solid volume. The same applies to surface area. The measured values in the columns "diameter" and 'height' were obtained by taking a sample of thirty PCDs of the same type and averaging the values. The value for weight was obtained by weighing 100 PCDs and dividing the measurement by 100. This approach was chosen since the blanks did not have critical deviations dimension-wise. Assuming a homogeneous density, this average is sufficiently accurate. The same applies to weighing the batches as a whole after treatment. Figure 3 shows the deviations in the weights of individual PCDs. All the weight measurements were taken into a stock plot to depict actual variations and uncertainties. The deviations of surface-to-volume ratio were obtained by relating the largest surface to the smallest volume and vice versa.

Table 2. Dimensions of PCD samples with grain size of 5 µm.

Blank Type	Symbol	Diameter [mm]	Height [mm]	Weight [g]	Volume [mm^3]	Surface Area [mm^2]	Surface/Volume [mm^{-1}]
Mant$^®$ MSD-06-005	DO6	2.97 ± 0.01	1.10 ± 0.02	0.029	7.62 ± 0.18	24.09 ± 0.28	3.16 ± 0.11
Mant$^®$ MSD-14-005	D14	4.05 ± 0.09	2.00 ± 0.04	0.099	25.71 ± 1.40	51.14 ± 1.97	1.99 (+0.20|−0.18)
Mant$^®$ MSD-15-005	D15	5.2	2.5	0.241	53.093	83.315	1.569
Mant$^®$ MSD-18-005	D18	5.22 ± 0.02	3.50 ± 0.02	0.299	74.90 ± 0.77	100.22 ± 0.68	1.338 ± 0.023

2.2. Procedure

The experiments were carried out in two glass reactors simultaneously set up in a fume cabinet, using argon gas (Linde Gas AG, Höllriegelskreuth, Germany) a flow meter, type Rota Yokogawa (Yokogawa Deutschland GmbH, Ratingen, Germany), and a bottle with sodium hydroxide (Merck KGaA, Darmstadt, Germany), as shown at Figure 4. The reactor vessels were three-necked round bottom flasks with a capacity of 500 mL, the necks with standard ground joints 29/32 served as couplings for a stirrer seal, two gas hose couplers, and as access points for sampling, respectively, as shown at Figure 5. For sampling, the gas inlet coupler had to be removed temporarily. The stirrer unit consisted of a motor unit with a drill chuck, type "IKA Eurostar digital" (IKA$^®$-Werke GmbH & Co.KG, Staufen, Germany) IKA$^®$-Werke GmbH & Co. KGIKA$^®$-Werke GmbH & Co. KGIKA and a Polytetrafluorethylen (PTFE) coated impeller including a PTFE stirrer seal, as shown at Figure 4. The standard ground joints on the sides were sealed with a high-viscosity, silicone-based lubricant.

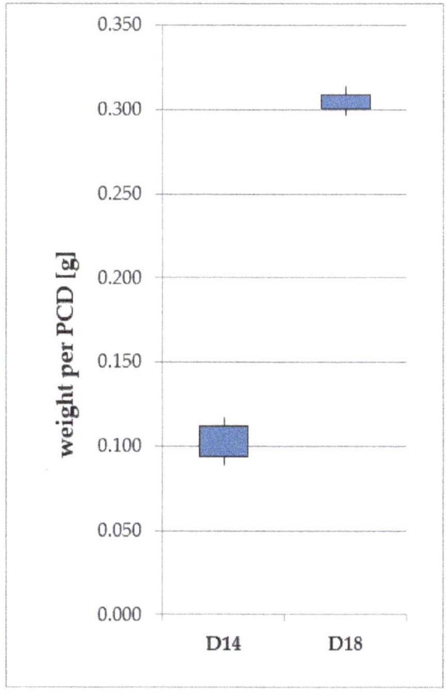

Figure 3. Deviations in weight of individual D14 and D18 PCD blanks.

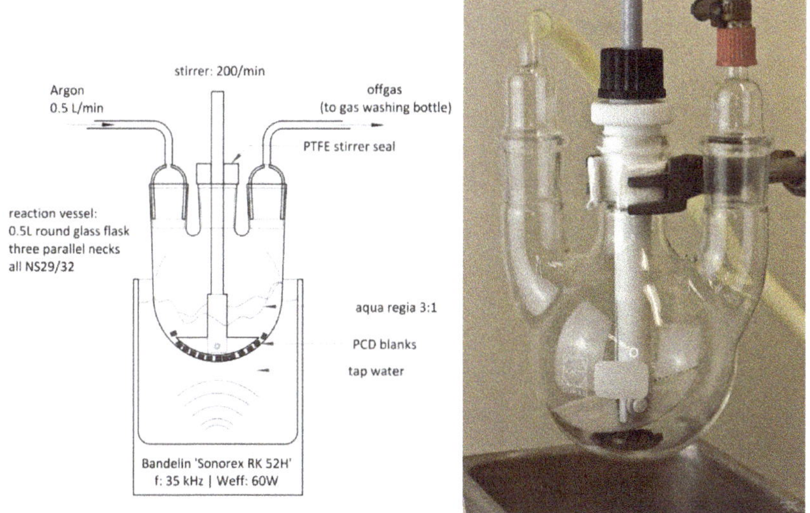

Figure 4. The description and picture of the leaching reactor.

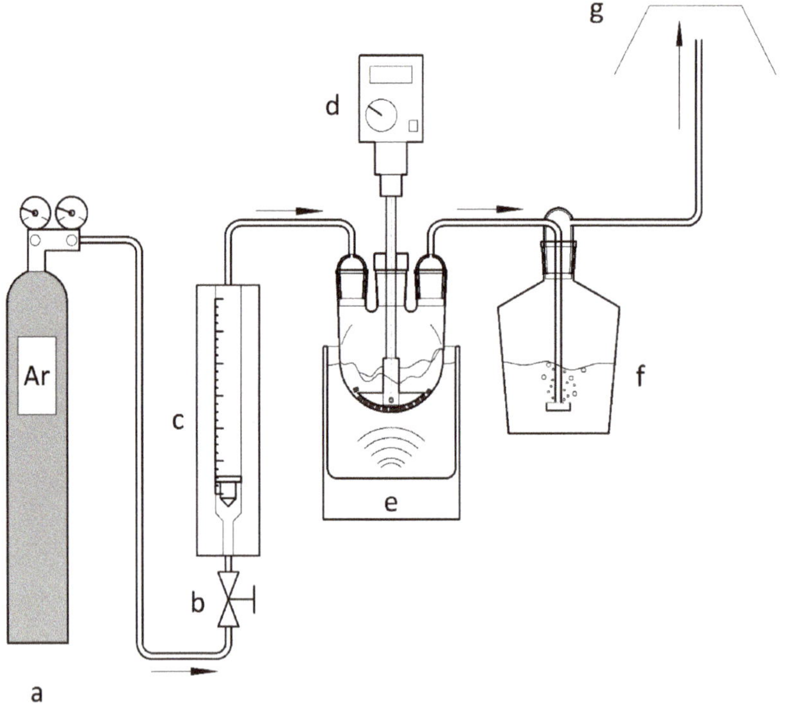

Figure 5. Innovative experimental set-up for leaching of PCD samples where: a—argon, b—valve; c—flow meter; d—mixer, e—reactor, f—bottle with dissolved sodium hydroxide, e—exhaust system.

The depth of immersion was chosen so that the surface level of the stirred liquid was as high as the water level in the heated ultrasound bath. The means of hindering evaporation can also be seen in this image. First, styrofoam beads were added to minimize the surface area available for evaporation, and secondly an acrylic lid made of two parts and an improvised cable tie hinge was used. For additional safety, each connection was clamped. For the purpose of temperature value measurement a "Testo 720" digital thermometer (testo SE & Co. KGaA, Lenzkirch, Germany) with "PT100" thermocouple (Temperatur Messelemente TMH, Hettstedt GmbH, Maintal, Germany) was used.

Below the aforementioned setup ultrasonic baths, "Bandelin Sonorex RK 52H" (BANDELIN electronic GmbH &Co. KG, Berlin, Germany) types were placed on lab jacks so they could be lowered for sampling and batch changes. This arrangement made disassembly easier and did not require readjusting the upper structure with every batch change. These ultrasound baths have a nominal frequency of 35 kHz and put out 60 W effectively, while output peaks can occur up to 240 W. The output level was fixed and ultrasound irradiation was altered by using it intermittently. In this case, they were filled with tap water to maximum capacity. The water volume was about 1.1 L due to the volume displaced by the reaction vessels.

As shown in Table 3, the parameters for the leaching experiments were proposed using our previous hydrometallurgical experience and previously performed experiments, reported in the literature [20].

Table 3. Parameters for the leaching of cobalt from polycrystalline diamond blanks.

ID	T_{Bath} [K]	PCD Type	Leaching Time [h/d] in the Presence of Ultrasound	S/L [g/L]
W1R1	333	D14	0	15
W1R2	353	D14	0	15
W2R1	333	D14 + D18	0	30
W2R2	353	D14 + D18	0	30
W3R1	333	D14 + D18	0	45
W3R2	353	D14 + D18	0	45
W4R1	333	D14 + D18	8 h/d	15
W4R2	353	D14 + D18	8 h/d	15
W5R1	333	D14 + D18	8 h/d	30
W5R2	353	D14 + D18	8 h/d	30
W6R1	333	D15 + D18	8 h/d	45
W6R2	353	D15 + D18	8 h/d	45
W7R1	333	D14 + D18	24 h/d	15
W7R2	353	D14 + D18	24 h/d	15
W8R1	333	D14 + D18	24 h/d	30
W8R2	353	D14 + D18	24 h/d	30
W9R1	333	D14 + D18	24 h/d	45
W9R2	353	D14 + D18	24 h/d	45

The abbreviations in the first column are read as successive week number and R1 and R2 representing reactors 1 and 2, respectively. These codes also served as stems for sample identification. The column header S/L is short for solid-to-liquid ratio in units of grams per liter. Constant parameters were stirring speed and batch time. The duration of each batch was planned to be between 90 and 100 h. Sample names D06, D14, D15 and D18 are abbreviated product names of Mant® MSD-06-005, MSD-14-005, MSD-15-005, and MSD-18-005, all self-supported PCD blanks with diamond grain sizes of 5 µm.

2.2.1. Preparation of Samples

The PCD blanks were weighed and measured with the teslameter (Projekt Elektronik GmbH, Berlin, Germany), beforehand to obtain the important '100%' reference value for evaluation. The aqua regia was mixed from three parts fuming hydrochloric acid 37% Emsure ACS/ISO quality (Merck KGaA, Darmstadt, Germany) and one part nitric acid 65% ISO analysis quality PanReac ApplicChem (Chicago, IL, USA). 240 mL of hydrochloric acid and 80 mL nitric acid were prepared in covered beakers for each reactor. Meanwhile, the ultrasound baths were filled with tap water and their heaters were set to 333 K and 353 K, respectively, to ensure that the bath temperature was nominal from the beginning of the experiment. Gas tightness of the apparatus was checked daily.

2.2.2. Conduct of Experiments

In weeks 1 through 3, the ultrasound baths were only used as heated water baths. In weeks 4–6 the ultrasound was intended to be switched on for eight hours per day. In addition to refilling water, the time of ultrasound irradiation had to be noted. In the remaining three weeks the ultrasound was switched on permanently. Concerning the forced gas flow, a current of around 0.25 L/min was sufficient to ensure the flow in one direction only. Argon was chosen over nitrogen or pressured air because oxygen and nitrogen might have skewed the equilibria with NO_2, N_xO_x or formed combustible mixtures with chlorine gas or hydrogen gas.

2.2.3. Sampling

Throughout the experiments, only cobalt content in the solution and changes in magnetic properties of the PCD were sampled each day. First, a few milliliters of the solution were pumped from each reactor using a plastic syringe and PTFE tube and transferred into a small beaker. From there,

1 mL of solution was taken with a pipette and added to a 50 mL round glass flask which was then filled with deionized water up to the 50 mL mark, resulting in a 1 in 50 dilution. This sample solution was transferred again into a 50 mL sample vial, and analyzed by the chemistry department at the IME, RWTH Aachen University (Aachen, Germany) by inductively coupled plasma–optical emission spectrometry (ICP-OES) (SPECTRO ARCOS, SPECTRO Analytical Instruments GmbH, Kleve, Germany). The solid sample was analyzed by X-ray fluorescence (Axios FAST, Malvern Panalytical GmbH, Germany).

The first indication that $CoCl_2$ was formed could be seen when taking samples from the solution. Especially towards Thursday and Friday of any experimental week, the solution taken from the reactor had a dark greenish teal color that changed to pink after a few seconds in the beaker, indicating the typical drying salt color change from the dihydrate ($CoCl_2 \cdot 2H_2O$) to the hexahydrate ($CoCl_2 \cdot 6H_2O$) form of $CoCl_2$ when cooling below approximately 308 K. Liquid samples were analyzed with the method of inductively coupled plasma optical emission spectrometry (ICP-OES).

2.2.4. Weighing of Sample

For the purpose of determining the mass balances, PCDs were weighed before and after each batch on the same scale in the laboratory, type LA620P (Sartorius AG, Göttingen, Germany). Before the experiments, the PCDs were weighed as they were delivered. After an experiment they were wet with aqua regia, so they had to be rinsed with distilled water at least twice. In between stages, the PCDs were left in fresh distilled water for about 15 min. After the last rinse they were shaken with a little ethanol to assist in the drying process. Then, after two days at 353 K in the laboratory dryer. They were weighed while still warm. This was to ensure that no humidity would skew the results of weighing. According to Redies GmbH & Co. KG (Aachen, Germany), Aachen, a loss of around 20 weight percent due to extraction of cobalt from PCD is to be expected.

2.2.5. Observation of Changes in Magnetic Properties of PCD

Changes in magnetic properties were observed using a teslameter, type FM 205 (Projekt Elektronik GmbH, Berlin, Germany), with a reference neodymium magnet. The apparatus for measuring of magnetic properties is shown in Figure 6.

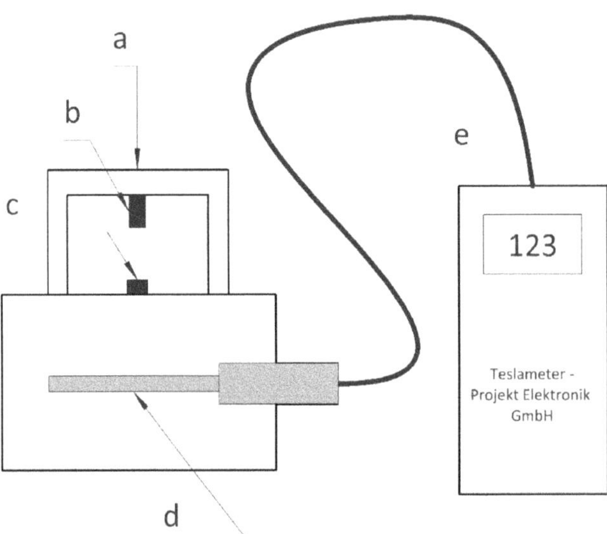

Figure 6. Measuring magnetic properties of a PCD. a: Magnet support, b: ∅ 5 mm by 8 mm Nd alloy magnet, c: PCD blank, d: probe, e: handheld teslameter.

It consists of a teslametric probe that is kept at a fixed distance from the reference magnet, such as Nd alloy. The probe will display a value at any time representing the current magnetic situation. To diminish disrupting effects from the surroundings, the probe was offset by the nearby reference magnet. The actual measurement was always a sum of the magnetic surroundings, reference magnet and the subject in between. With only air and plastic between magnet and probe, the reading on the display was 583. A measurement was taken before every run with readings varying above 600. The value displayed for air as a subject is taken as 0% as a reference value. The value for the unleached PCD marks the 100% value for each batch and reactor, respectively. This method allows for a normalization of values and a plot of inferred Co content in the PCD relative to its initial content. In the course of the experiment, this value dropped towards the value of normal air, enabling an estimate of the progress of relative cobalt content in the PCD.

Since all measurements were always compared against the offset value and normalized to the pre-experiment value being 100%, units cancel out. It has to be noted that this offset value was registered before each individual PCD measurement because it changed between 583 and 584—perhaps due to other magnetic influences, or the magnet may have been placed in a way that resulted in a measurement on the threshold between 583 and 584. To ensure consistency, the magnet was not moved until after the experiments. An important caveat is the fact that the relative Co content is inferred, not measured. The change in the magnetic field of the probe consists of more than just the effect due to the presence of cobalt. In this case, analyses have shown that there are oxygen and iron impurities present in the raw PCD. Fe(II), Fe(III), Co(II), and Co(III) oxides have magnetic properties that naturally differ from pure Co. The overall effect of magnetic metals is measured and the Co content is concluded from these values.

3. Results and Discussion

3.1. Mass Balance

The measured concentration of cobalt in the samples was plotted against the time when the samples were taken to visualize the accumulation in solution, as shown in Figure 7 (from 1 and 9 week).

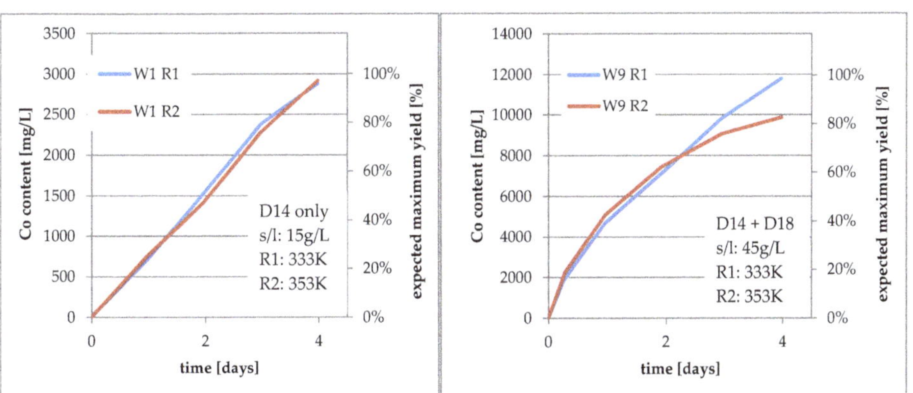

Figure 7. Cobalt concentration versus time in weeks 1 and 9.

These plots were chosen as the extreme cases, with the other results very similar and within their range. Without ultrasound and at the low end of solid-to-liquid ratios, the concentration of cobalt developed very similarly regardless of temperature, as shown at Figure 8. However, towards the end of week nine, at a high solid-to-liquid ratio and with full ultrasound, a difference emerged where the Co content in reactor 2 seemed to go into saturation. It appeared from week 4 onwards, so it may be linked to the higher temperature and use of ultrasound.

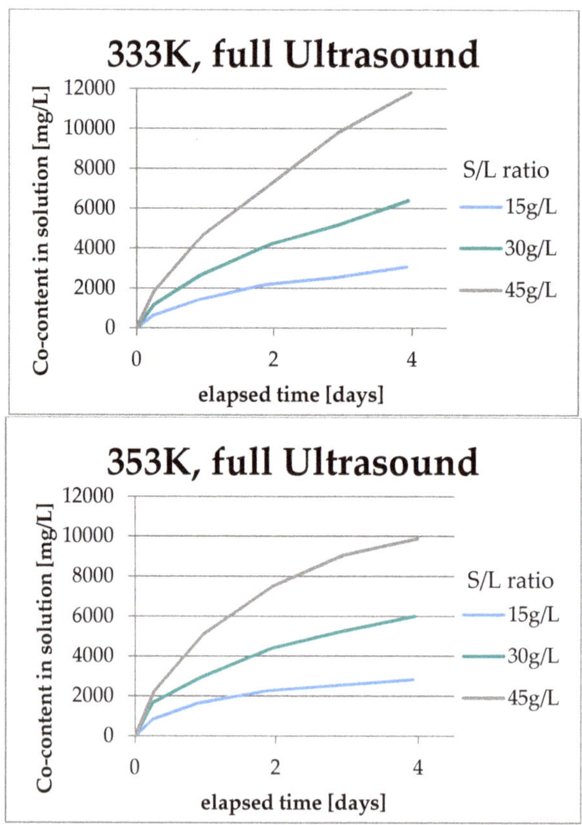

Figure 8. Concentration of cobalt in solution in time between 333 K and 353 K with full-time ultrasound.

When comparing the two reactors independently at Figure 8, this discrepancy becomes more obvious. With ultrasound used full-time, the solution at 353 K accumulated several percent less cobalt in total than the cooler reactor. However, the initial increase happened faster. To see the effect of ultrasound itself, the concentration curves from low and high solid-to-liquid ratios were plotted for both reactors, as shown at Figure 9. Interestingly, full-time ultrasound did not seem to achieve the highest cobalt yields. At high solid-to-liquid ratios, it did not even seem to do any better than the experiments without ultrasound.

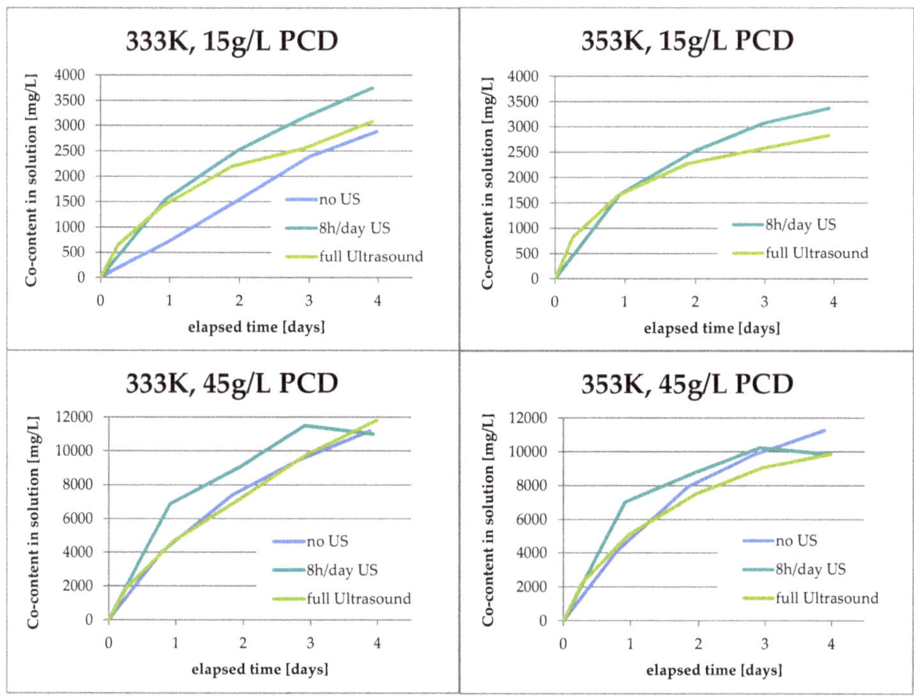

Figure 9. Influence of ultrasound on concentration of cobalt in the solution over time.

As stated above, the PCDs were weighed to gain insight into whether the chosen parameters, especially ultrasound, aided the leaching process. These measurements were used to compare the results of all runs, shown individually for D14 and D18 blanks in Figure 10. At first glance, the most obvious fact is that the larger PCDs were not leached to completion within the 90 to 100 h timeframe. The maximum values for lost PCD weight were 21.04% and 13.86% for D14 and D18, respectively. For D14, the most influential parameter appears to be ultrasound. There is a strong influence of bath temperature without it. However, already intermittent ultrasound is enough to drive the leaching efficiency towards the expected value. In the runs with ultrasound, higher temperature only has a slight effect. Solid-to-liquid ratio does not have a positive or negative impact. For D18, the influence of ultrasound is measurable, but is not as strong as it is for the smaller D14. The difference between intermittent and permanent ultrasound irradiation seems negligible, but higher solid-to-liquid ratios have an adverse effect in the runs with intermittent ultrasound. With permanent ultrasound, this inhibition is apparently gone. Higher bath temperature, on the other hand, influenced leaching in a positive way. The secondary axis "expected leaching efficiency" has to be viewed with caution. The 100% mark refers to the expected 20% weight percent of leachable substance in the PCD. In fact, this value has only been orally confirmed by the manufacturer and the only analysis is the surface SEM and EDS, as shown at Figures 1 and 2. The leaching efficiency was calculated using Equation (1). The obtained results with D14 show that this value is reasonably accurate.

$$\text{Leaching efficiency [\%]} = (\Delta m_{PCD} \text{ or } c(Co)_{SOL})/(0.2(m_{0,PCD})) \tag{1}$$

where Δm_{PCD} is the lost PCD weight after the experiment, $c(Co)_{SOL}$ is the cobalt content in solution after the experiment, 0.2 is the given factor of initial cobalt content in PCD and $m_{0,PCD}$ is the initial PCD weight.

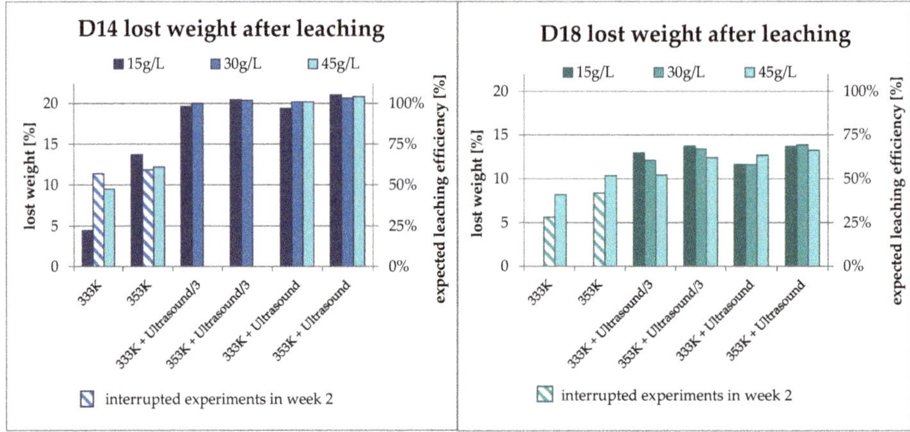

Figure 10. Lost weight for different PCD samples after leaching with different parameters described in the plot. 35 kHz ultrasound varied from "none", "ultrasound on" for a third of the run time, to "full-time ultrasound", as shown from left to right.

Weight data was used to calculate the average PCD weight lost per day, as shown at Figures 10 and 11. The difference between initial weights and final dry weights was divided by the actual batch time in hours. This was done to renormalize the values to a certain time interval because the batches had different run times. Though differences in the diagrams above may only be small, the diagrams below are truly adequate for a comparison. The negative effect of the solid-to-liquid ratio on leaching efficiency of cobalt from D18 blanks with intermittent ultrasound became clearer, as shown at Figure 11. A less obvious difference is the fact that the highest columns in the diagrams below now truly are the runs with the most extracted weight percentage per time.

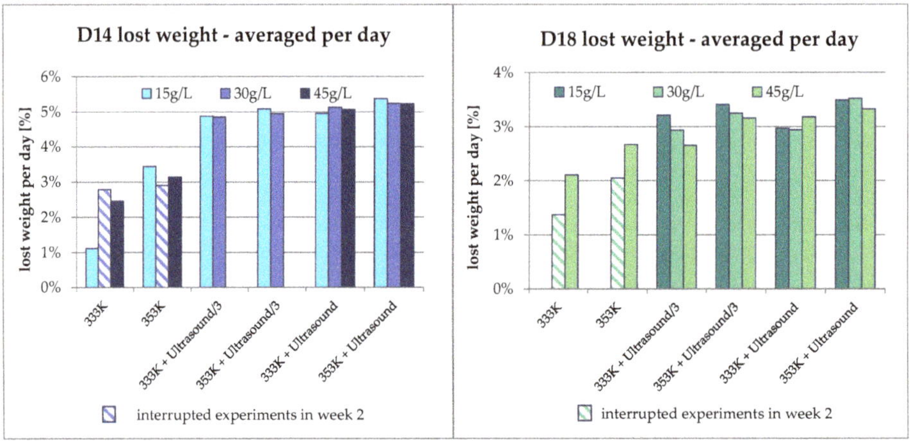

Figure 11. Lost weight for different PCD samples averaged per day. 35 kHz ultrasound varied from "none", "ultrasound on" for a third of the run time, to "full-time ultrasound", as shown from left to right.

In weeks 5, 8, and 9 the volume of remaining liquid was measured. The weight of dissolved cobalt was calculated by combining these volumes with the Co (II) concentrations from chemical analysis. A comparison between the weight of dissolved Co (II) versus lost PCD weight during leaching is presented in Figure 12. If pure metallic cobalt was used and remained in its metallic state during the making of the PCD, there should be no difference between these values. All the material that is leached from the PCD should be cobalt and end up in the solution as dissolved ions.

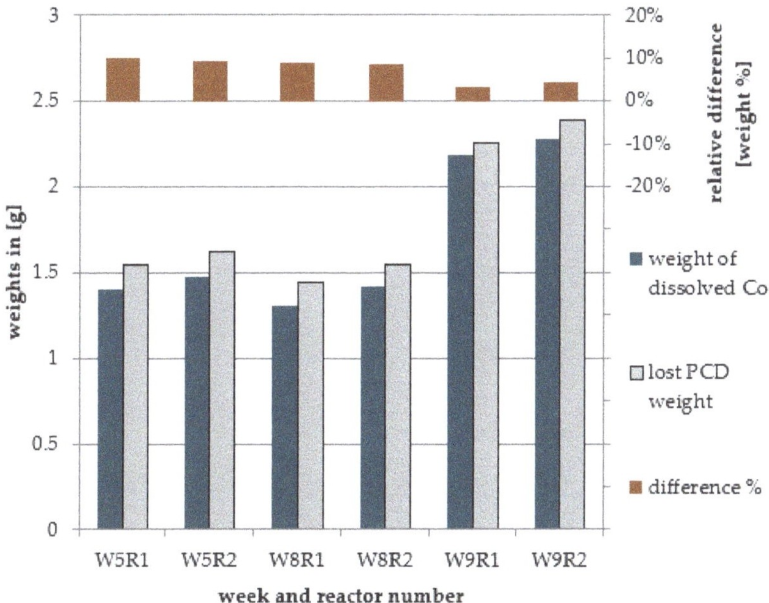

Figure 12. Weight of dissolved Co (II) versus lost PCD weight.

As can be clearly seen, there is a significant difference, ranging between +3% and +10%, among the compared values. On account of the EDS data (Table 1), this difference probably stems from cobalt oxides and iron impurities rather than fluctuations or measuring errors and uncertainties. Finally, the obtained solution using an aqua regia (Figure 13—left) from polycrystalline drawing die blanks via the ultrasound-assisted leaching process is shown in Figure 13—right. Cobalt powder was obtained from this solution using the precipitation method.

(left)   **(right)**

Figure 13. (**Left**): Aqua regia approximately 9 min after mixing and stirring. (**Right**): View of the reactor during the experiment, containing the PCD and cobalt bearing solution. In this case, it was running at 333 K, with full-time ultrasound and 45 g/L solid-to-liquid ratio.

3.2. Results Regarding Process Optimization

The results from this study suggest that the leaching of D14 does not require a 353 K bath temperature but can be done at 333 K. Ultrasound can accelerate the leaching process to the extent that the PCD can reach a desaturated state with less than 10% of metallic inclusions remaining after three to four days if they are leached at low solid-to-liquid ratios close to 15 g/L. The depth of ultrasound penetration into PCD, meaning the depth at which the disruptive effects of ultrasound are no longer able to outrun diffusion, was determined to be in the region of 1.4–1.7mm.

If the emphasis were put on just shortening the leaching time, one way could be to leach at 353 K, replacing the solution after three days to reset the concentration gradient and refresh the active compounds in solution. As there is more research needed regarding the kinetics and mechanism of the dissolution process, the same applies to process safety and possible replacement of aqua regia with other less harmful leaching agents. The potentiometric aspects of leaching of cobalt from PCD also deserve considerable attention in order to ensure controlled potential cobalt leaching as a selective way for total cobalt removal in a short time. The kinetics and mechanisms of the studied ultrasound-assisted leaching process from polycrystalline diamond blanks will be reported in Part 2 [24] of this research in detail.

3.3. Possible Recycling Routes for Cosolution in Order to Produce Cobalt Powder and Its Compounds

After the experiments, a highly acidic aqua regia solution laden with divalent cobalt remains. This cobalt content is very valuable and should not be discarded. One simple (but hardly elegant) way to reuse the cobalt chloride would be the complete evaporation of liquids using the remaining cobalt chloride as drying salt. Instead, there are ways to selectively extract Co from the solution with DEHPA2, for example, and then precipitating or electrolytic winning of the metal powder. As stated earlier, there also is the possibility to make cobalt nanopowder using the ultrasonic spray pyrolysis method and the chemical reduction method in the aqueous solution, which would be a very versatile substance to be used in battery technology as well as catalyst applications. The production of cobalt hydroxide shall be reached using sodium hydroxide as precipitation agent. A goal-oriented refining process such as solvent extraction as a traditional hydrometallurgical method could be imagined depending on the desired metal powder and its compounds.

4. Conclusions

This study was designed for the recovery of pure demetallized PCD and cobalt from raw PCD. In nine experimental runs with a 5 day duration, cobalt containing PCD was leached in aqua regia at atmospheric pressure between 333 K and 353 K. Using two reactors in parallel, temperature, ultrasound irradiation time, solid-to-liquid ratio, and PCD size were varied to find out which parameters are beneficial and could possibly accelerate this process. PCD weights and cobalt content in solution were also monitored. It was found that aqua regia accumulated more dissolved cobalt at 333 K than at 353 K probably due to volatile reagents being less available over time. The ultrasound treatment increases the leaching efficiency. With added ultrasound (even at just a third of total run time) and at a low S/L ratios close to 15 g/L, the leaching time for D14 to reach the 90% leached mark was reduced to three days, which is a significant shortening of leaching time. PCD type D18 with a thickness of 3.5 mm was not leached to completion within five days. The leaching temperature had more impact on the results than ultrasound. These findings were reinforced by the mass balance in which a small discrepancy was found. The PCD lost a fraction of weight that could not be explained by the weight of dissolved cobalt. From EDS data and the nature of PCD, this fraction probably consisted of oxygen from oxides in the PCD or single diamond grains that were broken off by the impact of ultrasound. Advances in synthesis of metallic powders using the ultrasound-assisted leaching process from polycrystalline diamond blanks can be used for the cemented tungsten carbide in order to estimate a scale-up of this process in future.

Recovery of Diamond and Cobalt Powders from Polycrystalline Drawing Die Scraps via Ultrasound-Assisted Leaching Process—Part 2: Kinetics and Mechanisms

The kinetic models were used for the study of cobalt dissolution from polycrystalline diamond blanks via a measurement of declining ferromagnetic properties over time. For a better understanding of this leaching process, thermochemical aspects were included in this work. The lowest free Gibbs energy corresponds to a low solid/liquid ratio and fully used ultrasound in the process. A transition from a reaction-controlled to a diffusion-controlled shrinking core model was found for PCD with a thickness larger than 2.8–3.4 mm. Intermittent ultrasound doubles the reaction rate constant and fully using of ultrasound causes a further increase with a factor of 1.5. The obtained activation energy between 333 K and 353 K is 20 kJ/mol, and small for all diamond blanks with a diameter size of 5 μm, which corresponds to the diffusion-controlled process.

Author Contributions: F.K. and S.S. conceptualized and managed the research. S.S. cowrote the paper. S.G. contributed the SEM and EDS analysis of the PCD surface. B.F. supervised personnel, coordinated resources, and co-wrote the paper. F.K. performed the experiments and wrote the paper. All authors have read and agreed to the published version of the manuscript.

Funding: This research was funded by Projektträger Jülich (PtJ), Grant Number 005-1902-0147.

Acknowledgments: We would like to thank Redies Deutschland GmbH & Co. KG (Aachen, Germany) for providing PCD samples as well as additional equipment.

Conflicts of Interest: The authors declare no conflict of interest.

References

1. Moskalyk, R.R.; Alfantazi, A.M. Nickel laterite processing and electrowinning practice. *Miner. Eng.* **2002**, *15*, 593–605. [CrossRef]
2. Olanipekun, E.O. Kinetics of leaching laterite. *Int. J. Miner. Process.* **2000**, *60*, 9–14. [CrossRef]
3. Narasimhan, K.S.; Bhima Rao, R.; Das, B. Technical note—Characterisation and concentration of laterites. *Miner. Eng.* **1989**, *2*, 425–429. [CrossRef]
4. Anokhin, A.S.; Strel'nikova, S.S.; Andrianov, M.A.; Tkachenko, V.V.; Shipkov, A.N.; Kukueva, E.V.; Gol'dt, A.E.; Zaremba, O.T. Formation of the Structure and Phase Composition of Polycrystalline Diamond with Cobalt Infiltration in the System Diamond—Hard Alloy. *Glass Ceram.* **2019**, *75*, 475–478. [CrossRef]
5. Wentorf, R.H.; DeVries, R.C.; Bundy, F.P. Sintered superhard materials. *Science* **1980**, *208*, 873–880. [CrossRef] [PubMed]
6. Scott, T.A. The influence of microstructure on the mechanical properties of polycrystalline diamond: A literature review. *Adv. Appl. Ceram.* **2018**, *117*, 161–176. [CrossRef]
7. Wang, S. Cobalt—Its recovery, recycling, and application. *Metals* **2006**, *58*, 47–50. [CrossRef]
8. General Electric Research Laboratory. Man-made diamonds. *Nature* **1955**, *176*, 50–56.
9. Rao, A.S.; Minango, R.; Nkhoma, J.; Singh, H.P. Process developments in the cobalt purification circuit at Chambishi RLE Cobalt Plant of ZCCM, Zambia. In *Queneau International Symposium Extractive Metallurgy of Copper, Nickel and Cobalt: Fundamental Aspects*; Paul, E., Ed.; The Minerals, Metals & Materials Society: Warrendale, PA, USA, 1993; Volume 1, p. 853.
10. Wakenge, C.I.; Dijkzeul, D.; Vlassenroot, K. Regulating the old game of smuggling? Coltan mining, trade and reforms in the Democratic Republic of the Congo. *J. Mod. Afr. Stud.* **2018**, *56*, 497–522. [CrossRef]
11. Slaczka, A.S. Effect of ultrasound on ammonium leaching of zinc from galmei ore. *Ultrasonics* **1986**, *24*, 53–55. [CrossRef]
12. Oncel, M.S.; Ince, M.; Bayramoglu, M. Leaching of silver from solid waste using ultrasound assisted thiourea method. *Ultrason. Sonochem.* **2005**, *12*, 237–242. [CrossRef] [PubMed]
13. Chang, J.; Zhang, E.-D.; Zhang, L.-B.; Peng, J.-H.; Zhou, J.-W.; Srinivasakannan, C.; Yang, C.-J. A comparison of ultrasound-augmented and conventional leaching of silver from sintering dust using acidic thiourea. *Ultrason. Sonochem.* **2017**, *34*, 222–231. [CrossRef] [PubMed]
14. Blake, J.R.; Gibson, D.C. Cavitation Bubbles near Boundaries. *Ann. Rev. Fluid Mech.* **1987**, *19*, 99–123. [CrossRef]

15. Herbert, E.; Balibar, S.; Caupin, F. Cavitation pressure in water. *Phys. Rev.* **2006**, *74*, 041603. [CrossRef] [PubMed]

16. OMFRS, L.R. On the pressure developed in a liquid during the collapse of a spherical cavity. *Philos. Mag.* **2009**, *34*, 94–98.

17. Hixson, A.W.; Baum, S.J. Mass Transfer and Chemical Reaction in Liquid-solid Agitation. *Ind. Eng. Chem.* **1944**, *36*, 528–531. [CrossRef]

18. Gürmen, S.; Stopić, S.; Friedrich, B. Synthesis of nanosized spherical cobalt powder by ultrasonic spray pyrolysis method. *Mater. Res. Bull.* **2006**, *41*, 1882–1890. [CrossRef]

19. Gürmen, S.; Güven, A.; Ebin, B.; Stopic, S.; Friedrich, B. Synthesis of nanocrystalline spherical cobalt-iron (Co-Fe) alloy particles by ultrasonic spray pyrolysis and hydrogen reduction. *J. Alloys Compd.* **2009**, *481*, 600–604.

20. Baghalha, M.; Gh, H.K.; Mortaheb, H.R. Kinetics of platinum extraction from spent reforming catalysts in aqua-regia solutions. *Hydrometallurgy* **2009**, *95*, 247–253. [CrossRef]

21. Gogoleva, E.M. The leaching kinetics of brannerite ore in sulfate solutions with iron (III). *J Radioanal. Nucl. Chem.* **2012**, *293*, 185–191. [CrossRef]

22. Wen, C.Y. Noncatalytic heterogeneous solid-fluid reaction models. *Ind. Eng. Chem.* **1968**, *60*, 34–54. [CrossRef]

23. Massucci, M.; Clegg, S.L.; Brimblecombe, P. Equilibrium Partial Pressures, Thermodynamic Properties of Aqueous and Solid Phases, and Cl_2 Production from Aqueous HCl and HNO_3 and Their Mixtures. *J. Phys. Chem. A* **1999**, *103*, 4209–4226. [CrossRef]

24. Kiessling, F.; Gürmen, S.; Stopic, S.; Friedrich, B. Advances in Synthesis of Metallic Powders Using Ultrasound Assisted Leaching Process from Polycrystalline Diamond Blanks—Thermochemistry and Kinetics of Ultrasound Leached Process—(Second Part). *Metals.* in press.

Article

Advances in Thermochemical Synthesis and Characterization of the Prepared Copper/Alumina Nanocomposites

Marija Korać [1], Željko Kamberović [1], Zoran Anđić [2] and Srećko Stopić [3,*]

1   Department of Metallurgical Engineering, Faculty of Technology and Metallurgy, University of Belgrade, Karnegijeva 4, 11120 Belgrade, Serbia; marijakorac@tmf.bg.ac.rs (M.K.); kamber@tmf.bg.ac.rs (Ž.K.)
2   Innovation center of Faculty of Chemistry Ltd., University of Belgrade, 12-16 Studentski trg, 11000 Belgrade, Serbia; zoranandjic@yahoo.com
3   IME Process Metallurgy and Metal Recycling, RWTH Aachen University, Intzestrasse 3, 52056 Aachen, Germany
*   Correspondence: sstopic@ime-aachen.de; Tel.: +49-241-80-95-860

Received: 30 April 2020; Accepted: 26 May 2020; Published: 28 May 2020

Abstract: This paper presents thermochemical synthesis of copper/alumina nanocomposites in a Cu-Al$_2$O$_3$ system with 1–2.5 wt.% of alumina and their characterization, which included: transmission electron microscopy: focused ion beam (FIB), analytical electron microscopy (AEM) and high resolution transmission electron microscopy (HRTEM). Thermodynamic analysis was used to study the formation mechanism of desirable products during drying, thermal decomposition and reduction processes. Upon synthesis of powders, samples were cold pressed (2 GPa) in tools dimension $8 \times 32 \times 2$ mm and sintered at temperatures within the range 800–1000 °C for 15 to 120 min in a hydrogen atmosphere. Results of characterization showed that dispersion-strengthened compacts could be produced by sintering of thermo-chemically prepared Cu-Al$_2$O$_3$ powders with properties suitable for material application, such as a contact material exhibiting high strength and high electrical conductivity at the same time. Additional research was carried out in order to analyze the application of the obtained nanocomposite powders for the synthesis of copper/alumina nanocomposites by a new method, which is a combination of a thermochemical procedure and mechanical alloying. The measured values of an electric conductivity and hardness were compared with ones in literature, confirming an advantage of the proposed combined strategy.

Keywords: synthesis; oxide; nanocomposites; characterization; copper; alumina; thermochemistry

---

1. Introduction

Research of metal matrix composite (MMC) materials has considerably intensified since its first mention in the 1950s [1]. Various combinations of base (Al, Cu, Ni, Mg, Ti, Fe, Co, etc.) and reinforcing material (e.g., oxides, borides, carbides, fibers, tubes) have been studied [2]. Through selection of appropriate combinations and ratios of materials, a wide spectrum of properties can be achieved, followed by extensive industrial applications.

Copper is considered the most significant base material for industrial applications, due to its good electrical and heat conductivity. Disadvantages of copper are its mechanical properties, such as high ductility, low wear resistance and thermodynamic instability at elevated temperatures. One of the possibilities for overcoming poor mechanical properties is reinforcement by dispersion strengthening; i.e., the introduction of fine ceramic particles. By the dispersion strengthening of copper, significant increases in mechanical properties can be achieved, with low adverse impacts on its electrical and heat conductivity. The main requirements for dispersed particles are higher thermodynamic stability at elevated temperatures; higher hardness, strength and wear resistance; and low solubility in base metal. Appropriate size and even distribution of dispersed particles also contribute to a positive effect via dispersion strengthening [3]. Finer particles with homogenous distribution and low volume fraction of dispersed particles in the total volume of the base metal will act as obstacles to dislocation motion, even at elevated temperatures without significant effects on conductivities, both thermal and electrical [4–7].

One of the most widely used oxides is alumina, which fulfills all the requirements for the dispersed particles and has low cost at the same time [8]. Additionally, alumina can increase the temperature of recrystallization of the copper matrix and demonstrates excellent strength at elevated temperature by pinning grain and sub-grain boundaries of the matrix. Finally, alumina particles add to strengthening by blocking the movement of dislocations [7,9,10]. The usual amount of alumina used for dispersion strengthening is 0.5–5.0 wt.% [11], but significant results regarding particle size can be achieved even with higher amounts, such as 50 wt.% of Al_2O_3 [12].

There are numerous routes for the synthesis of metal–matrix composites, but nowadays two main routes are the mechanical alloying and thermochemical route. Mechanical alloying is extensively used method for synthesizing of nanocrystalline materials by severe plastic deformation on the powder using high-energy ball milling technique [13–16]. This technique commonly employed for the prevention of formation of clusters and agglomerates enables the production of uniformly dispersed fine particles in a metal–matrix. On the other hand, using the thermochemical method [17] where input materials are in a liquid state enables production of finer particles and much more homogeneous structure of the final powder, which further contributes to the increase in the mechanical properties for the final product through various strengthening mechanisms.

Authors have also developed a new synthesis route based on the combination of routes mentioned above [18]. This route may be regarded as a new strategy for materials in the Cu-Al_2O_3 system, even though some phases of this process have been previously investigated by the authors [19–21]. Additionally, previous attempts have been made by authors for application of similar process in the system Cu-Ag-Al_2O_3, where a three-component system was produced by mechanically alloying the thermo-chemically-synthesized Cu-Al_2O_3 and Cu-Ag powder [22].

The main aim of this work was to investigate a thermochemical synthesis of metallic particles and nanocomposite with a microstructure and strengthening mechanism of copper with finely dispersed alumina particles. A novelty of this synthesis is a decreased reduction temperature for chemical reaction of the powder in a hydrogen atmosphere at 350 °C, which is an advantage in contrast to 820 °C for 1 h to produce the final Cu–Al_2O_3 nanocomposite powder, as described by Seyedraoufi et al. [23]. It can be very important point for decreasing production costs. A thermodynamic analysis of the reduction, spray drying and synthesis reactions was performed in order to predict a chemical behavior of the compounds. Amirjan et al. [24] have used artificial neural networks to predict Cu-Al_2O_3 properties. In order to prepare copper based composites, copper powder with four different amounts of Al_2O_3 reinforcement (1, 1.5, 2, 2.5 wt%) were mechanically alloyed, and the consolidated compacts of prepared powders were sintered in five different temperatures of 725–925 °C at seven several sintering times of 15–180 min. Guevara et al [25] have studied the synthesis of copper-alumina composites by mechanical milling via an analysis of materials and manufacturing processes. Ha et al. [26] studied the fabrication of Al_2O_3 dispersion strengthened copper alloy by spray in-situ synthesis casting process above 1250 °C as a new method. Mohammadi, E. et al. [27] used a combustion method for the synthesis of Cu-Al_2O_3, which take place in a short time at temperatures higher than 1000 °C. Generally, our synthesis method offers a cost-friendly process for the synthesis of Cu-Al_2O_3 in comparison to other processes [27].

These powders could be used for production of sintered materials with properties suitable for material applications, such as contact material exhibiting high strength and high electrical conductivity at the same time.

Some comparative results for different synthesis methods are presented, indicating that by mechanical alloying of atomized copper powders with produced composites, followed by thermo-mechanical treatment, sintered materials with improved properties could be produced.

2. Experimental

Water soluble copper and aluminum nitrates, $Cu(NO_3)_2 \cdot 3H_2O$ and $Al(NO_3)_3 \cdot 9H_2O$, were used to synthesize a two-component nanocomposite Cu-Al_2O_3 powder by the thermochemical procedure. The synthesis was carried out through four stages, as presented in Figure 1.

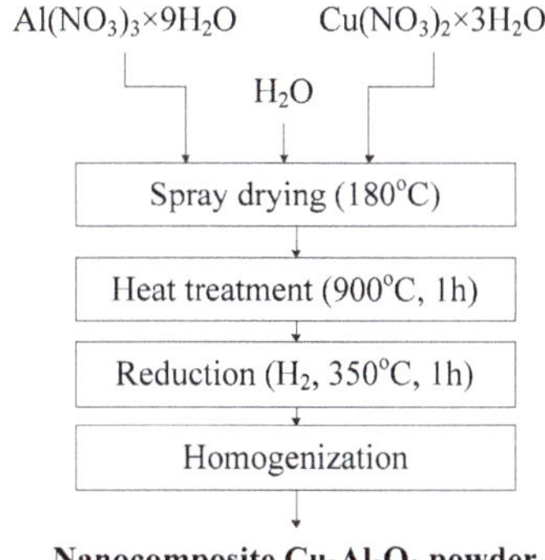

Figure 1. Flowsheet of the synthesis of Cu-Al_2O_3 nanocomposite powder by the thermochemical procedure [28].

Process temperatures are derived from thermodynamic consideration of the process and the following six chemical reactions:

Spray drying:

$$Cu(NO_3)_2 \cdot 6H_2O = Cu(NO_3)_2 + 6H_2O \tag{1}$$

$$Al(NO_3)_3 \cdot 6H_2O = Al(NO_3)_3 + 6H_2O \tag{2}$$

Heat treatment:

$$Cu(NO_3)_2 = CuO + N_2O_5 \tag{3}$$

$$2Al(NO_3)_3 = Al_2O_3 + 3\,N_2O_5 \tag{4}$$

Reduction:

$$CuO + H_2 = Cu + H_2O \tag{5}$$

$$Al_2O_3 + 3H_2 = 2Al + 3H_2O \tag{6}$$

Using HSC Chemistry® software package 6.12 (Outotec, Espoo, Finland), chemical and thermodynamic parameters of the processes for synthesis Cu-Al_2O_3 composites were analyzed. As shown at Figure 2, the calculated values of Gibbs energy of reactions versus temperature (up 1000 °C) for reactions (1)–(6) have positive and negative values; negative values confirmed the possibility for the beginning of these chemical reactions at the studied temperature. Because of the high positive values of Gibbs energy (more than 800 kJ/mol), reduction of aluminum oxide with hydrogen (as shown with Equation (6)) is not possible between 25 and 1000 °C.

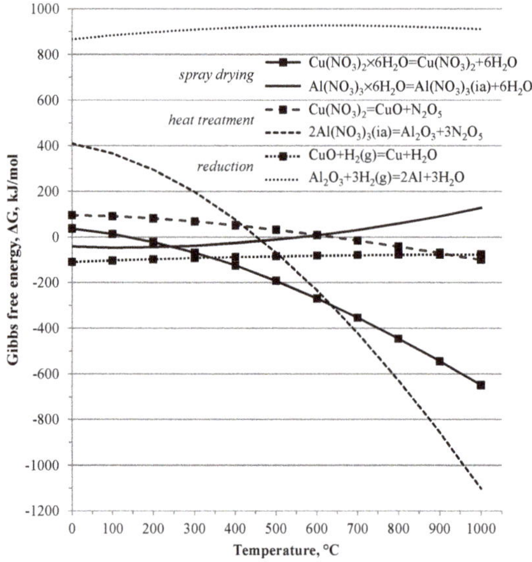

Figure 2. Gibbs energy of reactions versus temperature.

The first stage is the preparation of 50 wt.% aqueous solutions of $Cu(NO_3)_2 \cdot 3H_2O$ and $Al(NO_3)_3 \cdot 9H_2O$ (the quantities of salt were set so that the requested composition of a $Cu\text{-}Al_2O_3$ nanocomposite system with 5 wt.% of alumina could be produced). The second phase is spray drying of nitrate solution using Mini Spray Dryer B-290 Advance (BÜCHI Labortechnik GmbH, Essen, Germany) for producing the precursor powder, with inlet/outlet temperature 190/143 °C and a solution flow rate of 10% pump power. The third stage is oxidative calcination of the precursor powder in an air atmosphere at 900 °C for 1 h to form copper oxide and the phase transformation of Al_2O_3 up to the thermodynamically stable $\alpha\text{-}Al_2O_3$ phase. Final fourth stage was the reduction of thermally treated powders in hydrogen atmosphere flow rate 20 L/h at 350 °C for one hour, where copper oxide was transformed into elementary copper, while Al_2O_3 remained unchanged.

All temperatures were below the temperatures the melting temperature of Cu (1085 °C) [29] and Al_2O_3 (2072 °C) [30].

In previous work of authors procedures [22,29] and process parameters [21,31] are fully described for the synthesis of two-component nanostructured composite materials.

The obtained powders were cold-pressed (force 500 kN, calculated pressure 2 GPa) in tools with dimensions of $8 \times 32 \times 2$ mm and sintered at temperatures within 800–1000 °C for 15 to 120 min in a hydrogen atmosphere. Kinetics of the sintering process were determined and presented elsewhere [21].

Characterization of compacted powders after sintering at 900 °C for 2h included transmission electron microscopy: analytical electron microscopy (AEM), high resolution transmission electron microscopy (HRTEM) and focused ion beam (FIB) at e-beam 5.00 kV. The powder to be tested is suspended in a liquid (water, ethanol or butanol) with the aid of an ultrasonic device. Depending on particle size, requirements and type of examination, the powder is first ground. By means of a pipette a drop of suspension is taken up and placed on a carbon carrier net. The liquid is then allowed to evaporate (dry) under a lamp, resulting in a C-carrier net with the powder on top. After this powder preparation, our sample was studied by TEM Analysis. HRTEM analysis was performed using Philips CM200/FEG (FEI Company, Hillsboro, OR, USA).

Mechanical properties of sintered samples were also investigated and are presented in authors' previous research [28].

Ames Portable Hardness Tester was employed for hardness measurements using a 1/16″ ball with an applied load of 60 kg. For electrical conductivity measurement, SIGMATEST 2.069 (FOERSTER, Pittsburgh, PA, USA) operating at 120 kHz and with an 8 mm electrode diameter was used.

Values of hardness and electrical conductivity represent the mean values of at least six measurements conducted on the same composite.

3. Results and Discussion

In previous work of the authors [22], the results of determination of fluidness, pouring density and specific area of the obtained nanostructured composites with different amounts of Al_2O_3 dispersed in the copper matrix showed that all the investigated powders are not fluid and that mean values of pouring density and specific area are the same for different contents of Al_2O_3 up to the 5% investigated.

Additionally, in some previous studies of the authors [21,22,32] the results of differential thermal and thermogravimetric analysis (DTA-TGA) and scanning electron microscopy can be found, which show the flow of phase transformations during the process of oxidation, and the morphologies of the obtained powders.

Only peaks corresponding to the nitrates of copper and aluminum were identified in the structure during XRD examination of the precursor powder produced by spray drying an aqueous solution of copper and aluminum nitrates, which is in accordance with the experiment set-up [20–22]. X-ray diffraction analysis after annealing of dried powder exhibited peaks corresponding to CuO and Al_2O_3, and one unidentified peak. According to Lee [33] this peak corresponds to a third phase, $Cu_xAl_yO_z$, which appears in the structure due to the eutectic reaction of (Cu + Cu_2O) with Al_2O_3.

The produced powders were analyzed by AEM with corresponding EDX, as shown in the previous works of authors [34]. Based on AEM analysis, particles 20–50 nm in size, are clearly noticeable, as is the presence of agglomerates >100 nm. Particles are irregularly shaped; there are nodular individual particles with rough surface morphology. EDS analysis of marked spot show that the identified peaks correspond to Cu, Al and O. The intensities of peaks correspond to demanded compositions of the examined systems; therefore, the peak corresponding to copper is considerably higher than the peaks corresponding to aluminum and oxygen.

In order to identify the third phase, the authors performed additional research through the synthesis of Cu-50 wt.% by a thermochemical procedure. X-ray diffraction analysis of the obtained sample, presented in [18], shows the presence of copper peaks and $CuAl_2O_4$ compounds, which may represent a metastable phase that developed in the microstructure during the process of powder synthesis, thermal treatment and reduction on the surface of the contact between Cu and Al_2O_3 and is a seed for the development of the third phase during the sintering process.

FIB analysis of the sintered Cu-Al_2O_3 system based on the powders obtained by the thermochemical procedure, as shown in Figure 3, is characteristic for the final stage of sintering. FIB analysis did not indicate even at considerably higher magnifications, the existence of a phase rich with alumina. The bright fields are identified, i.e., a phase rich with copper, and gray fields, which can lead to a possible existence of the third $Cu_xAl_yO_z$ phase identified by X-ray diffraction analysis [18]. The formation of this phase is thermodynamically possible on Cu-Al contact surfaces. During eutectic joining of copper and Al_2O_3, the eutecticum formed by heating up to the eutectic temperature expands and reacts with Al_2O_3 creating $Cu_xAl_yO_z$, which is compatible with both phases on the inter-surface. According to [35,36], the process of formation of the third phase is developed through the following reactions: $2CuO + H_2 \rightarrow Cu_2O + H_2O$, $Cu_2O + Al_2O_3 \rightarrow 2CuAlO_2$ and/or $CuO + Al_2O_3 \rightarrow CuAl_2O_4$. $CuAlO_2$ is stable in air with the temperature range from 800 °C to 1000 °C, while $CuAl_2O_4$ is transformed into $CuAlO_2$ at the temperature of approximately 1000 °C. However, the presence of $Cu_xAl_yO_z$ phase demands a detailed characterization by using high resolution apparatus. Additionally, from micrographs of the examined samples, homogenous distribution of the present phase is clearly noticeable and the size of microstructural constituents in the range of 50–250 nm (Figure 3).

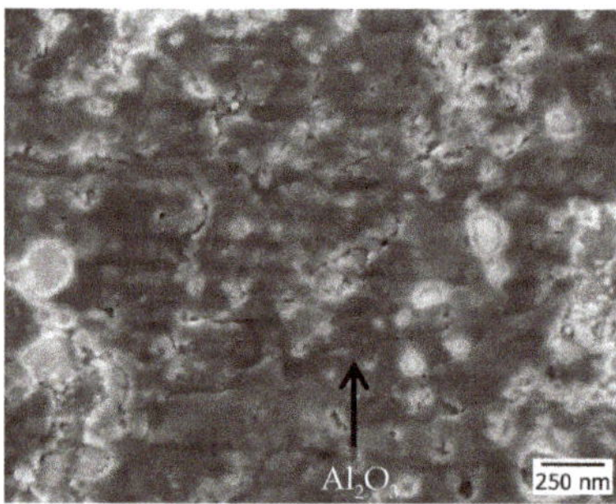

Figure 3. FIB image of compacted Cu-5wt.% Al$_2$O$_3$ composite sinter.

Additionally, the FIB analysis confirmed the analysis of structural stabilization of the system based on the values of the specific electric resistance of sintered samples ($\rho = 0.061 \times 10^{-6}$ Ω·m). Results show also that the hardness of sintered samples (HRB 10/40 (average = 124.7)) was very high for the achieved density of the sample yet lower than expected. The results of examining density, relative change of volume and the electrical and mechanical properties of sintered systems based on nanocomposite Cu-Al$_2$O$_3$ powders synthesized by the thermochemical process have been presented in previous papers by the authors [21,32].

Typical microstructure of Cu-Al$_2$O$_3$ 5 wt.% is presented in Figure 4. In BF (bright field)–DF (dark field) pair, it can be seen that a copper crystal exhibits annealing twins. Twins are slightly curved, a typical feature of deformation twinning, but in the presented case, it could be a consequence of a high temperature sintering stage.

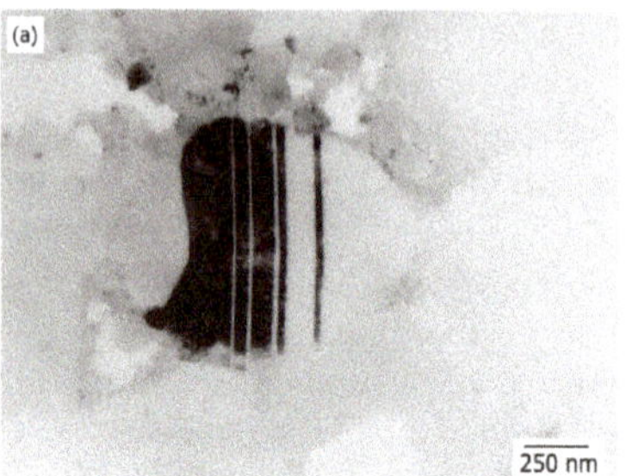

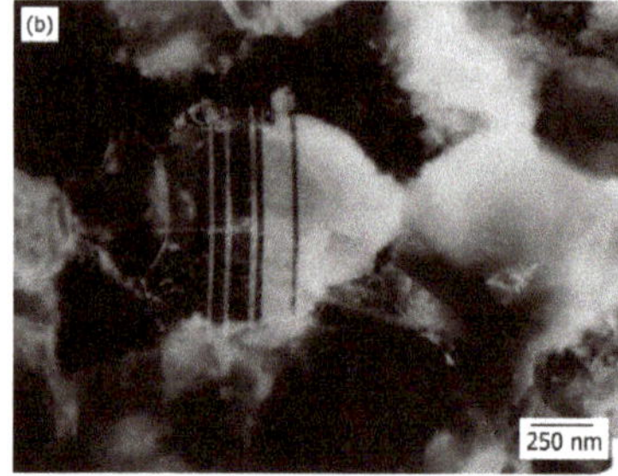

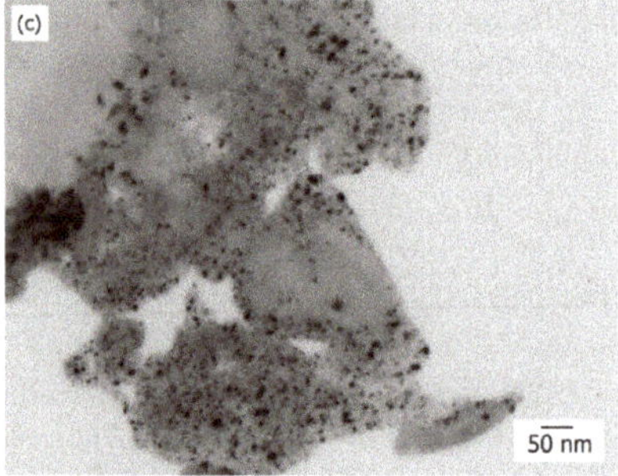

Figure 4. TEM analysis of sample after sintering: BF (bright field) and DF (dark field) images showing nano-twinning on Cu crystal (**a**,**b**), homogenous distribution of Al$_2$O$_3$ particles (**c**)

In Figure 4a,b, a typical TEM pair bright field (BF)-centered dark field (CDF) of nanocomposite Cu-5wt.% Al_2O_3 sintered system is shown, where the well-developed crystals of copper are exposed to twinning, despite their small size. In addition, detailed analysis indicates the detection of fine Al_2O_3 individual particles or aggregates. Conditions for twinning are accomplished when a great number of obstacles, such as homogeneously distributed Al_2O_3 particles, are created in the crystal which hamper dislocation mobility, dislocation plaits or already present twins. Since dislocations are piled up at the obstacles, in such local regions internal tension is increased, which, along with external tension, provokes creation of twins. Decreasing of dislocation mobility represents a condition for creating twin embryos; therefore, in Figure 4a,b, the clearly noticeable presence of twins indicates a decreased mobility of dislocations, i.e., stabilization of dislocation substructure, which is an elementary precondition for improving mechanical properties; i.e., reinforcing of metal materials.

Fine dark spots noticeable in the BF image (Figure 4c) present Al_2O_3 particles, size range 5–20 nm, dispersed in the copper matrix. Additionally, Figure 4c shows a homogenous distribution of Al_2O_3 particles, which is one of the requirements for dispersion of strengthened copper composites, to retain electrical conductivity of the base metal.

In a second set of TEM BF-DF pair images (Figure 5a,b) dispersion of alumina particles is also visible. Furthermore, TEM results in Figure 5a show the presence of dislocation density (the upper-right region of the grain) in a copper matrix surrounding the alumina particles, additionally increasing the strength of the material. Additionally, in Figure 5b, Moire fringes could be observed. According to [37] inside a single copper crystal, the clusters of Al_2O_3 particles could considerably alter the surrounding lattice structure, enough to prompt formation of Moire fringes.

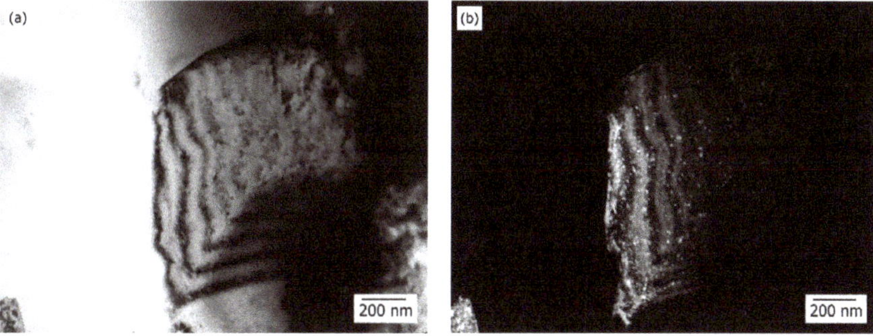

Figure 5. TEM analysis of a single copper grain containing a fine dispersion of alumina particles, dislocations and Moire fringes.

Figure 6a,b shows selected area diffraction patterns (SADPs), where both single spots and a Debye–Scherrer ring pattern can be observed. Single spots in SADP correspond to crystalline copper along the [111] axis, while the Debye–Scherrer rings in Figure 6b correspond to alumina Al_2O_3 particles.

In Figure 6a, besides the spots corresponding to copper, additional diffraction spots are visible, indicating presence of a solid solution. During sintering stage, formation of a third phase is possible, during eutectic reaction under suitable thermodynamic conditions at the Cu-Al_2O_3 interphase containing all three elements in a very narrow region. Existence of the third phase in the structure remains to be proven by further indexing and calculations. Composition of this phase could be, according to the literature, $CuAlO_2$ or $CuAl_2O_4$ [11,38] due to presence of an O-rich interface with larger adhesive energy [39]. Presence of this phase additionally reinforces the copper matrix by blocking the grain and sub-grain boundaries.

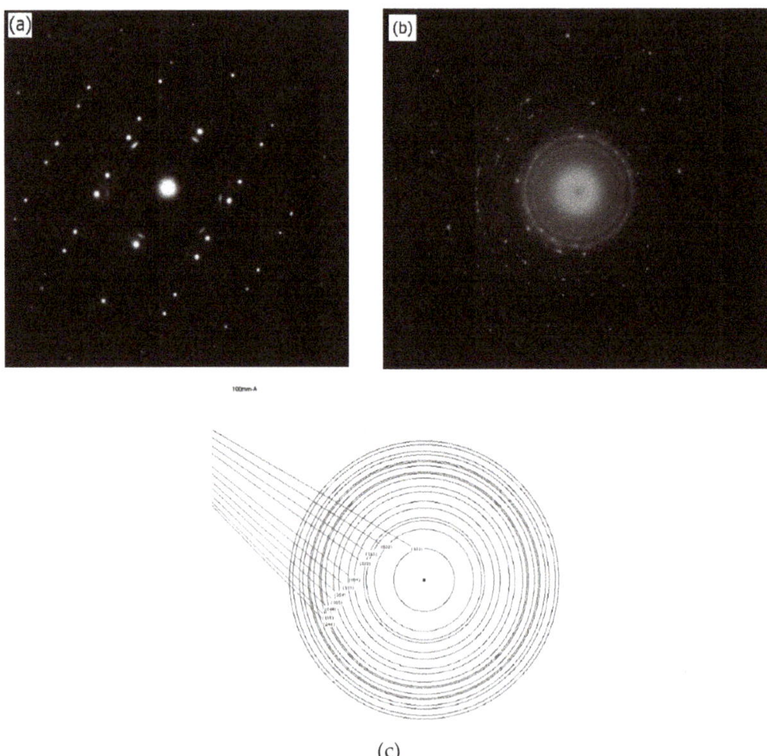

(c)

Figure 6. Selected area diffraction pattern (SADP) taken inside the grain in Figure 4: (**a**) single diffraction spots for copper along the [111] axis, (**b**) Debye–Scherrer rings for the pure Al_2O_3 particles and (**c**) inset showing the simulated Debye–Scherrer rings for $\gamma\text{-}Al_2O_3$.

Ratios of measured D-values from the ring diffraction pattern are in good agreement with the calculated ratios of the corresponding g-vectors for the γ-alumina (Figure 6c).

There is a subtle difference between the crystal structure of γ-alumina and μ-alumina, yet the performed characterization via TEM did not provided a definite proof of that. It is more likely that the structure is that of γ-alumina as the original structure is boehmite and its transformation sequence does not include μ-alumina in accordance with [40].

Successful application of synthesized of the nanocomposite $Cu\text{-}Al_2O_3$ powders obtained by the thermochemical procedure in mechanical alloying of atomized copper powders is in detail presented in [18]. Because of high strength and electrical properties, this material can be used as electrode material for lead wires, relay blades, different contact materials and various switches, and especially for electrode materials for spot welding due to high conductivity of copper and high hardness and excellent thermal stability of aluminum.

The obtained nanocomposite powders, with structure basically preserved with the final product, provided a significant reinforcement effect in the produced sintered system. This is a consequence of homogenous distribution of the elements in the structure, accomplished during synthesis of powder and presence of the third phase which causes stabilization of dislocation substructure, accomplishing a relevant reinforcing effect and achievement of a good combination of mechanical–electric properties of the sintered systems. Comparative analyses of mechanical properties of produced composites were derived from the previous work of the authors [18–22,28,34] regarding thermo-chemical synthesis, mechanical alloying and the new synthesis method. Sintering of copper/alumina nanocomposite powders was performed at 875 °C in 60 min in laboratory electro resistant furnace in a hydrogen atmosphere in order to avoid oxidation of samples. From the presented results in Table 1, it could be concluded that use of obtained powders for mechanical alloying followed by plastic deformation have the same level of hardness with a much lower amount of Al_2O_3, which has a direct consequence through higher values of electrical conductivity. The maximal values of electrical conductivity and hardness were obtained for the sample based on 1.0 wt.% percent of Al_2O_3 in structure. Regarding to the same chemical composition of copper/alumina nanocomposite in comparison to Amirjan [24], the values of electrical conductivity are higher in all cases, what confirms an advantage for our studied combined strategy.

Table 1. Comparison of the electrical conductivity and hardness values (HRF) after sintering at 875 °C and 60 min in our work with values of Amirjan [24].

Temperature (°C)	Time (min)	Electrical Conductivity ($MS \cdot m^{-1}$)	Electrical Conductivity, (% IACS *)		Rockvell Hardness Values (HRF)	
		1.0 wt.% Al_2O_3				
875	60	33.20	57.24	55.93 [24]	50	49 [24]
		1.5 wt.% Al_2O_3				
875	60	28.92	49.86	48.61 [24]	40	38 [24]
		2.0 wt.% Al_2O_3				
875	60	21.95	37.84	34.80 [24]	26	25 [24]

* IACS (International Annealed Copper Standard).

After annealing at 800 °C, hardness and electrical conductivity amounted to 58 HRF and 61.78% IACS, respectively.

The proposed strengthening mechanism is presented in Figure 7. The mechanism combines the strengthening in thermo-chemically synthesized composites and strengthening during mechanical alloying.

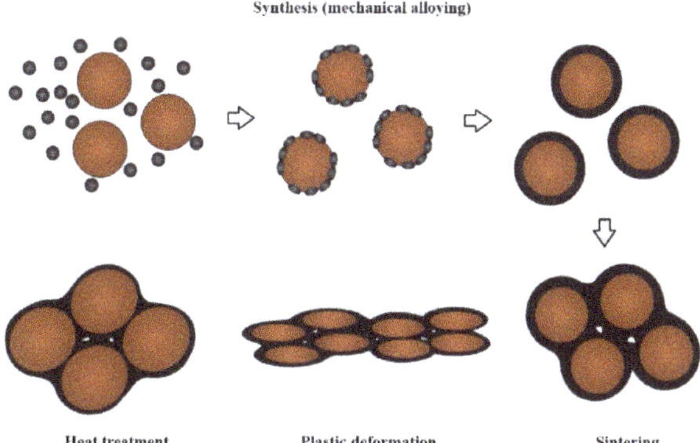

Figure 7. Microstructure transformation induced by following technological steps: mechanical alloying, heat treatment, plastic deformation and sintering.

Throughout thermo-chemical synthesis, copper base strengthening is achieved by dispersion of fine particles of Al_2O_3, and strengthening by grain boundaries, as presented in this paper. As reported by Amirjan [24] with respect to strengthening mechanism of Orawan, with increasing reinforcement amount, the distances between particles in the microstructure will decrease. Therefore, the dislocations can encompass the particles easily and lead to lower values of hardness. We assume the grain size of the composite matrixes microstructure increases with increasing sintering time. According to Hall–Petch effect, larger grain size in microstructure leads to a decrease in hardness values.

During mechanical alloying of atomized copper particles, copper-alumina composites are built into its surface. Along with the process of sintering occurs the formation of the compact structure and formation of the third phase on the grain boundary, which causes strengthening on the grain boundaries. Due to the plastic deformation, the deformation strengthening occurs, and after heat treatment, the strengthening by annealing occurs. The annealing treatment increases the system's strength by reducing dislocation emission sources and improves material ductility through strengthening grain boundaries' resistance to intergranular cracks.

4. Conclusions

Characterization of produced nanostructured composites in system Cu-Al$_2$O$_3$ showed the possibility of their synthesis via a thermochemical route. The mechanism of formation of copper-alumina nanocomposite was studied using thermodynamic analysis of drying, thermal decomposition, reduction step and homogenization. The reduction of thermally treated powders was performed in hydrogen with flow rate 20 L/h at 350 °C for one hour, where copper oxide was transformed into elementary copper, while Al$_2$O$_3$ remained unchanged.

By AEM analysis it was confirmed that homogenous distribution of Al$_2$O$_3$ particles was achieved by the thermochemical route followed by cold pressing and sintering, a necessary requirement for retaining electrical conductivity of the base metal.

Increasing of the strength of the material was achieved by presence of dislocation density in a copper matrix surrounding the alumina particles, confirmed by TEM analysis.

Additionally, the selected area diffraction pattern showed the possible presence of a third phase formed during the sintering stage at interphase containing all three elements in a very narrow region, which additionally reinforces the copper matrix by blocking the grain and sub-grain boundaries. The existence of the third phase in the structure remains to be proven by further indexing and calculations. The proposed strengthening mechanism combines the strengthening in thermo-chemically synthesized composites and strengthening during mechanical alloying. The maximal values of electrical conductivity and hardness were obtained for the sample based on 1.0 wt.% percent of Al$_2$O$_3$ in structure. Regarding the same chemical composition of copper/alumina nanocomposite in comparison to literature values by Amirjan [24], the values of electrical conductivity and hardness are higher in all cases, which confirms an advantage for our studied combined strategy.

The future study can be focused on the kinetics of the thermochemical synthesis of the studied nanocomposites. The economic size (cost effect) of this method shall be calculated via partial operations. This is a practical way for manufacturing these composites for powder metallurgy based on different applications.

Author Contributions: M.K. conceptualized and managed the research, and co-wrote the paper together with the other co-authors. Ž.K. ensured financial support, supervised M.K. and co-wrote the paper. Z.A. participated in analysis and discussion of the obtained results and co-wrote this paper. S.S. helped in discussion of the results and co-wrote this paper. All authors have read and agreed to the published version of the manuscript.

Funding: This research received no external funding.

Acknowledgments: Ministry of Education, Science and Technological development of Republic of Serbia through project TR34033 financially supported the research presented within this paper.

Conflicts of Interest: The authors declare no conflict of interest.

References

1. Poole, C.P.; Owens, F.J. *Introduction to Nanotechnology*; John Wiley & Sons: Hoboken, NJ, USA, 2003.
2. Eckert, J.; Reger-Leonhard, A.; Weiß, B.; Heilmaier, M. Nanostructured materials in multicomponent alloy systems. *Mater. Sci. Eng. A* **2001**, *301*, 1–11. [CrossRef]
3. Tian, B.H.; Liu, P.; Song, K.X.; Li, Y.; Liu, Y.; Ren, F.Z.; Su, J.H. Microstructure and properties at elevated temperature of a nano-Al$_2$O$_3$ particles dispersion-strengthened copper base composite. *Mater. Sci. Eng. A* **2006**, *435–436*, 705–710. [CrossRef]
4. Naser, J.; Ferkel, H.; Riehemann, W. Grain stabilisation of copper with nanoscaled Al$_2$O$_3$-powder. *Mater. Sci. Eng. A* **1997**, *234–236*, 470–473. [CrossRef]
5. Trojanová, Z.; Ferkel, H.; Luká, P.; Naser, J.; Riehemann, W. Thermal stability of copper reinforced by nanoscaled and microscaled alumina particles investigated by internal friction. *Scr. Mater.* **1990**, *40*, 1063–1069. [CrossRef]
6. Song, J.; Koch, V.; Wang, l.; Stopic, S.; Bogovic, J.; Friedrich, B.; Möbius, A.; Fuhrmann, A. Nanoscale Particles enhanced Gold Plating. *Adv. Mater. Res.* **2011**, *320*, 210–215. [CrossRef]
7. Casati, R.; Vedani, M. Metal Matrix Composites Reinforced by Nano-Particles—A Review. *Metals* **2014**, *4*, 65–83. [CrossRef]
8. Fathy, A.; Shehata, F.; Abdelhameed, M.; Elmahdy, M. Compressive and wear resistance of nanometric alumina reinforced copper matrix composites. *Mater. Des.* **2012**, *36*, 100–107. [CrossRef]
9. Plascencia, G.; Utigard, T.A. High temperature oxidation mechanism of dilute copper aluminium alloys. *Corros. Sci.* **2005**, *47*, 1149–1163. [CrossRef]
10. Lianga, S.; Fana, Z.; Xua, L.; Fangb, L. Kinetic analysis on Al$_2$O$_3$/Cu composite prepared by mechanical activation and internal oxidation. *Compos. Part A-Appl. Sci. Manuf.* **2004**, *35*, 1441–1446. [CrossRef]
11. Jena, P.K.; Brocchi, E.A.; Solórzano, I.G.; Motta, M.S. Identification of a third phase in Cu–Al$_2$O$_3$ nanocomposites prepared by chemical routes. *Mater. Sci. Eng. A* **2004**, *371*, 72–78. [CrossRef]
12. Brocchi, E.A.; Motta, M.S.; Solorzano, I.G.; Jena, P.K.; Moura, F.J. Alternative chemical-based synthesis routes and characterization of nano-scale particles. *Mater. Sci. Eng. B* **2004**, *112*, 200–205. [CrossRef]
13. Lü, L.; Lai, M.O.; Zhang, S. Modeling of the mechanical-alloying process. *Mater. Process. Technol.* **1995**, *52*, 539–546. [CrossRef]
14. Fathy, A.; Wagih, A.; El-Hamid, M.A.; Hassan, A.A. The effect of Mg add on morphology and mechanical properties of Al–xMg/10Al$_2$O$_3$ nanocomposite produced by mechanical alloying. *Adv. Powder Technol.* **2014**, *25*, 1345–1350. [CrossRef]
15. Zebarjad, S.M.; Sajjadi, S.A. Microstructure evaluation of Al–Al$_2$O$_3$ nanocomposite produced by mechanical alloying method. *Mater. Des.* **2006**, *27*, 684–688. [CrossRef]
16. Dash, P.K.; Murty, B.S.; Karthik Aamanchi, R.B. Synthesis and Mechanical Characterisation of Aluminium-Copper-Alumina Nano Composites Powder Embedded in Glass/Epoxy Laminates. *Am. J. Nanomaterials* **2015**, *3*, 28–39.
17. Shehata, F.; Abdelhameed, M.; Fathy, A.; Elmahdy, M. Preparation and characteristics of Cu-Al$_2$O$_3$ nanocomposite. *Open J. Metal* **2011**, *1*, 25–35. [CrossRef]
18. Korać, M.; Kamberović, Ž.; Anđić, Z.; Filipović, M. Nanocomposites and Polymers with Analytical Methods. In *Sintered Materials Based on Copper and Alumina Powders Synthesized by a Novel Method*; InTech: London, UK, 2011; pp. 181–198.
19. Korać, M.; Kamberović, Ž.; Filipović, M. Determination of Al$_2$O$_3$ particle size in Cu-Al$_2$O$_3$ nanocomposite materials using UV spectrophotometry. *J. Metall. Metal.* **2008**, *14*, 279–284. (In Serbian)
20. Korać, M.; Kamberović, Ž.; Tasić, M.; Gavrilovski, M. Nanocomposite powders for new contact materials based on copper and alumina. *Chem. Ind. Chem. Eng. Q* **2008**, *14*, 215–218. [CrossRef]
21. Anđić, Z.; Korać, M.; Kamberović, Ž.; Vujović, A.; Tasić, M. Analysis of the Properties of a Cu-Al$_2$O$_3$ Sintered System based on Ultra Fine and Nanocomposite Powders. *Sci. Sinter.* **2007**, *39*, 145–152. [CrossRef]
22. Anđić, Z.; Korać, M.; Tasić, M.; Raić, K.; Kamberović, Ž. The synthesis of ultrafine and nanocomposite powders based on copper, silver and alumina. *Kovove Mater.* **2006**, *44*, 145–150.
23. Seyedraoufi, Z.A.; Saghafian, H.; Shabestari, S.G. Thermochemical synthesis of nanostructured Cu-Al$_2$O$_3$ composite powder. *Chem. Ind. Chem. Eng. Q* **2014**, *20*, 339–344. [CrossRef]

24. Amirjan, M.; Khorsand, H.; Siadati, M.H.; Farsani, R.E. Artificial Neural Network prediction of Cu–Al$_2$O$_3$ composite properties prepared by powder metallurgy method. *J. Mater. Res. Technol.* **2013**, *2*, 351–355. [CrossRef]

25. Lara-Guevara, A.; Rojas-Rodriguez, I.; Hernandez, R.V.; Bernal-Correa, R.A.; Sierra-Gutierrez, A.; Herrera-Ramos, A.; Rodriguez-Garcia, M.E. Synthesis of copper-alumina composites by mechanical milling: An analysis. *Mater. Manuf. Processes* **2013**, *28*, 157–162. [CrossRef]

26. Han, S.J.; Seo, J.; Choe, K.H.; Kim, M.H. Fabrication of Al$_2$O$_3$ dispersion strengthened copper alloy by spray, in-situ synthesis casting process. *Met. Mater. Int.* **2015**, *21*, 270–275. [CrossRef]

27. Mohammadi, E.; Nasiri, H.; Khaki, V.J.; Zebarjad, S.M. Copper-alumina nanocomposite coating on copper substrate through solution combustion. *Ceram. Inter.* **2018**, *44*, 3226–3230. [CrossRef]

28. Anđić, Z.; Korać, M.; Tasić, M.; Kamberović, Ž.; Raić, K. Synthesis and sintering of Cu-Al$_2$O$_3$ nanocomposite powders produced by a thermochemical route. *Metal. J. Metall.* **2007**, *13*, 71–82.

29. Errandonea, D. High-pressure melting curves of the transition metals Cu, Ni, Pd, and Pt. *Phys. Rev. B* **2013**, *87*, 054108. [CrossRef]

30. Zhang, S. Melting Temperature of Al$_2$O$_3$ under Pressure. *Adv. Mater. Res.* **2012**, *549*, 745–748.

31. Anđić, Z.; Tasić, M.; Korać, M.; Jordović, B.; Maričić, A. Influence of alumina content on the sinterability of the Cu-Al$_2$O$_3$ pseudo alloy (composite). *Mater. Technol.* **2004**, *38*, 245–248.

32. Korać, M.; Anđić, Z.; Tasić, M.; Kamberović, Ž. Sintering of Cu–Al$_2$O$_3$ nanocomposite powders produced by a thermochemical route. *J. Serb. Chem. Soc.* **2007**, *72*, 1115–1125. [CrossRef]

33. Lee, D.W.; Ha, G.H.; Kim, B.K. Synthesis of Cu-Al$_2$O$_3$ nano composite powder. *Scr. Mater.* **2001**, *44*, 2137–2140. [CrossRef]

34. Anđić, Z.; Vujović, A.; Tasić, M.; Korać, M.; Kamberović, Ž. Synthesis and Characterization of Dispersion Reinforced Sintered System Based on Ultra Fineand Nanocomposite Cu-Al$_2$O$_3$ Powders. In *Nanocrystal*; InTech: London, UK, 2011; pp. 217–236.

35. Entezarian, M.; Drew, R.A.L. Direct bonding of copper to aluminum nitride. *Mater. Sci. Eng.* **1992**, *A212*, 206. [CrossRef]

36. Jena, P.K.; Brocchi, E.A.; Motta, M.S. Characterization of Cu–Al$_2$O$_3$ nano-scale composites synthesized by in situ reduction. *Mater. Sci. Eng.* **2001**, *C15*, 175–177.

37. Tu, J.F. TEM Nano-Moiré Pattern Analysis of a Copper/Single Walled Carbon Nanotube Nanocomposite Synthesized by Laser Surface Implanting. *C* **2018**, *4*, 19. [CrossRef]

38. Yi, S.; Trumble, K.P.; Gaskell, D.R. Thermodynamic analysis of aluminate stability in the eutectic bonding of copper with alumina. *Acta Mater.* **1999**, *47*, 3221–3226. [CrossRef]

39. Tanaka, S.; Yang, R.; Kohyama, M.; Sasaki, T.; Matsunaga, K.; Ikuhara, Y. First-Principles Characterization of Atomic Structure of Al$_2$O$_3$(0001)/Cu Nano-Hetero Interface. *Mater. Trans. JIM* **2004**, *45*, 1973–1977. [CrossRef]

40. Zhou, R.S.; Snyder, R.L. Structures and transformation mechanisms of the η, γ and θ transition aluminas. *Acta Crystallogr. Sect. B: Struct. Sci.* **1991**, *47*, 617–630. [CrossRef]

Article

Reaction Mechanism and Process Control of Hydrogen Reduction of Ammonium Perrhenate

Junjie Tang [1,2], Yuan Sun [2,*], Chunwei Zhang [2,3], Long Wang [2], Yizhou Zhou [2,*], Dawei Fang [3] and Yan Liu [4]

[1] School of Biomedical & Chemical Engineering, Liaoning Institute of Science and Technology, Benxi 117004, China; tangjunjieyh@lnist.edu.cn

[2] Superalloys research department, Institute of Metal Research, Chinese Academy of Sciences, Shenyang 110016, China; 13842034917@163.com (C.Z.); wanglong@imr.ac.cn (L.W.)

[3] College of chemistry, Liaoning University, Shenyang 110036, China; davidfine@163.com

[4] Key Laboratory for Ecological Utilization of Multimetallic Mineral, Ministry of Education, Northeastern University, Shenyang 110819, China; liuyan@smm.neu.edu.cn

* Correspondence: yuansun@imr.ac.cn (Y.S.); yzzhou@imr.ac.cn (Y.Z.); Tel.: +86-024-23971767 (Y.S.); +86-024-83978068 (Y.Z.)

Received: 17 April 2020; Accepted: 13 May 2020; Published: 15 May 2020

Abstract: The preparation of rhenium powder by a hydrogen reduction of ammonium perrhenate is the only industrial production method. However, due to the uneven particle size distribution and large particle size of rhenium powder, it is difficult to prepare high-density rhenium ingot. Moreover, the existing process requires a secondary high-temperature reduction and the deoxidization process is complex and requires a high-temperature resistance of the equipment. Attempting to tackle the difficulties, this paper described a novel process to improve the particle size distribution uniformity and reduce the particle size of rhenium powder, aiming to produce a high-density rhenium ingot, and ammonium perrhenate is completely reduced by hydrogen at a low temperature. When the particle size of the rhenium powder was 19.74 μm, the density of the pressed rhenium ingot was 20.106 g/cm^3, which was close to the theoretical density of rhenium. In addition, the hydrogen reduction mechanism of ammonium perrhenate was investigated in this paper. The results showed that the disproportionation of ReO_3 decreased the rate of the reduction reaction, and the XRD and XPS patterns showed that the increase in the reduction temperature was conducive to increasing the reduction reaction rate and reducing the influence of disproportionation on the reduction process. At the same reduction temperature, reducing the particle sizes of ammonium perrhenate was conducive to increasing the hydrogen reduction rate and reducing the influence of the disproportionation.

Keywords: ammonium perrhenate; rhenium; disproportionation reaction; hydrogen reduction

---

1. Introduction

Rhenium as an important rare metal is widely used in metallurgy and the aerospace industry [1]. The plasma method, electrolysis method, vapor deposition method, and powder metallurgy are the main processes for the preparation of rhenium [2,3]. Jurewicz et al. [4] prepared a high-purity nanometer rhenium powder by the plasma method. Leonhardt et al. [5] used plasma spray spheroidization to control the microstructure of rhenium and obtained spherical rhenium powders. Schrebler et al. [6] also prepared spherical rhenium powder from a mixture of rhenic acid and sodium sulfate by electrolysis. Liu et al. [7] prepared small particles of superfine spherical rhenium powder by vapor deposition. The rhenium powders prepared by the above processes have a uniform particle size distribution and large specific surface area, and the pure rhenium materials prepared from these rhenium powders have high densities. However, due to the complexity of the above processes and high equipment requirements, none of the above processes have been industrialized. The hydrogen reduction of ammonium perrhenate is a commonly used process to prepare rhenium powder in industry, which has the characteristics of a simple process flow, low production cost, and low equipment requirements [8,9]. The preparation process flow chart of rhenium ingot is shown in Figure 1. However, the rhenium powders that are produced by this preparation technology have an uneven particle size distribution, small specific surface area, and the rhenium ingots produced have poor compactness. Moreover, due to the low efficiency of the mass and heat transfer in the traditional process, the reaction with hydrogen is not complete and the second reduction step at a high temperature is required [10–12].

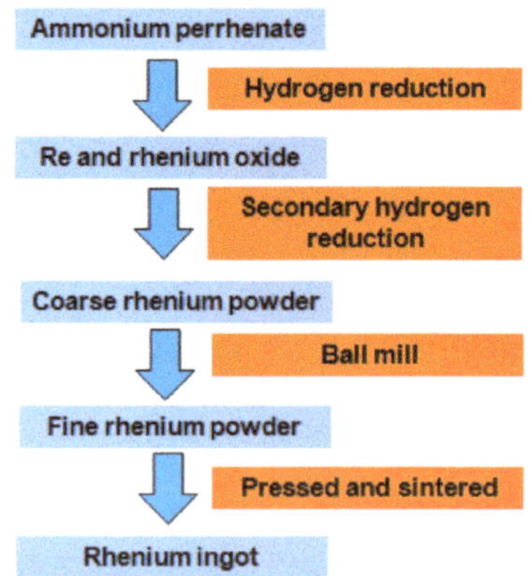

Figure 1. Industrial preparation of rhenium ingots.

In order to optimize the process of preparing rhenium by a high-temperature reduction, we can look into work dealing with the preparation of other metal powders by a high-temperature reduction. In recent years, there have been many studies on the preparation of metal powders by a high-temperature hydrogen reduction [13–21]. For instance, Wang et al. [22] proposed a novel route to synthesize Mo powders via a carbothermic prereduction of molybdenum oxide followed by a reduction by hydrogen; they removed oxygen from the samples by a secondary reduction. Kanga et al. [23] prepared nanosized W and W-Ni powders by applying ball milling and a hydrogen reduction of oxide powders. Gua et al. [24] prepared Mo nanopowders through a hydrogen reduction of a combustion-synthesized foam-like MoO_2 precursor. All of the above studies are based on a reducing substance pretreatment, which provides a certain reference experience for this study. However, the hydrogen reduction reaction of ammonium perrhenate is a complicated process. This process not only involves reduction, but also a disproportionation reaction. Colton [25] pointed out that the disproportionation of ReO_3 occurred above 300 °C to produce Re_2O_7 and ReO_2. This disproportionation reaction is the main reason why ammonium perrhenate cannot be completely reduced to rhenium at a low temperature. There are few reports on the preparation of a high-quality Re powder by a hydrogen reduction at present. Bai et al. [26] reduced volatile rhenium oxide to prepare Re powder. However, due to the high equipment requirements of this process, it cannot be used for industrial production.

Chemical Reaction Considerations

Ammonium perrhenate decomposes into oxides with different valence states when reduced at a high temperature, and the main oxides are Re_2O_7, ReO_3, and ReO_2 [27]. The total equation for the reduction of ammonium perrhenate in hydrogen is represented by Equation (1). Ammonium perrhenate is decomposed by heat to Re_2O_7, which is gradually recombined with hydrogen and finally reduced to Re, as shown in Equations (2)–(4). ReO_3 is very reactive; ReO_3 is disproportionated at a high temperature to produce ReO_2 and Re_2O_7, as shown in Equation (5).

$$2NH_4ReO_4 \text{ (s)} + 7H_2 \text{ (g)} = 2Re \text{ (s)} + 8H_2O \text{ (g)} + 2NH_3 \text{ (g)} \tag{1}$$

$$Re_2O_7 \text{ (s)} + H_2 \text{ (g)} = 2ReO_3 \text{ (s)} + H_2O \text{ (g)} \tag{2}$$

$$ReO_3 \text{ (s)} + H_2 \text{ (g)} = ReO_2 \text{ (s)} + H_2O \text{ (g)} \tag{3}$$

$$ReO_2 \text{ (s)} + 2H_2 \text{ (g)} = Re \text{ (s)} + 2H_2O \text{ (g)} \tag{4}$$

$$3ReO_3 \text{ (s)} = Re_2O_7 \text{ (s)} + ReO_2 \text{ (s)} \tag{5}$$

In the present work, we determined the reduction mechanism of ammonium perrhenate through a differential thermal analysis, and innovatively proposed that the disproportionation reaction in the reduction process was the main reason affecting the complete reduction of ammonium perrhenate. This work also determined an innovative process for reducing the particle size and reduction temperature of rhenium powder, aiming to produce a high-density rhenium ingot, and ammonium perrhenate is completely reduced by hydrogen at a low temperature. This is the technological innovation point of this paper. Moreover, the optimization and innovation of this process is based on the already industrialized hydrogen reduction process to produce rhenium ingot, which makes it easy to realize as an industrialized production process.

2. Materials and Methods

2.1. Instrument

Instrument: The following instruments were used herein: RE-2000A rotary evaporator, Qiqiang instrument manufacturing co. LTD, Shanghai, China; SSX-550 scanning electron microscope, Shimadzu, Osaka, Japan; PW3040/60 X-ray diffractometer (XRD), Panalytical Company, Almelo, Netherlands; Escalab250 250 X-ray photoelectron spectroscopy (XPS), Hewlett-Packard Company, Palo alto, CA, USA; VEP223 high-temperature vacuum sintering furnace, Beizhen Vacuum Technology Co. Ltd., Shenyang, China; HYL-1076 laser particle size analyzer, Haoyu Technology Co. Ltd., Dandong, China; the self-developed recrystallization reactor, Institute of metals, Chinese academy of sciences, Shenyang, China; STA PT1600 Differential Thermal Analysis (DTA), Linseis Co. Ltd., Selb, Germany.

2.2. Materials

Materials: NH_4ReO_4 (99.99%, Re $\geq$ 69.4%) from Halin Chemical Co. LTD, Weifang, China.

2.3. Analytical Methods

XRD detection: The light tube type was a Cu target, ceramic X light tube. λ = 0.15406 nm, scan range was 10–90 degrees, the scanning speed was 2 degrees/min.

DTA detection: The temperature ranged from 300 to 700 °C, and the gas atmosphere was nitrogen.

Particle size distribution detection: the test medium was ethanol, the optical model was Mie, and the distribution type was volume distribution.

The parameters of SEM: The electron acceleration voltage was 20.0 KV, the working distance was 21.8 mm, and the magnification was 15,000 times.

XPS analysis parameters: The fitting software was XPSPEAK (XPSPEAK.41, Hewlett-Packard Company, Palo alto, CA, USA). The read base pressure was 2.4×10^{-8} Pa, utilizing monochromatic Al Kα radiation operating at 1486.6 eV. At the pass energy of 50 eV, with a 0.1 eV step, the high-resolution scans were performed. At the pass energy of 100 eV and a step size of 1 eV, the survey spectra were acquired. The reproducible C (1 s) binding energy of all samples was 284.6 eV and the charge neutralization was achieved using low-energy electrons and argon ions. The spin–orbit splitting was 2.4 eV, and the spin orbital split intensity ratio of $4f_{7/2}$ and $4f_{5/2}$ was 4:3.

2.4. Experimental Procedure

As illustrated by Figure 2 was the experimental flow chart of a hydrogen reduction of ammonium perrhenate in this study. Ammonium perrhenate (99.99%, Re ≥ 69.4%) was prepared with deionized water into a saturated solution at room temperature, the room temperature was about 25 °C. Then the saturated solution of ammonium perrhenate at room temperature was placed in an RE-2000A rotary evaporator, and part of the water was evaporated to form a hot saturated solution at 120 °C [28]. The thermally saturated solution of ammonium perrhenate was introduced into the recrystallization condensation reactor; the stirring speed and cooling temperature were adjusted and recrystallized at 5 °C for 3 h. Finally, the cooled solid–liquid mixture was filtered and dried to obtain ammonium perrhenate crystals. The SEM diagrams of the recrystallized ammonium perrhenate are shown in Figure 3 [29], and the median diameters (D_{50}) and specific surface area are shown in Table 1. The recrystallization ammonium perrhenate particles (60 g) were reduced with the different temperatures (300–900 °C) in the high-temperature vacuum sintering furnace; the hydrogen flow rate was 500 mL/min, and the heating rate was 10 °C/min. After 3 h of reduction, the reduction product of ammonium perrhenate was obtained.

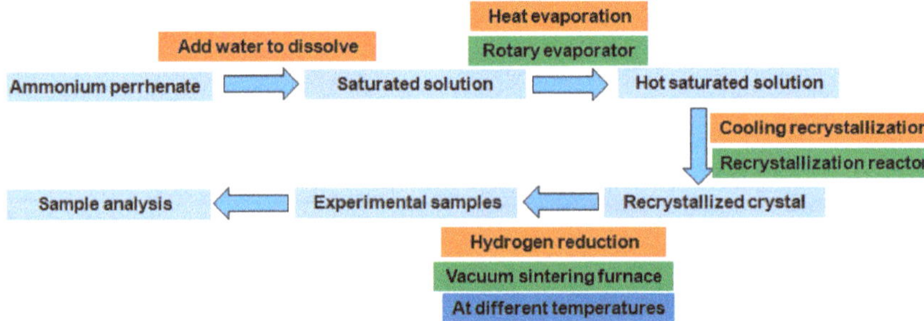

Figure 2. The experimental working procedure diagram.

Figure 3. SEM images of recrystallized ammonium perrhenate particles at different agitation intensities and ammonium perrhenate raw material particles (**a**) ammonium perrhenate raw material particles, (**b**) 100 rpm stirring strength recrystallized particles, and (**c**) 200 rpm stirring strength recrystallized particles) [29].

Table 1. Specific surfaces and median diameters (D_{50}s) of the NH4ReO4 particles.

Ammonium Perrhenate	Specific Surface (m^2/kg)	D_{50} (μm)
100 rpm stirring strength recrystallized	21.72	81.05
200 rpm stirring strength recrystallized	26.93	71.17
Raw material particles	14.79	123.90

Preparation of rhenium ingots: 20 g rhenium powder was put into the powder press mold (the height was 32.00 mm and the inner diameter of the mould was 16.60 mm) to press. The pressure of the powder press was 74,000 N, the sintering furnace temperature was 2300 °C, and the sintering time was 3 h. The theoretical density of rhenium is 21.04 g/cm^3.

3. Results and Discussion

3.1. Reduction Mechanism Analysis

In order to research the reaction mechanism of ammonium perrhenate in the reduction process, the decomposition products of ammonium perrhenate (Re, ReO_2, and ReO_3) were analyzed by a differential thermal analysis (DTA), and the differential thermal analysis curve is shown in Figure 4. It can be seen that an obvious endothermic peak appeared at 350 to 400 °C. Re is stable at a high temperature, and the decomposition temperature of ReO_2 is 700 °C. Therefore, the generation of this endothermic peak can only be due to the disproportionation of ReO_3. The disproportionation reaction products of ReO_3 are ReO_2 and Re_2O_7, and the reaction equation is shown in Equation (5). In the hydrogen reduction process, the ammonium perrhenate is firstly decomposed into Re_2O_7, the Re_2O_7 reacts with hydrogen to form ReO_3, and ReO_3 reacts with hydrogen to form ReO_2 until they are reduced to Re. In the process of a hydrogen reduction of ammonium perrhenate, if the disproportionation reaction and reduction reaction exist simultaneously, the disproportionation reaction will be the main reason affecting the complete reduction of ammonium perrhenate at a low temperature.

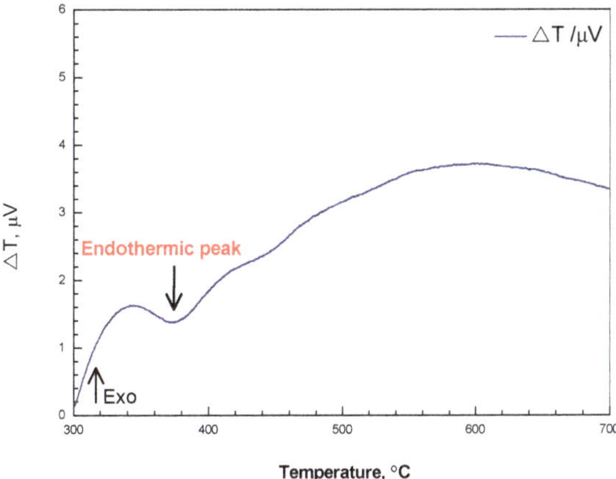

Figure 4. The differential thermal analysis curve of Re, ReO_2, and ReO_3.

In order to research the effect of disproportionation on the hydrogen reduction of ammonium perrhenate, the hydrogen reduction experiments of ammonium perrhenate were carried out at the same reduction time and at different reduction temperatures. The reducing substance was recrystallized ammonium perrhenate at a 200 rpm stirring strength (D_{50} was 71.17 μm, specific surface was 26.93 m^2/kg). The XRD patterns of the reduction products of ammonium perrhenate at different temperatures are shown in Figure 5. The characteristic peaks of the reduction products were complex at lower temperatures (300–600 °C), and the diffraction peaks of the reduction products indicated Re and ReO_2, and ReO_3. The characteristic peaks of Re did not change obviously from 400 to 600 °C. However, in the range of 300 to 400 °C, the characteristic peaks of ReO_3 were enhanced. In the range of 400 to 600 °C, the characteristic peaks of ReO_2 were enhanced, while that of ReO_3 were weakened. When the temperature reached 700 °C, the characteristic peaks of the reduction products were Re, and other crystal peaks were not observed. This result suggested that the contents of ReO_3 in the reduction products increased in the range of 300 to 400 °C. However, within the temperature range of 400 to

600 °C, the content of Re in the reduction products did not increase significantly, while the content of ReO_2 increased significantly.

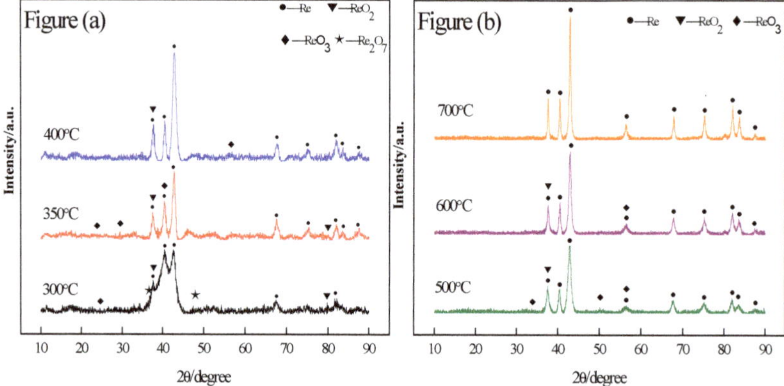

Figure 5. XRD patterns of the reduction products of ammonium perrhenate (D_{50} was 71.17 µm, specific surface was 26.93 m^2/kg) at different temperatures. Figure (**a**) was the XRD diffraction pattern of the reduced product at 300 to 400 °C; Figure (**b**) was the XRD diffraction pattern of the reduced product at 500 to 700 °C.

In order to further clarify the influences of reduction temperatures on an ammonium perrhenate hydrogen reduction, X-ray photoelectron spectroscopy was used for the rhenium atomic quantitative analysis in different valence states. According to the references [30–32], a rhenium atom has a split energy level (f), where the spin–orbit splitting was 2.4 eV, and the spin orbital split intensity ratio of Re $4f_{7/2}$ and Re $4f_{5/2}$ was 4:3. The background was a mixed Shirley background (Shirley + straight line) [33], and the slope of the line was eight. The peak positions and fit paramters for all samples are given in Table 2. The X-ray photoelectron spectroscopy of the reduced products at the different reduction temperatures are shown in Figure A1.

Table 2. The peak positions and fit paramters for Re $4f_{7/2}$ and Re $4f_{5/2}$.

Fitting Parameters	Re	Re^{4+}	Re^{6+}	Re^{7+}
B.E (4f7/2)	40.31	42.7	45.2	46.1
B.E (4f5/2)	42.71	45.1	47.6	48.5
FWHM	1.3	1.7	1.75	1.8

The X-ray photoelectron spectroscopy of the reduced products at different reduction temperatures are shown in Figure A1. According to the peak areas of the different valence states of rhenium, the content of each valence state was calculated, as shown in Table 2. The percent of rhenium atom in each valence state was calculated, as shown in Table 3.

Table 3. The percent of rhenium atom in each valence state of the reduction products of ammonium perrhenate (D_{50} was 71.17 µm, specific surface was 26.93 m^2/kg) at different temperatures.

Reduction Temperature (°C)	Re (%)	Re^{4+} (%)	Re^{6+} (%)	Re^{7+} (%)
300	75.9	20.0	3.0	1.1
350	69.6	16.4	24.0	0.0
400	59.3	14.1	26.6	0.0
500	65.9	16.2	17.9	0.0
600	76.1	18.1	7.8	0.0
700	100	0.0	0.0	0.0

According to the results in Table 3, the percents of rhenium atom in each valence state of the reduced products at different temperatures were plotted, as shown in Figure 6. Combined with Table 2 and Figure 6, it can be concluded that the percent of Re^{6+} in the reduction products increased in the range of 300 to 400 °C, and within the temperature range of 400 to 600 °C, the percent of Re in the reduction products did not increase significantly, while the content of Re^{4+} increased significantly. These results showed that the ReO_3 content in the reduction products increased in the temperature range of 300–400 °C and decreased with the increase in the reduction temperature; the Re content decreased in the temperature range of 300–400 °C and increased with the increase in the reduction temperature. In order to explain this rule, the following conclusions were obtained by combining the XRD and DTA detection results: the disproportionation reaction of ReO_3 occurred between 350 and 400 °C, ReO_3 decomposed into ReO_2 and Re_2O_7, and Re_2O_7 was converted to ReO_3 by a hydrogen reduction. Therefore, in the temperature range of 350 to 400 °C, the effect of disproportionation hindered the normal reduction process, and this is why ammonium perrhenate cannot be reduced to more rhenium at 350 to 400 °C in the same reduction time. This phenomenon shows that the disproportionation of ReO_3 hindered the normal reduction reaction in a certain temperature range, but with the increase in the temperature, the reduction reaction rate gradually increased, and the content of rhenium in the reduction products also increased. At the reduction temperature of 300 °C, the disproportionation of ReO_3 did not occur, and only the reduction reaction was carried out. Therefore, the content of Re in the reduction products was higher. As discussed above, in the process of a hydrogen reduction of ammonium perrhenate, the disproportionation of ReO_3 decreased the rate of the reduction reaction, and the increase in the temperature can increase the reduction reaction rate and reduce the effect of disproportionation on the reduction process. Increasing the reduction time and lowering the reduction temperature below the temperature at which the disproportionation of ReO_3 occurs can also increase the content of Re in the reduction products.

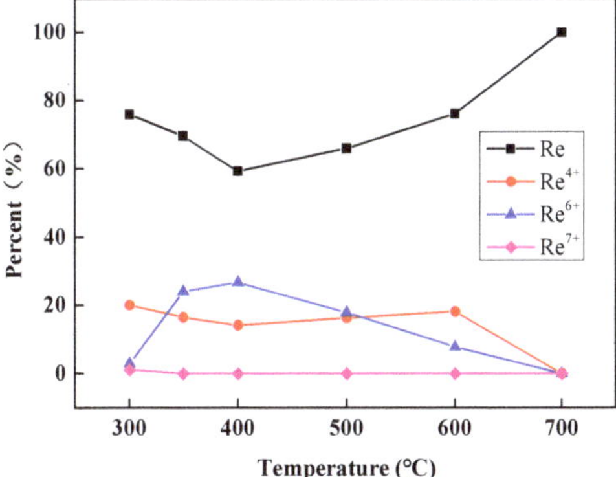

Figure 6. Oxidation state atomic content changes of ammonium perrhenate (D_{50} was 71.17 μm, specific surface was 26.93 m^2/kg) hydrogen reduction reaction at different temperatures.

3.2. Influence of Particle Size on Reduction

In the solid-state reaction system, the reaction rate is related not only to the temperature but also to the diffusion rate of the reactant [34]. The particle size of the reactant decreases, which is conducive to increasing the diffusion rate [35]. In order to research the effects of particle sizes of ammonium perrhenate on the reduction effect, the recrystallized ammonium perrhenate (D_{50} 81.05 μm) was used for the reduction experiments under the same operating conditions. The XRD patterns of the reduction products of ammonium perrhenate at different temperatures are shown in Figure 7. The characteristic peaks of the reduction products were complex at lower temperatures (300–700 °C), and the diffraction peaks of the reduction products indicated Re and ReO_2, and ReO_3. The characteristic peaks of Re did not change obviously from 500 to 800 °C. However, in the range of 300 to 400 °C, the characteristic peaks of ReO_3 were enhanced. In the range of 500 to 800 °C, the characteristic peaks of ReO_2 were enhanced, while that of ReO_3 were weakened. When the temperature reached 900 °C, the characteristic peaks of the reduction products were Re and ReO_2, and other crystal peaks were not observed. This result suggested that the contents of ReO_3 in the reduction products increased in the range of 300 to 400 °C. However, within the temperature range of 500 to 800 °C, the content of Re in the reduction products did not increase significantly, while the content of ReO_2 increased significantly.

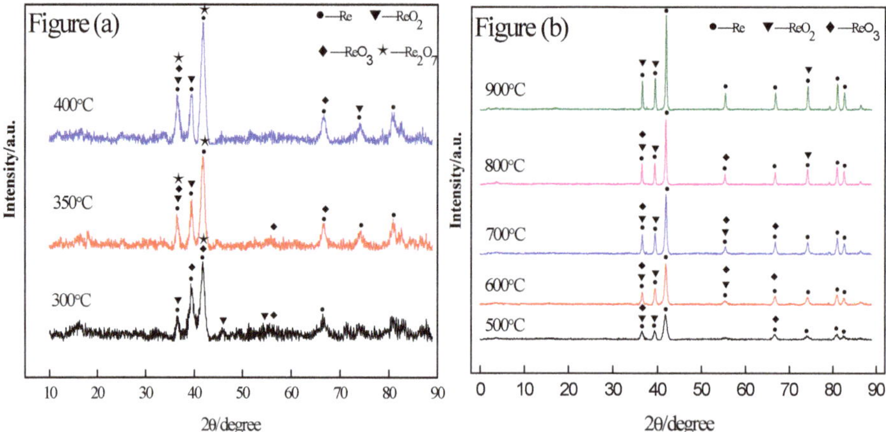

Figure 7. XRD patterns of the reduction products of ammonium perrhenate (D_{50} was 81.05 μm, specific surface was 21.72 m^2/kg) at different temperatures. Figure (**a**) was the XRD diffraction pattern of the reduced product at 300 to 400 °C; Figure (**b**) was the XRD diffraction pattern of the reduced product at 500 to 900 °C.

In order to further clarify the influence of particle sizes on an ammonium perrhenate hydrogen reduction, X-ray photoelectron spectroscopy was used to quantitatively analyze this feature. The X-ray photoelectron spectroscopy of the reduced products at different reduction temperatures are shown in Figure A2. The percent of rhenium atom in each valence state was calculated, as shown in Table 4. According to the results in Table 4, the percents of rhenium atom in each valence state of the reduced products at different temperatures were plotted, as shown in Figure 8. Combined with Table 4 and Figure 8, it can be concluded that the content of Re in the reduction products increased with the decrease of ReO_3 between 350 and 800 °C. At the reduction temperature of 300 °C, the disproportionation of rhenium trioxide did not occur and the content of rhenium was higher. The composition law of the reduction products was the same as that of previous reduction experiments, which proved the accuracy of the reduction mechanism analysis. When the reduction temperature reached 800 °C, rhenium trioxide still existd. It can be concluded that the disproportionation of ReO_3 existed in the range of 350 to 800 °C. Compared with the previous reduction experiment, the temperature range at which the disproportionation reaction occurs had increased. This was because the particle size of ammonium perrhenate increased, the solid reaction diffusion rate decreased, and the contact between the reactant and hydrogen was not ideal. Therefore, under the same operating conditions, reducing the particle size of ammonium perrhenate can increase the hydrogen reduction diffusion rate and reduce the influence of the disproportionation on the reduction process.

Table 4. The percents of rhenium atom in each valence state of the reduction products of ammonium perrhenate (D_{50} was 81.05 μm, specific surface was 21.72 m^2/kg) at different temperatures.

Reduction Temperature (°C)	Re (%)	Re^{4+} (%)	Re^{6+} (%)	Re^{7+} (%)
300	65.7	20.0	10.8	3.5
350	58.9	21.8	19.1	0.2
400	36.3	10.7	45.9	7.1
500	68.1	15.7	16.2	0.0
600	72.2	18.2	9.6	0.0
700	72.7	22.0	5.3	0.0
800	79.0	17.2	3.8	0.0
900	80.1	19.9	0.0	0.0

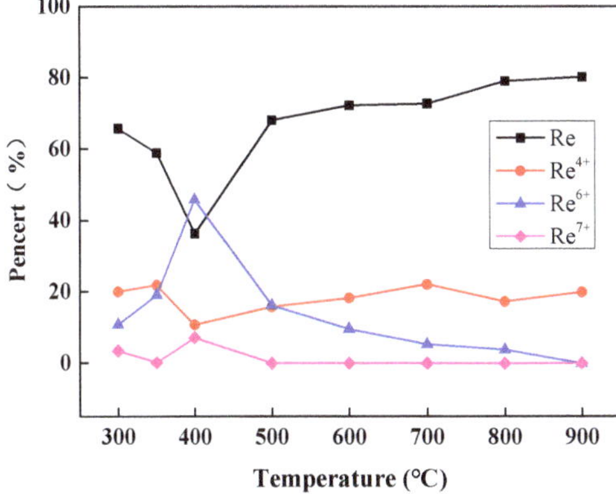

Figure 8. Oxidation state atomic content changes of ammonium perrhenate (D_{50} was 81.05 μm, specific surface was 21.72 m^2/kg) hydrogen reduction reaction at different temperatures.

3.3. Analysis of the Products of the Rhenium Ingots

The Re powders were prepared by unrecrystallized ammonium perrhenate and recrystallized ammonium perrhenate with a different D_{50}, and the rhenium ingots were prepared from these rhenium powders. The relevant physical parameters of the products are shown in Table 5. The particle size distribution of the rhenium powder is shown in Figure 9. It can be seen from Table 5 and Figure 9 that the rhenium powder prepared by ammonium perrhenate with a small particle size is also of relatively small particle size; the particle size distribution uniformity of rhenium powder prepared by recrystallized ammonium perrhenate was improved obviously, and the rhenium ingot prepared has a higher density. The theoretical density of rhenium is 21.04 g/cm^3, and the density of the rhenium ingot prepared by D_{50} 71.17 μm ammonium perrhenate reached 20.106 g/cm^3. The rhenium ingot prepared under this condition was close to the theoretical density. The SEM images of the surface defects of the rhenium ingots are shown in Figure 10. It can be seen that the rhenium ingot prepared by the small particle size rhenium powder not only has a high density but also has a small surface hole defect. Therefore, reducing the particle size of the rhenium powder is a key factor in preparing a high-density rhenium ingot.

Table 5. The rhenium materials prepared from ammonium perrhenate of different particle sizes.

D_{50} of Ammonium Perrhenate (μm)	D_{50} of Re Powders (μm)	Specific Surface Area of Re (m^2/kg)	Density of Rhenium Ingots (g/cm^3)
81.05	34.04	97.54	18.779
71.17	19.74	163.70	20.106
123.90	52.15	72.73	17.726

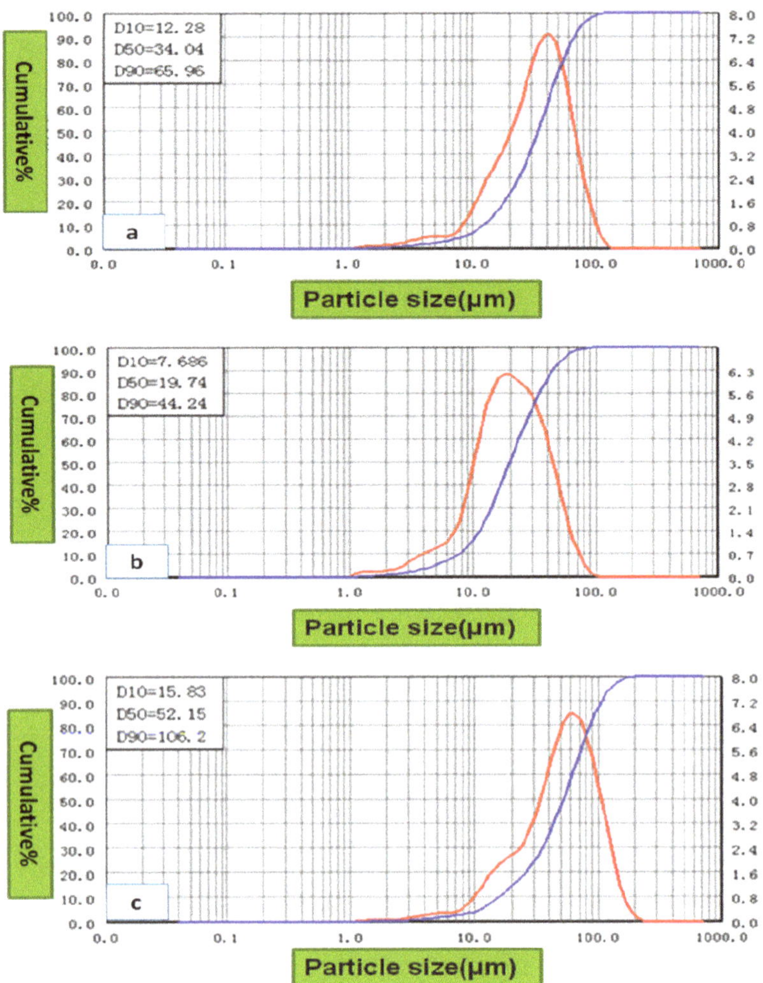

Figure 9. The particle size distribution of rhenium powder prepared by ammonium perrhenate with different particle sizes ((**a**) prepared by 81.05 μm D_{50} ammonium perrhenate, (**b**) prepared by 71.17 μm D_{50} ammonium perrhenate, (**c**) prepared by 123.90 μm D_{50} ammonium perrhenate).

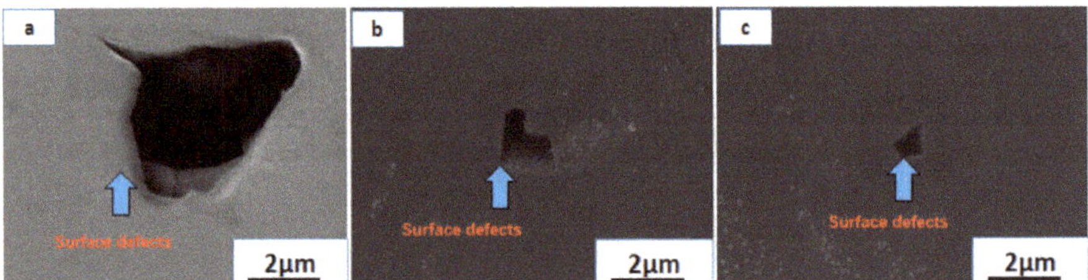

Figure 10. SEM images of the surface defects of the rhenium ingots ((**a**) prepared by 52.15 μm D_{50} Re powders, (**b**) prepared by 34.04 μm D_{50} Re powders, (**c**) prepared by 19.74 μm D_{50} Re powders).

3.4. Proposed Flow Sheet

Based on the experimental results, the flow sheet for the processing of rhenium ingots by ammonium perrhenate was tentatively suggested, which is shown in Figure 11. The ammonium perrhenate particles were refined by homogeneous recrystallization, and the D_{50} of the ammonium perrhenate particles was refined from 123.90 to 71.17 μm. The recrystallized ammonium perrhenate was completely reduced by hydrogen at 700 °C for 3 h, and the D_{50} of 19.74 μm rhenium powder was obtained. The density of the rhenium ingots pressed by these rhenium powders was 20.106 g/cm^3. The theoretical density of rhenium is 21.04 g/cm^3, and a rhenium ingot that reaches the theoretical density of more than 90% is a high-quality product. In this study, the rhenium ingot density was 95.56% of the theoretical density, which reached the high-quality product standard. The optimized production process of rhenium ingot not only realized a low-temperature reduction, but also increased the density of the rhenium ingot, which can provide a theoretical basis and practical experience for industrial production.

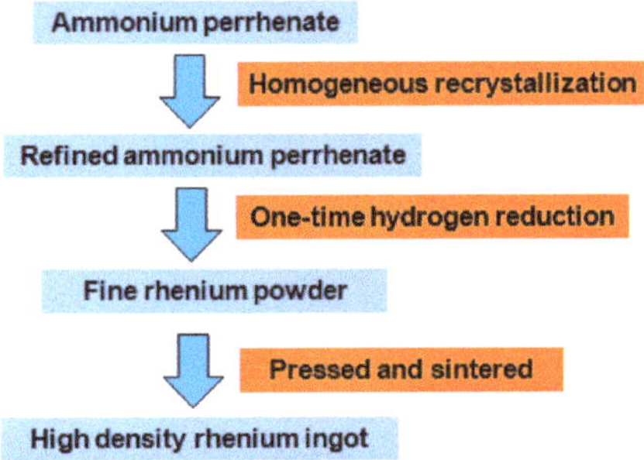

Figure 11. Proposed flow sheet for the preparation of high-density rhenium ingot by homogeneous recrystallization of ammonium perrhenate by a hydrogen reduction.

4. Conclusions

In this study, the influences of disproportionation on the hydrogen reduction of ammonium perrhenate were investigated and the following conclusions were drawn:

(1) In the process of a hydrogen reduction of ammonium perrhenate, the disproportionation of ReO_3 decreased the rate of the reduction reaction, and the increase in the reduction temperature was conducive to increasing the reduction reaction rate and reducing the influence of disproportionation on the reduction process.

(2) At the same reduction temperature, reducing the particle sizes of ammonium perrhenate was conducive to increasing the hydrogen reduction rate and reducing the influence of the disproportionation on the reduction process.

(3) The rhenium ingot prepared by the small particle size rhenium powder not only has a high density but also has a small surface hole defect. Therefore, reducing the particle size of the rhenium powder is a key factor in preparing high-density rhenium ingot.

(4) It is feasible to increase the density of rhenium ingot by reducing the particle size of the rhenium powder. The particle size of ammonium perrhenate was reduced to a rhenium powder with a D_{50} of 19.74 μm and a specific surface area of 163.70 m^2/kg, which was pressed into a rhenium ingot with a density of 20.106 g/cm^3, close to the theoretical density of rhenium.

Author Contributions: J.T.: Writing—original draft; Data curation; Formal analysis; Investigation; Funding acquisition. Y.S.: Data curation; Formal analysis; Funding acquisition; Resources. C.Z.: Data curation; Software. L.W.: Resources; Writing—review and editing. Y.Z.: Investigation; Supervision; Funding acquisition; Resources. D.F.: Writing—review and editing; Resources. Y.L.: Writing—review and editing; Funding acquisition. All authors have read and agreed to the published version of the manuscript.

Funding: Liaoning provincial department of science and technology doctoral research initiation fund, grant number 2019-BS-130; Open project fund of key laboratory of ecological metallurgy of polymetallic symbiosis in ministry of education, Northeastern University, grant number NEMM2019003.

Acknowledgments: School of Biomedical & Chemical Engineering, Liaoning Institute of Science and Technology and the analysis and test center of institute of metals, Chinese academy of sciences undertook the sample test for this study. The authors are grateful for these supports.

Conflicts of Interest: The authors declare no conflict of interest.

Appendix A

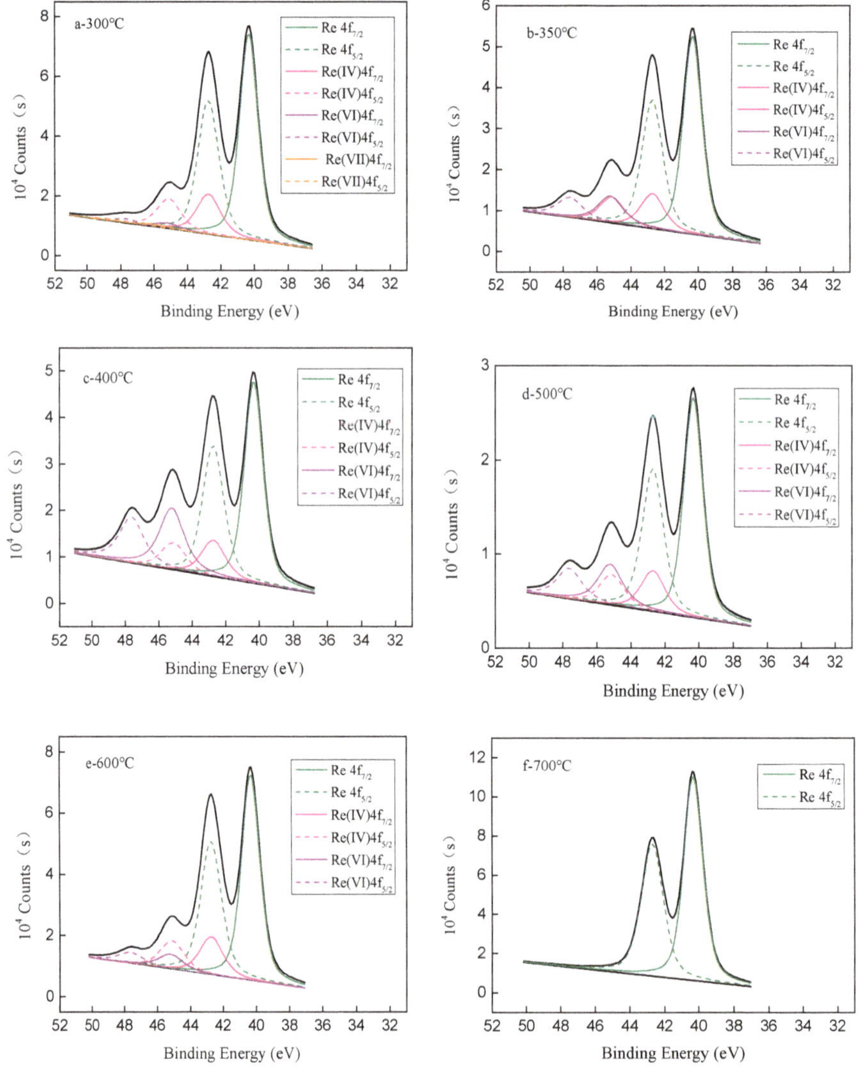

Figure A1. X-ray photoelectron spectroscopy of the reduction products of ammonium perrhenate (D_{50} was 71.17 µm, specific surface was 26.93 m^2/kg) at different temperatures.

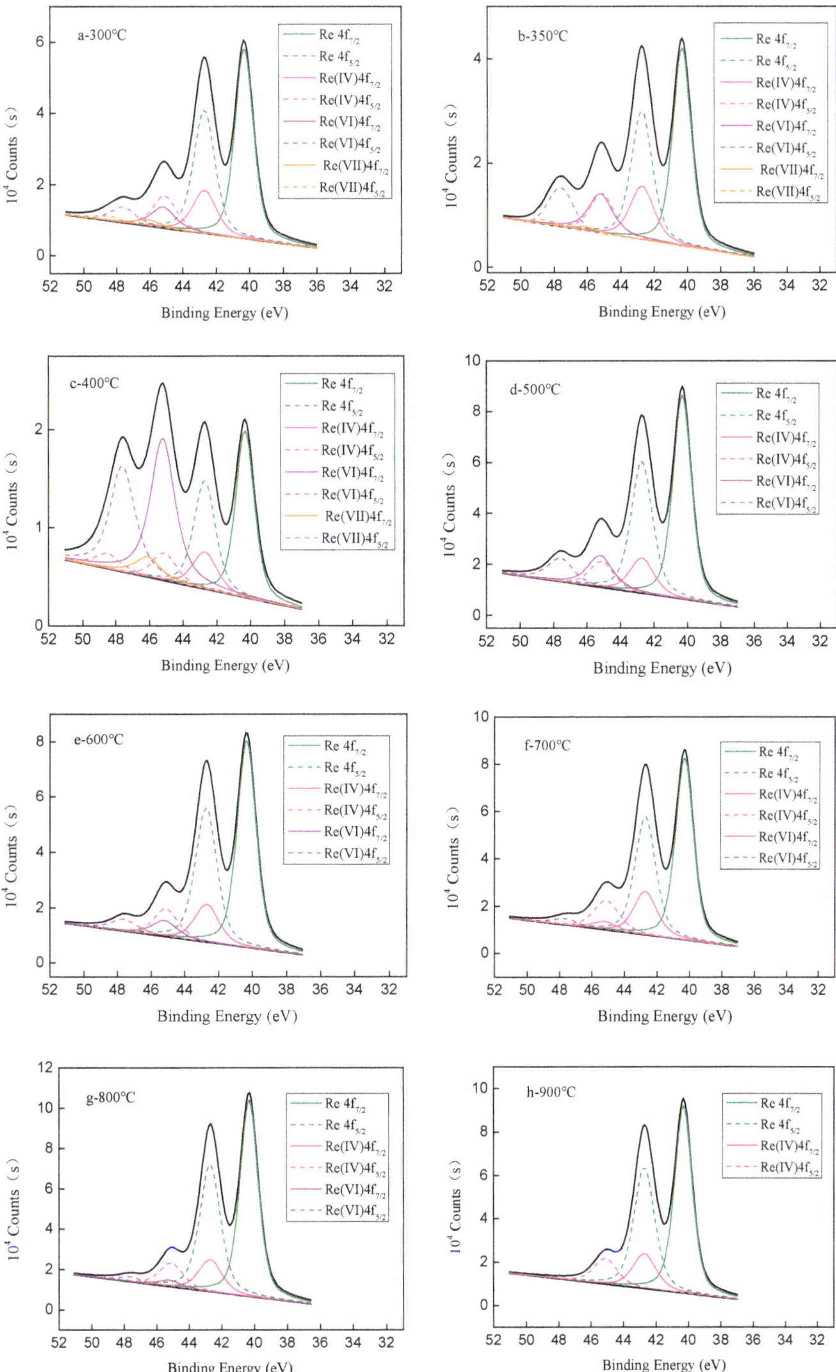

Figure A2. X-ray photoelectron spectroscopy of the reduction products of ammonium perrhenate (D_{50} was 81.05 μm, specific surface was 21.72 m^2/kg) at different temperatures.

References

1. Li, L.P.; Liu, Y.; Zhang, W.; Jiang, L.J.; Zhang, W.Z. Recent Development of Rhenium Technology. *China Molybdenum Ind.* **2016**, *40*, 1–6.
2. Noar, A.; Eliaz, N.; Gileadi, E.; Taylor, S.R. Properties and applications of rhenium and its alloys. *AMMTIAC Q.* **2010**, *5*, 11–14.
3. Li, H.M.; He, X.T.; Zhou, Y.; Guo, J.M.; Han, S.L.; Wang, H.; Li, Y.; Tan, M.L. Resources, Application and Extraction Status of Rhenium. *Precious Met.* **2014**, *35*, 77–81.
4. Jurewicz, J.W.; Guo, J.Y. Process for Plasma Synthesis of Rhenium Nano and Micro Powders, and for Coating and New Net Shape Deposits Thereof and Apparatus Therefor. U.S. Patent 2005/0211018A1, 29 September 2005.
5. Leonhardt, T.; Trybus, C.; Hickman, R. Consolidation Methods for Spherical Rhenium and Rhenium Alloys. *Powder Metall.* **2003**, *46*, 148–153. [CrossRef]
6. Schrebler, R.; Cury, P.; Orellana, M.; Gómez, H.; Córdova, R.; Dalchiele, E.A. Electrochemical and Nanoelectrogravimetric Studies of the Nucleation and Growth Mechanisms of Rhenium on Polycrystalline Gold Electrode. *Electrochim. Acta* **2001**, *46*, 4309–4318. [CrossRef]
7. Liu, Z.H.; Zhang, S.Y.; Liu, Z.Y.; Li, Y.H.; Wang, J. Principle and Research Development of Powder Materials Prepared by Chemical Vapor Deposition. *Mater. Sci. Eng. Powder Metall.* **2009**, *14*, 359–364.
8. Stefan, L.; Helumt, A. Hydrogen as a reducing agent: State-of-the-art science and technology. *JOM* **2007**, *59*, 20–26.
9. Mannheim, R.L.; Garin, J.L. Microstructural characterization of rhenium powder obtained at various temperatures. *Adv. Powder Technol.* **2003**, *416*, 273–282. [CrossRef]
10. Trybus, C.L.; Wang, C.M.; Pandheeradi, M.; Meglio, C.A. Powder metallurgical processing of rhenium. *Adv. Mater. Process.* **2002**, *160*, 23–26.
11. Shen, Y.Y.; Yi, X.M.; Liao, B.B. A Preparation Method of High Purity Rhenium Powders. CN Patent NO. 1396027, 12 February 2003.
12. Yang, H.B.; Peng, J.Y.; Cheng, T.Y.; Xiong, N.; Yin, J.C.; Chen, F.X. A Manufacturing Method of Pure Rhenium Flakes with Difficulty in Deformation. CN Patent NO. 101177748A, 14 May 2008.
13. Bartosz, T.; Beata, B.; Kusz, B. A study of a reduction of a micro-and nanometric bismuth oxide in hydrogen atmosphere. *Ther. Acta* **2018**, *669*, 99–108.
14. Dang, J.; Wu, Y.J.; Lv, Z.P.; You, Z.X.; Zhang, S.F.; Lv, X.W. A new kinetic model for hydrogen reduction of metal oxides under external gas diffusion controlling condition. *Int. J. Refract. Met. Hard Mater.* **2018**, *77*, 90–96. [CrossRef]
15. Liu, Q.P.; Bai, Y.; Yi, S.F.; An, P.F. The Effect of Hydrogen Dew Point on the Properties of Molybdenum Dioxide and Molybdenum Powder. *China Tungsten Ind.* **2018**, *33*, 42–46.
16. Zhuo, H.O.; Ye, N.; Zhou, Q.; Liu, W.S.; Tang, J.C. Comparative study of tungsten powders prepared by carbon-hydrogen co-reduction and common hydrogen reduction. *Chin. J. Nonferrous Met.* **2018**, *28*, 743–748.
17. Chen, L.J.; Xie, Z.H.; Wang, R.X.; Tian, L.; Nie, H.P. Reduction Carbonization Mechanism Analysis of Tungsten Oxide Powder and Preparation of Ultrafine Tungsten Carbide Powder. *Rare Met. Cem. Carbides* **2018**, *46*, 11–20.
18. Song, H.L.; Jiang, P.G.; Liu, W.J.; Wang, Z.B. Research progress on hydrogen reduction kinetics of tungsten oxide. *Nonferrous Met. Sci. Eng.* **2017**, *8*, 64–69.
19. Vesel, A.; MoZetic, M.; Marianne, B.P. Sequential oxidation and reduction of tungsten/tungsten oxide. *Thin Solid Film* **2015**, *591*, 174–181. [CrossRef]
20. Yu, K.N.; Mao, M.H.; Liang, H.Z.; Xu, J. Preparation of Ultrafine Co Powder from Basic $Co(OH)_2$ Slurry by Hydrogen Reduction. *Chin. J. Process Eng.* **2001**, *1*, 62–65.
21. Fang, Z.Z.; Wang, H.; Kumar, V. Coarsening, densification, and grain growth during sintering of nano-sized powders—A perspective. *Int. J. Refract. Met. Hard Mater.* **2017**, *62*, 110–117. [CrossRef]
22. Wang, D.H.; Sun, G.D.; Zhang, G.H. Preparation of ultrafine Mo powders via carbothermic pre-reduction of molybdenum oxide and deep reduction by hydrogen. *Int. J. Refract. Met. Hard Mater.* **2018**, *75*, 70–77. [CrossRef]
23. Kang, H.J.; Jeong, Y.K.; Oh, S.T. Hydrogen reduction behavior and microstructural characteristics of WO_3 and WO_3-NiO powders. *Int. J. Refract. Met. Hard Mater.* **2019**, *80*, 69–72. [CrossRef]

24. Gu, S.Y.; Qin, M.L.; Zhang, H.A.; Ma, J.D.; Qu, X.H. Preparation of Mo nanopowders through hydrogen reduction of a combustion synthesized foam-like MoO_2 precursor. *Int. J. Refract. Met. Hard Mater.* **2018**, *76*, 90–98. [CrossRef]

25. Colton, R. Some Complex anioks containing rhenium. *Aust. J. Chem.* **1965**, *18*, 435–439. [CrossRef]

26. Bai, M.; Liu, Z.H.; Zhou, L.J.; Liu, Z.Y.; Zhang, C.F. Preparation of ultrafine rhenium powders by CVD hydrogen reduction of volatile rhenium oxides. *Trans. Nonferrous Met. Soc. China* **2013**, *23*, 538–542. [CrossRef]

27. Zhou, L.J. Research on Preparation of Ultrafine Rhenium Powder by Chemical Vapor Deposition. Master's Thesis, Central South University, Hunan, China, 2010.

28. Tang, J.J.; Sun, Y.; Hou, G.C.; Ding, Y.T.; He, F.; Zhou, Y.Z. Studies on Influencing Factors of Ammonium Rhenate Recovery from Waste Superalloy. *Appl. Sci.* **2018**, *8*, 2016. [CrossRef]

29. Tang, J.J.; Feng, L.; Zhang, C.W.; Sun, Y.; Wang, L.; Zhou, Y.Z.; Fang, D.W.; Liu, Y. The Influences of Stirring on the Recrystallization of Ammonium Perrhenate. *Appl. Sci.* **2020**, *10*, 656. [CrossRef]

30. Okal, J.; Tylus, W.; Kepi´nski, L. XPS study of oxidation of rhenium metal on γ-Al_2O_3 support. *J. Catal.* **2004**, *225*, 498–509. [CrossRef]

31. Kuznetsov, V.V.; Gamburg, Y.D.; Zhulikov, V.V.; Batalov, R.S.; Filatova, E.A. Re-Ni cathodes obtained by electrodeposition as a promising electrode material for hydrogen evolution reaction in alkaline solutions. *Electrochim. Acta* **2019**, *317*, 358–366. [CrossRef]

32. Iqbal, S.; Shozi, M.L.; Morgan, D.J. X-ray induced reduction of rhenium salts and supported oxide catalysts. *Surf. Interface Anal.* **2017**, *49*, 223–226. [CrossRef]

33. Zhang, B.; Sun, Y.Z.; Wang, W.H. Line Shapes Used in XPS and Background Subtraction. *Phys. Exam. Test.* **2011**, *1*, 18–23.

34. Hillig, W.B.; Adjerid, S.; Flaherty, J.E.; Hudson, J.B. The effect of combined diffusion and kinetic transport barriers on multi-phase solid state reactions with a vapour reactant. *J. Mater. Sci.* **1996**, *31*, 5865–5871. [CrossRef]

35. Ler, K.M.K.; Götze, L.C.; Rybacki, E.; Dresen, G.; Abart, R. Enhancement of solid-state reaction rates by non-hydrostatic stress effects on polycrystalline diffusion kinetics. *Am. Mineral.* **2010**, *95*, 1399–1407.

Article

Electrochemical Deposition of Al-Ti Alloys from Equimolar AlCl₃ + NaCl Containing Electrochemically Dissolved Titanium

Vesna S. Cvetković [1],*, Nataša M. Vukićević [1], Ksenija Milićević-Neumann [2], Srećko Stopić [2], Bernd Friedrich [2] and Jovan N. Jovićević [1]

[1]   Department of Electrochemistry, Institute of Chemistry, Technology and Metallurgy, National Institute, University of Belgrade, Njegoševa 12, 110000 Belgrade, Serbia; vukicevic@ihtm.bg.ac.rs (N.M.V.); jovicevic@ihtm.bg.ac.rs (J.N.J.)

[2]   IME Process Metallurgy and Metal Recycling, RWTH Aachen University, Intzestrasse 3, 52056 Aachen, Germany; ksenija@maanaelectric.com (K.M.-N.); sstopic@metallurgie.rwth-aachen.de (S.S.); bfriedrich@metallurgie.rwth-aachen.de (B.F.)

*   Correspondence: v.cvetkovic@ihtm.bg.ac.rs; Tel.: +381-11-3640-228

Received: 14 November 2019; Accepted: 31 December 2019; Published: 4 January 2020

Abstract: Al-Ti alloys were electrodeposited from equimolar chloroaluminate molten salts containing up to 0.1 M of titanium ions, which were added to the electrolyte by potentiostatic dissolution of metallic Ti. Titanium dissolution and titanium and aluminium deposition were investigated by linear sweep voltammetry and chronoamperometry at 200 and 300 °C. Working electrodes used were titanium and glassy carbon. The voltammograms on Ti obtained in the electrolyte without added Ti ions indicated titanium deposition and dissolution proceeding in three reversible steps: $Ti^{4+} \rightleftarrows Ti^{3+}$, $Ti^{3+} \rightleftarrows Ti^{2+}$ and $Ti^{2+} \rightleftarrows Ti$. The voltammograms recorded with glassy carbon in the electrolyte containing added titanium ions did not always clearly register all of the three processes. However, peak currents, which were characteristics of Al, Ti and Al-Ti alloy deposition and dissolution, were evident in voltammograms on both working electrodes used. A constant potential electrodeposition regime was used to obtain deposits on the glassy carbon working electrode. The obtained deposits were characterized by SEM, energy-dispersive spectrometry and XRD. In the deposits on the glassy carbon electrode, the analysis identified an Al and AlTi₃ alloy formed at 200 °C and an Al₂Ti and Al₃Ti alloy obtained at 300 °C.

Keywords: Al-Ti alloy; electrochemical co-deposition; chloroaluminate melt; XRD

---

1. Introduction

Intermetallic materials based on a combination of aluminium and titanium, which possess high specific strength and low weight and required stiffness and excellent oxidation resistance at elevated temperatures (particularly over 600 °C), are of increasing importance as new structural materials in aerospace industry and medicine [1–3]. Due to its ability to increase the temperature of titanium allotropic transformation, aluminum is the main alloying element for titanium. The density of aluminum is less than the density of titanium, so the addition of aluminum increases the specific strength of Ti alloys. High functional properties make the Ti-Al system the foundation of many titanium alloys. The presence of thermodynamically stable intermetallic phases in titanium-aluminum composite materials allows for and significantly enhances physical and mechanical characteristics of these systems [4].

Over the last thirty years, various processing methods have been studied to fabricate these intermetallic materials [3]. The most prominent methods for fabricating Al-Ti alloys are rapid solidification [2], sputter deposition [5], ball milling [6], mechanical alloying [7], spark plasma sintering [8], reaction sintering of elemental powders, etc. [3].

In general, Al-Ti alloys could be electrodeposited from electrolytes containing Ti(II) species [2,3,9–11]. Electrodeposition synthesis of aluminium-titanium-based materials is a tempting process that shows the potential to replace processing methods identified earlier [2,7]. The fundamental aspects of chemistry and electrochemistry of titanium ions in molten salt electrolytes have been investigated, but data on the electrochemical behaviour of titanium ions in molten chloride/fluoride salt electrolytes are scarce and contradictory. The main barrier for successful development of an electrochemical route for aluminium-titanium alloy production is associated with the existence of different oxidation states of dissolved titanium species, namely Ti(II), Ti(III) and Ti(IV) [9,12–15].

Electrochemical deposition of aluminium-titanium intermetallics has been investigated from either an Lewis acidic chloroaluminate molten salts electrolyte made of 2:1 $AlCl_3$-NaCl or $AlCl_3$-1-ethyl-3-methylimidazolium chloride ionic liquid (IL) [1–3,7,10]. In Lewis acidic 2:1 $AlCl_3$-NaCl electrolyte systems, authors particularly studied the influence of Ti^{2+} concentration on the alloy composition and found that, with low Ti^{2+} concentrations, alloy composition depended on current density. For example, an Al_3Ti alloy containing 25% atomic fraction of titanium was deposited only at low current densities [1]. With an increase in current density, the titanium content in the alloys decreased [2]. The concentration limit of titanium in the alloy composition was proposed to be due to a mechanism, by which an Al-Ti alloy forms through the reductive decomposition of a divalent species—$Ti(AlCl_4)_2$. The electrochemical reduction of Ti^{2+} ions to metallic Ti was not observed at potentials more positive than that required for aluminium deposition, but an Al_3Ti alloy was deposited onto a copper working substrate under specific deposition conditions [2].

In comparison to other molten salt electrolytes systems, the electrochemical behaviour of titanium ions in chloride melts is different because of the stability of various oxidation states of titanium ions, which is caused by the influence of electrolyte composition [12–15] and temperature [14,15].

However, to our knowledge, there is no information published that addresses electrodeposition of Al-Ti alloys from an equimolar chloroaluminate $AlCl_3$-NaCl molten electrolyte on glassy carbon (GC) at temperatures below 300 °C. Equimolar $AlCl_3$-NaCl electrolytes have been characterised in the following ways: (a) lower vapour pressure above an equimolar melt at the same temperature applied than on an acidic $AlCl_3$-NaCl electrolyte [16,17]; (b) the aluminium deposition potential from $AlCl_4^-$ ions in an equimolar and acidic $AlCl_3$-NaCl electrolyte was more negative than the aluminium deposition potential from $Al_2Cl_7^-$ ions in an acidic melt, which provided a larger potential distance to the titanium deposition potential [18,19]; (c) the deposition current density of aluminium was greater for the reaction: $AlCl_4^- + 3e^- \rightarrow Al + 4Cl^-$, than for the reaction: $4Al_2Cl_7^- + 3e^- \rightarrow Al + 7AlCl_4^-$ for the same value of an overpotential (exceeding −60 to −80 mV) recorded with the same $AlCl_3$ concentration in the melt [1,3,18,19].

The aim of the present paper is to study titanium and aluminium co-deposition from an equimolar chloroaluminate molten salt containing Ti ions introduced by electrochemically dissolved Ti metal. This novel electrodeposition route consisting of anodic dissolution of Ti and co-deposition of Ti and Al may be a useful route for Al-Ti alloy production.

2. Materials and Methods

An equimolar mixture of $AlCl_3$ and NaCl served as a base electrolyte [20], and preparation of the electrolyte was identical to those described in previous articles [21,22].

Electrochemical measurements and electrodeposition processes were carried out at 200 and 300 °C in a three-electrode electrochemical cell. In the cell used for titanium ion introduction into the equimolar $AlCl_3$-NaCl molten salt, the working electrode (WE, an anode) was a titanium plate (Ti 99.99% Alfa Aesar, Haverhill, MA, USA), the counter electrode was titanium and the reference electrode was an aluminium rod with a diameter of 3 mm (Al 99.999% Haverhill, MA, USA). In the cells used for aluminium and titanium deposition and co-deposition, the cathode was a glassy carbon

(GC, Alfa Aesar, Haverhill, MA, USA) cylinder, a titanium plate was used as the counter electrode and the reference electrode was an aluminium rod with a diameter of 3 mm.

All the reported potentials of WEs in this work were measured relative to the equilibrium potential of the aluminium reference electrode in the melt used under given conditions.

Before the experiment, the GC WE was polished with 0.05 μm alumina powder (Merck & Co., Kenilworth, NJ, USA) and cleaned several times by sonication in ethanol and Milli-Q water with each duration of 3 min.

The aluminium electrodes were etched in solutions made of 50 vol% HF, 15 vol% H_2O, 25 vol% ammonia solution and 5 vol% H_2O_2. Thereafter, the electrodes were rinsed with deionised water and absolute ethyl alcohol and dried before use.

The Ti WEs were etched in a mixture of HF + H_2O + H_2O_2 (volume ratio: 1:20:1), rinsed with distilled water and dried before use.

Argon flow was maintained in the cell, and the electrolyte was not stirred during experiments. After the electrodes were introduced into the electrolyte, the system was left for 5 to 10 min to achieve thermal equilibrium.

The study started with recording potentiodynamic polarization curves for the Ti WE in an equimolar $AlCl_3$-NaCl melt without previously added titanium at 200 or 300 °C. The potential (measured relative to the aluminium reference electrode) was scanned from a starting value, $E_I = 0.0$ V, to a final value, $E_F = 1.200$ V, with a scan rate of 1 mV·s^{-1}.

In linear sweep voltammetry (LSV) experiments when Ti was used as a WE, and an electrolyte without previously added titanium, the potential was changed with a scan rate of 20 mV s^{-1} from a potential slightly more negative than the open-circuit potential (OCP) of Ti to a different cathodic end potential (E_C), then back to anodic potential (E_A) and finally to the starting potential. In order to examine the anodic part of the voltammograms, LSV experiments were performed, starting from the open-circuit potential to the final anodic end potential (E_A), and back to slightly more negative potential than the OCP.

Ti ions were introduced into the electrolyte by electrochemical dissolution of titanium metal with constant potentials: at 200 °C, the potential was maintained at 0.500 V; and at 300 °C, the potential was maintained at 0.450 V.

The voltammograms obtained on the GC WE in the equimolar chloroaluminate molten salts with Ti ions present started from a potential E_I, usually 0.050 V more negative than the GC OCP (measured against the aluminium reference electrode), changed to a cathodic potential limit, E_F, and back to E_I with various sweep rates.

Controlled electrodeposition onto the GC electrode in the electrolyte with previously added titanium ions was initiated 5 min. after insertion of the WE into the melt in order to allow for thermal equilibrium. Titanium and aluminium were electrodeposited at a constant overpotential at two different temperatures (200 and 300 °C). After potentiostatic deposition, the WE was taken out of the cell, washed thoroughly with absolute ethanol (Zorka-Pharma, Šabac, Serbia) in order to remove any melt residue and dried in a desiccator furnished with silica gel. The morphology and the composition of the samples deposited were explored by a scanning electron microscope (VEGA 3 model; TESCAN, Brno, the Czech Republic), equipped with an energy-dispersive spectrometer (Oxford INCA 3.2, Oxford Instruments, High Wycombe, UK) and an optical microscope (VH-Z100R model; Keyence, Osaka, Osaka Prefecture, Japan).

The deposit collected from the GC WE obtained at 200 °C was analyzed by XRD on a Philips PW1050 powder diffractometer at room temperature with Ni-filtered Cu Kα radiation ($\lambda = 1.54178$ Å) and a scintillation detector within a 2θ range of 20–85° in steps of 0.05° and a scanning time of 5 s per step, and the deposit was obtained at 300 °C by SmartLab® X-ray diffractometer (Rigaku Co., Tokyo, Japan) using Cu Kα radiation ($\lambda = 1.542$ Å). The patterns were collected within a 2θ range of 10–90° at a scan rate of 0.5°/min with a divergent slit of 0.5 mm, operated at 40 kV and 30 mA. The phases formed during the deposition were identified by a comparison of the recorded diffraction peaks with the references from the Joint Committee on Powder Diffraction Standards (JCPDS) database.

3. Results and Discussion

3.1. Dissolution of Titanium

The composition of a solvent-fused salt has a dramatic influence on electrodeposition process of titanium [14,15]. The published works on melts used for titanium deposition (inorganic and organic melts and ILs) emphasize problems encountered with the control of electrolytes made by titanium salts dissolution. These involve titanium salt sublimation at elevated temperatures, titanium ions unwanted disproportionation and titanium oxide deposition onto electrodes including passivation. To avoid most of the mentioned problems, it was decided to introduce titanium into an equimolar $AlCl_3$-NaCl melt by electrochemically controlled dissolution. Figure 1 exhibits voltammograms obtained with a titanium electrode in an equimolar $AlCl_3$-NaCl melt at 200 °C.

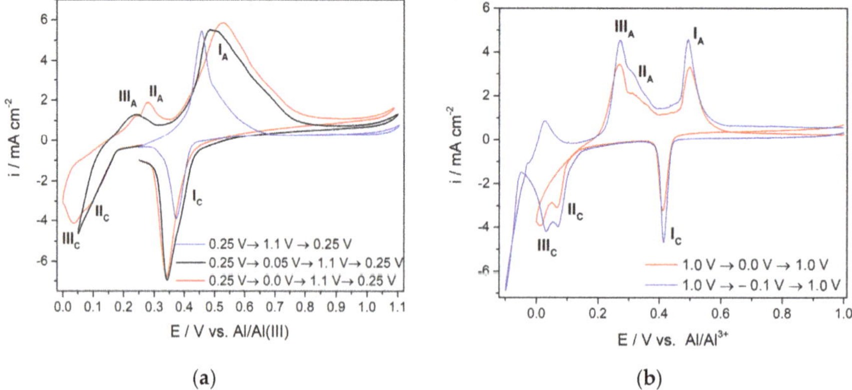

Figure 1. Voltammograms with a Ti working electrode in an equimolar $AlCl_3$-NaCl melt (T = 200 °C, v = 20 mV·s^{-1}): (**a**) potential changes during the first cycle from a starting point (E_I = 0.250 V) to an anodic potential limit (E_A = 1.100 V) and then back to 0.250 V and potential changes during the second and third cycles starting from E_I = 0.250 V to different cathodic potential limits E_C, then to E_A and finally to a value E_F = 0.250 V; (**b**) potential changes during cycles from a starting point (E_I = 1.000 V) to different cathodic potential limits and back to a final value (E_F = 1.000 V).

It was found that in the presence of aluminium ions in the equimolar chloroaluminate molten salt electrolyte, the electrochemical reduction of titanium ions to metallic titanium was complicated by the formation of intermediate oxidation states of Ti^{4+}, Ti^{3+} and Ti^{2+} [1,3]. These were recorded as cathodic peaks I_C (Ti^{4+}/Ti^{3+}), II_C (Ti^{3+}/Ti^{2+}) and III_C (Ti^{2+}/Ti) and their respective anodic counterparts I_A, II_A and III_A, shown in Figure 1. These observations were similar to the results reported on Pt and Ti electrodes from an $AlCl_3$ + N-(n-butyl)pyridinium chloride (mole ratio: 2:1) melt at 25 °C [23]. The potentials related to these processes can be read from voltammograms. Their values greatly depend on temperature, the composition of electrolytes and concentrations (amounts) of dissolved titanium ions in the melt [1,3,10,13–15,24]. In the equimolar $AlCl_3$-NaCl melt used, the average recorded values of the reversible potential for the pairs Ti^{4+}/Ti^{3+}, Ti^{3+}/Ti^{2+} and Ti^{2+}/Ti were approximately 0.410, 0.190 and 0.149 V. However, they can be identified also from the potentiodynamic polarization curve of titanium recorded in the used melt at 200 °C (Figure 2). It is apparent that the potentials designated as $E_{Ti^{2+}/Ti} \approx 0.200$ V, $E_{Ti^{3+}/Ti^{2+}} \approx 0.240$ V and $E_{Ti^{4+}/Ti^{3+}} \approx 0.370$ V are in reasonably good agreement with the reversible potentials determined by the peak pairs I_C/I_A, II_C/II_A and III_C/III_A from the voltammograms in Figure 1.

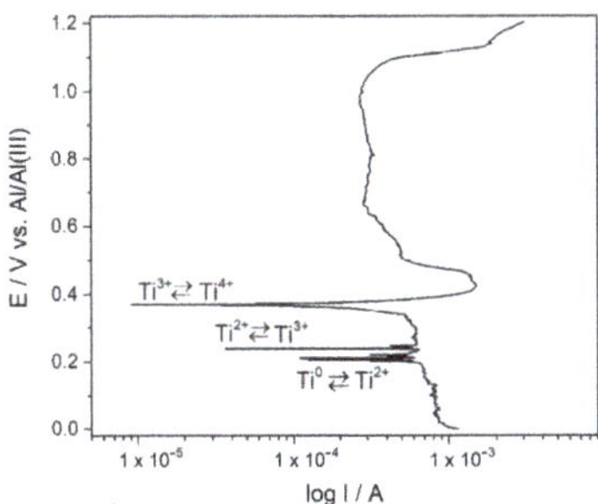

Figure 2. Potentiodynamic polarization curve of the Ti working electrode in the equimolar $AlCl_3$-NaCl melt at $T = 200\ °C$, $E_I = 0.0$ V and $E_F = 1.200$ V.

Two important features of titanium in the equimolar $AlCl_3$-NaCl melt at 200 °C should be mentioned:

(1) The changes of the peak shape that the reaction (I_C/I_A) exhibits when recorded with different sweep rates are presented in Figure 3. It was proposed [12–15] that, in all alkali chloride melts, this pair reflects redox reaction Ti^{4+}/Ti^{3+}. Using the analysis of the relationship between the peak maximum current densities (for both cathodic and anodic currents shown in Figure 3) and the square root of a scan rate used, it was found that the relationship is linear, which confirmed that the process is a simple diffusion-controlled reversible process [25]. The positions of the other cathodic and anodic peak currents on the voltammograms (namely Ti^{3+}/Ti^{2+} and Ti^{2+}/Ti) show the reversibility of the process as well. However, they were not defined well enough in order to conduct the same analysis.

(2) The reversible potential of $Ti^{2+} \rightleftarrows Ti$ in the equimolar $AlCl_3$ -NaCl melt was recorded at ≈ 0.200 V, which is a potential positive to that required for aluminium deposition. This finding was similar to the findings observed in different electrolytes [1,3].

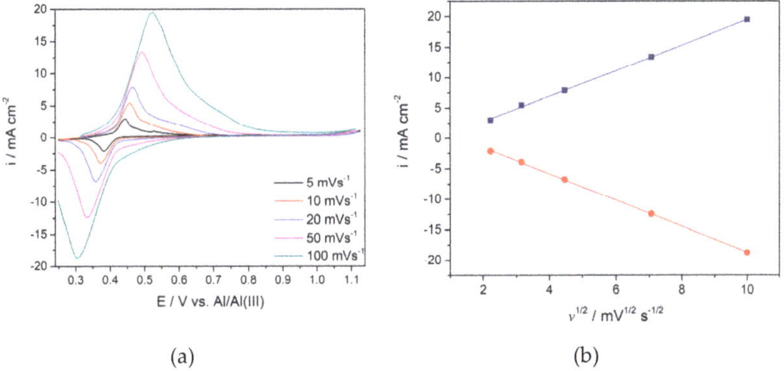

Figure 3. (**a**) Voltammgrams of the Ti working electrode in the equimolar $AlCl_3$-NaCl melt at T = 200 °C with different sweep rates; (**b**) plots of anodic and cathodic peak current densities vs. square root of scan rate calculated from (**a**).

In most molten chloride/fluoride electrolytes, there are equilibria between metallic titanium and Ti^{2+}, Ti^{3+} and Ti^{4+} ions [12–15]. According to some studies in chloride electrolyte systems, metallic titanium is usually in equilibrium with two different titanium species Ti^{2+} and Ti^{3+} [9,14]. The presence of different oxidation states of titanium ions in molten chloride electrolytes and the tendency for reoxidation or disproportionation reactions mostly cause poor current efficiency and deposited product quality [9,15]. Furthermore, the melt temperature has a significant influence on equilibrium and electrodeposition processes of titanium in aforementioned electrolytes [14,15].

Taking into account the voltammograms obtained in the system used (Figure 1), the electrodissolution of titanium was done potentiostatically at 0.500 V and 200 °C and at 0.450 V and 300 °C (Figure 4). The chosen anodic potentials were sufficient enough compared to the reversible potential of the Ti^{3+}/Ti^{4+} redox couple to sustain titanium dissolution at a current density of around 1 mA·cm^{-2}. The alternate rise and drop of the dissolution current recorded in Figure 4 is due to processes of dissolution-precipitation reactions on the electrode surface at a working potential applied. Similar effects at potentials, which were anodic but close to the Ti^{2+}/Ti^{3+} reversible potential, were addressed in the literature [1,3] and were attributed to the precipitation of a Ti^{3+} product, while the breakdown of a passive film was positioned at potentials, where Ti^{4+} species generation started, i.e., at potentials close to the Ti^{3+}/Ti^{4+} reversible potential.

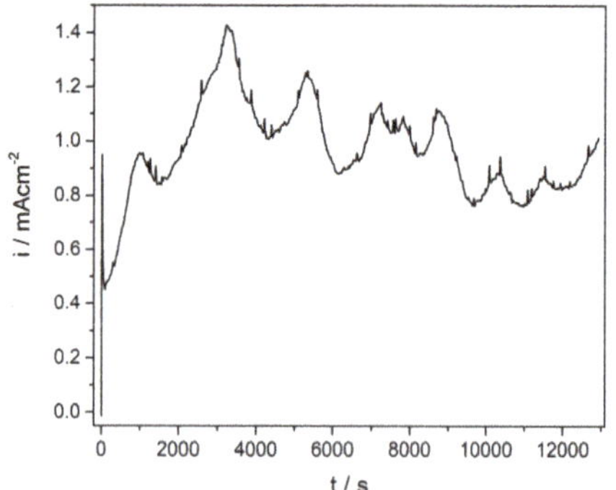

Figure 4. Anodic dissolution of the Ti working electrode at 0.500 V for 3.6 h in the equimolar AlCl$_3$–NaCl melt at $T = 200$ °C.

Thus, a melt that was equimolar in AlCl$_3$ and NaCl and contained ≈ 0.1 M titanium was prepared to be used in experiments involving titanium and aluminium electrodeposition on GC. The Ti molarity was calculated from the Ti anode mass lost during controlled potentiostatic dissolution (Faraday's law applied to Ti $\rightarrow$ Ti^{2+}) and from the slopes in Figure 3, following the procedure proposed in a similar system using Randles-Sevcik equation [3,26]:

$$i_p = 0.4463\left(\frac{F^3}{RT}\right)^{\frac{1}{2}} n^{\frac{3}{2}} AD_0^{\frac{1}{2}} C_0^* v^{\frac{1}{2}}, \tag{1}$$

where i_p is the peak current in amperes; v is the sweep rate in V·s^{-1}; C_0^* is the concentration in mol·cm^{-3}, D_0 is the diffusion coefficient in cm^2·s^{-1}; A is the area of an electrode in cm^2, n is the number of electrons, F is the Faraday's constant; R is the gas constant; T is the temperature. Both methods showed that the concentration of titanium in the melt used was around 0.1 M. A titanium anode was used to replace (by its dissolution) Ti ions reduced to titanium metal from the electrolyte. Thus, the Ti ions concentration was kept close to a wanted value during experiments.

3.2. Deposition of Titanium and Aluminium onto GC

The LSV results of the GC WE in the equimolar AlCl$_3$-NaCl melt used with anodically dissolved titanium and recorded at 200 and 300 °C are presented in Figures 5 and 6. The obtained voltammograms were very similar to those obtained on W, Cu, mild steel and Pt in chloride and fluoride/chloride inorganic melts and ILs with different titanium concentrations and working temperatures [1,3,10,15,23].

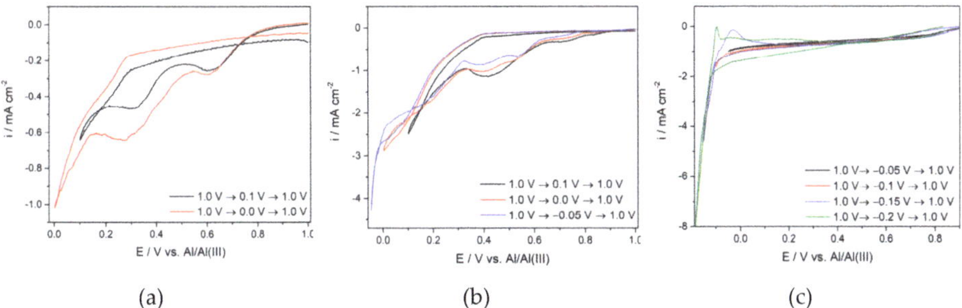

Figure 5. Voltammograms of the glassy carbon (GC) working electrode in the equimolar AlCl$_3$-NaCl melt containing anodically dissolved Ti, obtained at different cathodic potential limits and at T = 200 °C with different sweep rates: (**a**) ν = 5 mV·s^{-1}; (**b**) ν = 20 mV·s^{-1}; (**c**) ν = 5 mV·s^{-1}.

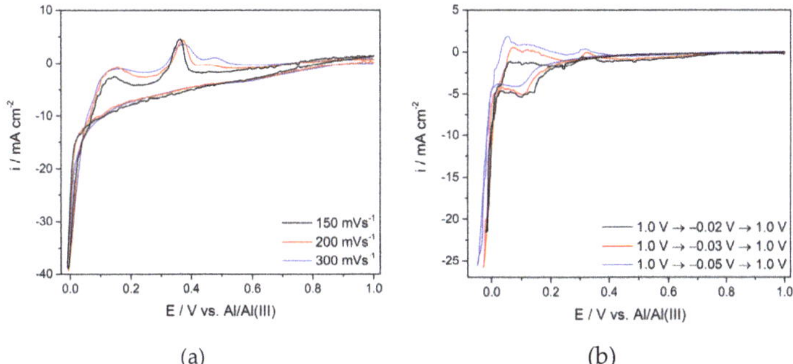

Figure 6. Voltammograms obtained on the GC working electrode in the equimolar AlCl$_3$-NaCl melt containing anodically dissolved Ti at T = 300 °C on different conditions: (**a**) potential change: E_I = 1.000 V to E_F = 0.0 V with different sweep rates; (**b**) at different cathodic potential limits at a constant sweep rate ν of 50 mV·s^{-1}.

The voltammograms in Figures 5 and 6 do not exhibit well-defined cathodic and anodic sides of the voltammograms, as was the case with the titanium WE in the same electrolyte with a much lower titanium concentration (Figure 1). At a lower temperature (Figure 5), the cathodic side of the voltammogram was better defined than its anodic counterpart and tentatively suggests three current increases reflecting all three steps of Ti^{4+} ions being reduced to Ti metal at potentials more positive than the aluminium reversible potential. It appears that the peak potentials, although not always easily identified, approached the values recorded for the same reactions on the Ti WE (see Figure 1). At a higher temperature, the peak current density structures were recognizable for both the cathodic and anodic sides of the voltammograms. However, the anodic side of the voltammograms was better defined, showing all three expected oxidation peaks after the applied cathodic potential limit was made more negative than −0.020 V.

When the cathodic potential limit was pushed further to an aluminium overpotential region, pronounced cathodic currents were recorded in the electrolyte with 0.1 M of titanium ions, independent of the temperature applied. The anodic response to the entrance into the aluminium overpotential region showed peaks, suggesting dissolution of Ti and Al and most probably dissolution of an Al-Ti alloy. The charge under the curve of the corresponding anodic peaks, however, did not always equal those under the curves of the cathodic counterparts.

The data obtained by LSV were used to define the potentials needed for the electrodeposition of an Al-Ti alloy, which was the primary goal of this work. The chronoamperometic response in the form of i = f(t) to an overpotential of −0.085 V applied to the GC WE for two hours at 200 °C in the equimolar

AlCl$_3$-NaCl melt containing 0.1 M of titanium ions is presented in Figure 7a. When the falling part of the transient in Figure 7a was transformed into the form of $i = f(t^{-1/2})$, a linear relationship became obvious [25], shown in Figure 7b, suggesting that after approximately 400 s, the Al and Ti deposition was proceeded under diffusion control. With all other conditions being the same, the chronopotentiograms recorded at 300 °C on the GC electrode were very similar in shape, with deposition current densities being about two times greater than the same potential applied at 200 °C.

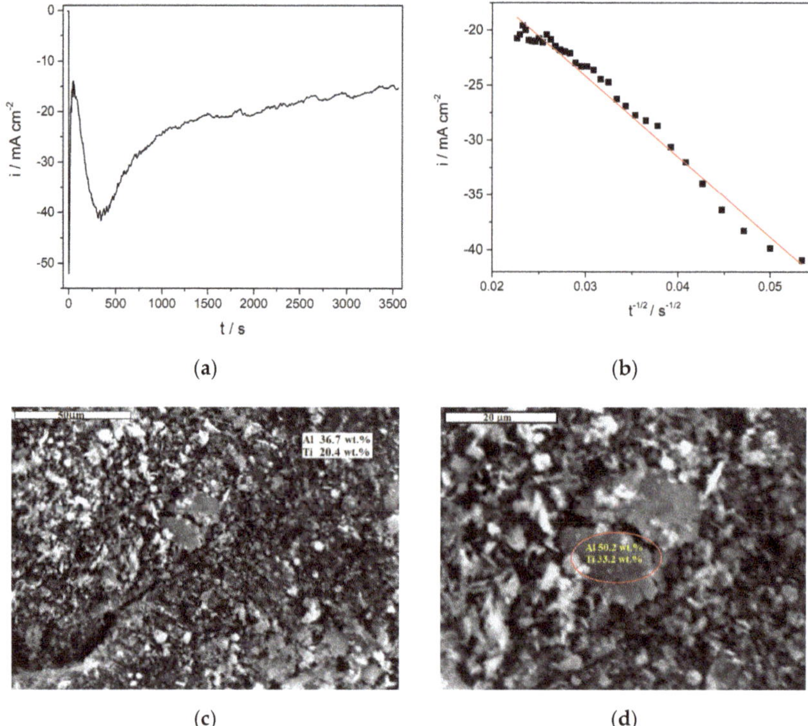

Figure 7. Potentiostatic deposition on the GC working electrode from the equimolar AlCl$_3$-NaCl melt containing anodically dissolved Ti at −0.085 V and 200 °C for two hours: (**a**) current–time transient of the deposition; (**b**) current as a function of $t^{-1/2}$ for the falling part of the transient in (a); (**c,d**) SEM photographs of the deposit obtained with energy-dispersive spectroscopy (EDS) results embedded.

For both temperatures applied, thick but nonuniform deposits were obtained (Figure 7c,d and Figure 8). At a higher magnification, grains of different sizes similar to those obtained from AlCl$_3$-BMIC ILs on mild steel published recently [27] can be observed. The energy-dispersive spectroscopy (EDS) analysis made from a larger portion of the same deposits (approximately 400 µm^2) reported 36.7 wt. % of Al and 20.4 wt. % of Ti (Figure 7c). In Figure 7d, a result of EDS analysis for one of the larger grains in the deposit is presented numerically (in the inserted circle), and it suggests presence of 50.2 wt. % of Al and 33.2 wt. % of Ti. The deposits obtained at 300 °C and −0.020 V after two hours showed a larger average grain size than the deposit obtained after two hours at −0.085 V and 200 °C (Figure 8).

Figure 8. Optical microscopy image of the deposit obtained on the GC electrode from the equimolar AlCl$_3$-NaCl melt containing anodically dissolved Ti at −0.020 V and 300 °C for two hours.

Data acquired from the XRD analysis of the deposits are shown in Figure 9.

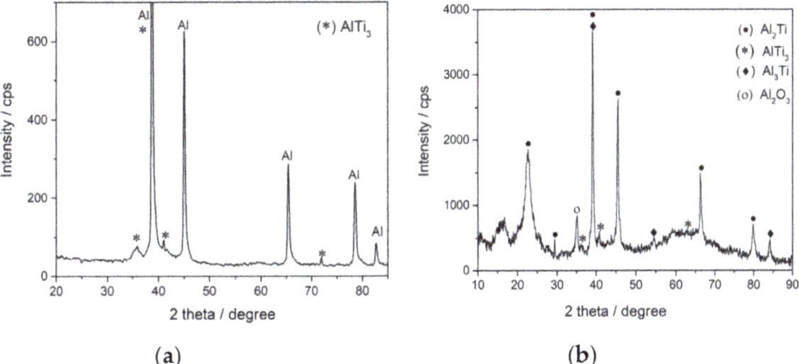

Figure 9. XRD patterns of the deposits obtained potentiostatically on the GC electrode on different conditions: (**a**) at −0.085 V and 200 °C for two hours; (**b**) at −0.020 V and 300 °C for two hours.

The analysis for the deposit obtained at 200 °C (Figure 9a) exhibited diffraction peaks at 2θ = 35.9° with reflection (200), 38.95° with reflection (002), 41.036° with reflection (201) and 71.97° with reflection (203), which are characteristics of a hexagonal AlTi$_3$ alloy (JCPDS No. 03-065-7534). Several stronger diffraction peaks at 2θ = 38.47°, 44.73°, 65.13°, 78.22° and 82.43° with the respective reflections (111), (200), (220), (311) and (222) should be attributed to face-centered cubic Al (JCPDS No. 00-004-0787). The spectrum indicates no evidence of additional peaks, implying that the deposit produced by the electrochemical deposition method in this study was relatively pure.

According to the data from XRD analysis of the deposit obtained on GC at a higher temperature (300 °C), dominating alloy appeared to be Al$_2$Ti (Figure 9b). The spectrum gives diffraction peaks at 2θ = 22.68°, 29.37°, 38.99°, 45.66°, 66.55° and 79.94°, with the respective reflections (101), (008), (116), (200), (220) and (316), which are characteristics of body-centered tetragonal Al$_2$Ti (JCPDS No. 00-052-0861). Peaks in the spectrum at 2θ = 39.15°, 54.32° and 84.15° with the respective reflections (112), (211) and (224) can be attributed to body-centered tetragonal Al$_3$Ti (JCPDS No.03-065-2667). The spectrum also indicates characteristics of a hexagonal AlTi$_3$ alloy, of which the peaks are at 2θ = 35.76°, 40.84° and 62.69° with the respective reflections (200), (201) and (103) (JCPDS No. 03-0-052-0859). The peak at 2θ = 35.15° can be associated with the trace of Al$_2$O$_3$ (JCPDS No: 00-046-1212), implying that it was not possible to handle a sample without exposing it to the atmosphere.

Each of the peaks attributed to a certain alloy were chosen from a group of five or seven highest peak intensities as defined by the JCPDS database for the mentioned alloy. However, due to the fact that the Al-Ti binary phase diagram is not fully understood [4,28,29] and the fact that we only presented initial experimental results, there is a space for improvement in attribution of the peak positions in XRD analysis of the obtained electrodeposits in the future.

The AlTi$_3$ alloy obtained on the GC WE from the equimolar AlCl$_3$-NaCl melt with titanium ions added by electrodissolution of metal Ti is a finding different from a TiAl$_3$ alloy predominantly produced electrochemically on Cu and mild steel from electrolytes reported in the literature [1,2,7,10,11]. The AlTi$_3$ alloy is hexagonal in structure, and in the Al-Ti phase diagram, it appears in the composition region of 13–25 wt. % Al (i.e., 75–87 wt. % Ti) at temperatures below 1200 °C [30]. The AlTi$_3$ intermetallic compound is largely accepted as having a variable composition, with a wide homogeneity domain and as an intermetallic compound formed by order-disorder transformation (α_{Ti}) $\leftrightarrow$ (AlTi$_3$). This aluminide is emerging as a revolutionary material for high-temperature applications and aeronautical industry [31, 32].

Al$_3$Ti is an intermediate phase of a tetragonal structure and appears in the binary phase diagram in a composition region, where the Al mass is between 75% and 100% [28]. The Al$_3$Ti alloy has a great potential application in aerospace and automobile as a high-temperature structural material, but it has poorest ductility among three typical Al-Ti alloys (AlTi$_3$, AlTi and Al$_3$Ti), which limits its engineering applications [33].

The Al$_2$Ti compound is considered stable up to 1216 °C, existing with a very narrow Al range between 60 and 67 at % [29]. Al$_2$Ti is one of the four intermetallic phases in the Ti-Al binary system that are stable below 1150 °C [34]. Al$_2$Ti is a very promising material for elevated-temperature applications.

However, although the Al-Ti binary phase diagram has been intensively studied, it still cannot be considered fully reliable [4,28,29]. It seems that there are 12 intermetallic compounds recognised [35]. According to the Ti-Al phase diagram, there can be up to seven stable intermetallic phases. The most stable intermetallic phases that increase the physico-mechanical properties of titanium aluminide are γ-TiAl, α2-Ti$_3$Al and γ-TiAl + α2-Ti$_3$Al. Lately, it was pointed out that AlTi$_3$ and AlTi intermetallic compounds are largely accepted as having variable compositions, with a wide homogeneity domain, while Al$_2$Ti and Al$_5$Ti$_2$ are accepted as having constant compositions [29]. According to the same authors, Al$_3$Ti is treated as an intermetallic compound with a variable composition or with a constant composition [29]. If thermal procedures are used, all of the abovementioned intermetallics can be produced at temperatures above 1110 °C.

When electrochemical co-deposition is used as a procedure for binary Al-Ti alloy generation, such high temperatures are not required. The mechanism of alloy formation in this case includes nanoscale relationships between adatoms of co-depositing metals on the substrate, such as interdiffusion in a solid state [18,36]. Interdiffusion phenomena were investigated in the Ti-Al system (a Ti region from 25 to 100 at %), but only at elevated temperatures (between 516 and 1200 °C). In a temperature range between 516 and 642 °C, metallic Ti, as well as Ti-Al alloys, was coated with a solid layer of metallic Al, and interdiffusion was studied. A well-adhered layer of TiAl$_3$ was formed, while no other intermetallic compounds were observed and no solid solution of Al in Ti was recorded [37]. In another study, interdiffusion phenomena were investigated in a Ti-Al system but at higher temperatures (between 768 and 972 °C) [38]. In the Al-rich part of the diagram, the Ti$_2$Al$_5$ phase was identified. It was found that, at temperatures between 768 and 865 °C, Ti was the more mobile element in the Ti$_3$Al phase whereas in the Al-richer compounds Al was the more mobile element at temperatures between 784 and 972 °C. The results of the studies in our work indicated appreciable interdiffusion between the co-deposited Al and Ti even at 200 and 300 °C, which led to the formation of Al$_3$Ti, Al$_2$Ti and AlTi$_3$.

4. Conclusions

The electrodeposition of Al-Ti alloys from an equimolar AlCl$_3$-NaCl melt on a GC electrode was successfully performed. It was shown that there is a novel way to obtain Al-Ti alloys, such as AlTi$_3$, Al$_2$Ti and Al$_3$Ti, in a very controlled manner under favorable and technologically suitable conditions.

The voltammograms generated from a system of a Ti WE in the equimolar AlCl$_3$-NaCl melt at 200 and 300 °C without introducing Ti ions indicated titanium deposition and dissolution proceeding in three reversible steps: $Ti^{4+} \rightleftarrows Ti^{3+}$, $Ti^{3+} \rightleftarrows Ti^{2+}$ and $Ti^{2+} \rightleftarrows Ti$, occurring at potentials more positive than the reversible potential of Al. The reversible potential of titanium in the equimolar $AlCl_3$-NaCl melt was identified as ≈ 0.200 V, and the starting deposition potential of titanium onto Ti was ≈ 0.020 V at 200 and 300 °C.

The titanium deposition starting potential on the GC electrode in the electrolyte made of an equimolar $AlCl_3$-NaCl melt containing ≈ 0.1 M of titanium ions appeared to be between 0.050 and 0.0 V for both temperatures applied (200 and 300 °C). However, we did not succeed in depositing pure titanium without aluminium, because their deposition potentials were very close.

The XRD analysis of the deposits revealed that $AlTi_3$, Al_2Ti and Al_3Ti alloys were generated on the GC electrode, with $AlTi_3$ dominating at a lower temperature and Al_2Ti dominating at a higher temperature.

The results obtained in this work suggest new possibilities of aluminium-titanium alloys formation (including $AlTi_3$) using low temperatures via a better controlled process.

Author Contributions: V.S.C. designed and managed the research and participated in the manuscript preparation; N.M.V. performed most of the experiments and participated in the manuscript preparation; K.M.-N., S.S. and B.F. helped with the corrections of the manuscript; J.N.J. supervised the experiments and the manuscript writing. S.S. and B.F. from RWTH Aachen University provided funding for publication. All authors discussed the results and commented on the manuscript. All authors have read and agreed to the published version of the manuscript.

Funding: Part of the research was supported by the funds of the bilateral research project (ID: 451-03-01971/2018-09/4) supported by the Ministry of Education, Science and Technological Development of the Republic of Serbia and German Academic Exchange Service (DAAD).

Acknowledgments: Vesna S. Cvetković and Nataša M. Vukićević acknowledge the financial support for the investigation received from the Ministry of Education, Science and Technological Development of the Republic of Serbia.

Conflicts of Interest: The authors declare no conflicts of interest.

References

1. Stafford, G.R. The Electrodeposition of Al_3Ti from Chloroaluminate Electrolytes. *J. Electrochem. Soc.* **1994**, *141*, 945–953. [CrossRef]
2. Tsuda, T.; Hussey, C.L.; Stafford, G.R.; Bonevich, J.E. Electrochemistry of titanium and the electrodeposition of Al-Ti alloys in the Lewis acidic aluminum chloride-1-ethyl-3-methylimidazolium chloride melt. *J. Electrochem. Soc.* **2003**, *150*, C234–C243. [CrossRef]
3. Stafford, G.R.; Moffort, T.P. Electrochemistry of Titanium in Molten $2AICI_3$-NaCI. *J. Electrochem. Soc.* **1995**, *142*, 3288–3296. [CrossRef]
4. Kosova, N.; Sachkov, V.; Kurzina, I.; Pichugina, A.; Vladimirov, A.; Kazantseva, L.; Sachkova, A. The preparation of the Ti-Al alloys based on intermetallic phases. *IOP Conf. Ser. Mater. Sci. Eng* **2016**, *112*, 012039. [CrossRef]
5. Yan, Q.; Yoshioka, H.; Habazaki, H.; Kawashima, A.; Asami, K.; Hashimoto, K. Passivity and its breakdown on sputter-deposited amorphous AlTi alloys in a neutral aqueous solution with Cl^-. *Corros. Sci.* **1990**, *31*, 401–406. [CrossRef]
6. Moon, K., II; Lee, K.S. Study of the microstructure of nanocrystalline Al-Ti alloys synthesized by ball milling in a hydrogen atmosphere and hot extrusion. *J. Alloys Compd.* **1999**, *291*, 312–321. [CrossRef]
7. Stafford, G.R.; Tsuda, T.; Hussey, C.L. Order/disorder in electrodeposited aluminum-titanium alloys. *J. Min. Metall.* **2003**, *39*, 23–42. [CrossRef]
8. Yang, Y.F.; Qian, M. Spark plasma sintering and hot pressing of titanium and titanium alloys. In *Titanium Powder Metallurgy: Science, Technology and Application*; Elsevier Inc.: Amsterdam, The Netherlands, 2015; pp. 219–236. [CrossRef]
9. Haarberg, G.M.; Kjos, O.S.; Martinez, A.M.; Osen, K.S.; Skybakmoen, E.; Dring, K. Electrochemical behaviour of dissolved titanium species in molten salts. *ECS Trans.* **2010**, *33*, 167–173. [CrossRef]
10. Tsuda, T.; Hussey, C.L.; Stafford, G.R. Electrodeposition of Titanium-aluminum alloys in the lewis acidic aluminum chloride-1-ethyl-3-methylimidazolium chloride molten salt. *ECS Proc. Vol.* **2002**, *2002*, 650–659. [CrossRef]

11. Uchida, J.; Seto, H.; Shibuya, A. Electrodeposition of Al-Ti alloy from chloroaluminate bath. *J. Surf. Finish. Soc. Jpn.* **1995**, *46*, 1167–1172. [CrossRef]

12. Uda, T.; Okabe, T.H.; Waseda, Y.; Awakura, Y. Electroplating of titanium on iron by galvanic contact deposition in NaCl-TiCl$_2$ molten salt. *Sci. Technol. Adv. Mater.* **2006**, *7*, 490–495. [CrossRef]

13. Zhu, X.; Wang, Q.; Song, J.; Hou, J.; Jiao, S.; Zhu, H. The equilibrium between metallic titanium and titanium ions in LiCl–KCl melts. *J. Alloys Compd.* **2014**, *587*, 349–353. [CrossRef]

14. Song, J.; Wang, Q.; Wu, J.; Jiao, S.; Zhu, H. The influence of fluoride ions on the equilibrium between titanium ions and titanium metal in fused alkali chloride melts. *Faraday Discuss.* **2016**, *190*, 421–432. [CrossRef] [PubMed]

15. Song, J.; Xiao, J.; Zhu, H. Electrochemical behavior of titanium ions in various molten alkali chlorides. *J. Electrochem. Soc.* **2017**, *164*, E321–E325. [CrossRef]

16. Trémillon, B.; Letisse, G. Proprietes en solution dans le tetrachloroaluminate de sodium fondu I. systemes "acide-base". *J. Electroanal. Chem. Interfacial Electrochem.* **1968**, *17*, 371–386. [CrossRef]

17. Del Duca, B.S. Electrochemical behavior of the aluminum electrode in molten salt electrolytes. *J. Electrochem. Soc.* **1971**, *118*, 405–411. [CrossRef]

18. Stafford, G.R.; Hussey, C.L. Electrodeposition of Transition Metal-Aluminum Alloys from Chloroaluminate Molten Salts. In *Advances in Electrochemical Science and Engineering*; Alkire, R.C., Kolb, D.M., Eds.; Wiley-VCH Verlag GmbH: Winheim, Germany, 2001; Volume 7, pp. 275–348.

19. Kan, H.M.; Wang, Z.W.; Wang, X.Y.; Zhang, N. Electrochemical deposition of aluminum on W electrode from AlCl$_3$-NaCl melts. *Trans. Nonferrous Met. Soc. China* **2010**, *20*, 158–164. [CrossRef]

20. Berg, W.R.; Hjuler, A.H.; Bjerrum, N. Phase diagram of the NaCl-AlCl$_3$ system near equimolar composition, with determination of the cryoscopic constant, the enthalpy of melting and oxid contaimination. *Inorg. Chem.* **1984**, *23*, 557–565. [CrossRef]

21. Radović, B.S.; Cvetković, V.S.; Edwards, R.A.H.; Jovićević, J.N. Al–Fe alloy formation by aluminium underpotential deposition from AlCl$_3$+NaCl melts on iron substrate. *Int. J. Mater. Res.* **2011**, *102*, 59–68. [CrossRef]

22. Vukićević, N.M.; Cvetković, V.S.; Jovanović, L.; Stevanović, S.I.; Jovićević, J.N. Alloy Formation by Electrodeposition of Niobium and Aluminium on Gold from Chloroaluminate Melts. *Int. J. Electrochem. Sci.* **2017**, *12*, 1075–1093. [CrossRef]

23. Ali, M.R.; Nishikata, A.; Tsuru, T. Electrodeposition of Al-Ti alloys from aluminum chloride-N-(n-butyl)pyridinium chloride room temperature molten salt. *Indian J. Chem. Technol.* **2003**, *10*, 14–20.

24. Endres, F.; Zein El Abedin, S.; Saad, A.Y.; Moustafa, E.M.; Borissenko, N.; Price, W.E.; Wallace, G.G.; MacFarlane, D.R.; Newman, P.J.; Bund, A. On the electrodeposition of titanium in ionic liquids. *Phys. Chem. Chem. Phys.* **2008**, *10*, 2189–2199. [CrossRef] [PubMed]

25. Greef, R.; Peat, R.; Peter, L.M.; Pletcher, D.; Robinson, J. *Instrumental Methods in Electrochemistry*, 1st ed.; Kemp, T.J., Ed.; Ellis Horwood Limited: Chichester, UK, 1985.

26. Bard, A.J.; Faulkner, L.R. *Electrochemical Methods. Fundamentals and Applications*, 2nd ed.; Jonh Wiley & Sons, Inc.: New York, NY, USA, 2001.

27. Xu, C.; Hua, Y.; Zhang, Q.; Li, J.; Lei, Z.; Lu, D. Electrodeposition of Al-Ti alloy on mild steel from AlCl$_3$ -BMIC ionic liquid. *J. Solid State Electrochem.* **2017**, *21*, 1349–1356. [CrossRef]

28. Landolt, H.; Börnstein, R. *Phase Equilibria, Crystallographic and Thermodynamic Data of Binary Alloys*; Springer: Berlin/Heidelberg, Germany, 1991. [CrossRef]

29. Batalu, D.; Coşmeleaţă, G.; Aloman, A. Critical analysis of the Ti-Al phase diagrams. *UPB Sci. Bull. Ser. B Chem. Mater. Sci.* **2006**, *68*, 77–90.

30. Massalski, T.B.; Okamoto, H.; Subramanian, P.R.; Kacprzak, L. *Binary Alloy. Phase Diagrams*, 2nd ed.; Massalski, T., Okamoto, H., Subramanian, P.R., Kacprzak, L., Eds.; ASM International, Materials Park: Geauga County, OH, USA, 1990; Volume 1–3.

31. Djanarthany, S.; Viala, J.C.; Bouix, J. An overview of intermetallic matrix composites based on Ti$_3$Al and TiAl. *Mater. Chem. Phys.* **2001**, *72*, 301–319. [CrossRef]

32. Froes, F.H.; Suryanarayana, C.; Eliezer, D. Synthesis, properties and applications of titanium aluminides. *J. Mater. Sci.* **1992**, *27*, 5113–5140. [CrossRef]

33. Wei, N.; Han, X.; Zhang, X.; Cao, Y.; Guo, C.; Lu, Z.; Jiang, F. Characterization and properties of intermetallic Al$_3$Ti alloy synthesized by reactive foil sintering in vacuum. *J. Mater. Res.* **2016**, *31*, 2706–2713. [CrossRef]

34. Benci, J.E.; Ma, J.C.; Feist, T.P. Evaluation of the intermetallic compound Al$_2$Ti for elevated-temperature applications. *Mater. Sci. Eng. A* **1995**, *192–193*, 38–44. [CrossRef]

35. Villars, P.; Calvert, L. *Pearson's Handbook of Crystallographic Data for Intermetallic Phases*, 2nd ed.; Villars, P., Calvert, L., Eds.; ASM International, Materials Park: Geauga, OH, USA, 1991.

36. Cvetković, V.S.; Vukićević, N.M.; Jovićević, J.N. Aluminium and Magnesium Alloy Synthesis by Means of Underpotential Deposition from Low Temperature Melts. In *Metals and Metal-Based Electrocatalytic Materials for Alternative Energy Sources and Electronics*; Stevanović, J., Ed.; Nova Science Publisher: New York, NY, USA, 2019; pp. 371–423.

37. Van Loo, F.J.J.; Rieck, G.D. Diffusion in the titanium-aluminium system-I. Interdiffusion between solid Al and Ti or Ti-Al alloys. *Acta Metall.* **1973**, *21*, 61–71. [CrossRef]

38. Van Loo, F.J.J.; Rieck, G.D. Diffusion in the titanium-aluminium system-II. Interdiffusion in the composition range between 25 and 100 at.% Ti. *Acta Metall.* **1973**, *21*, 73–84. [CrossRef]

Article

Application of the Flotation Tailings as an Alternative Material for an Acid Mine Drainage Remediation: A Case Study of the Extremely Acidic Lake Robule (Serbia)

Nela Petronijević [1,*], Srđan Stanković [1], Dragana Radovanović [2], Miroslav Sokić [1], Branislav Marković [1], Srećko R. Stopić [3] and Željko Kamberović [4]

[1] Center for Metallurgical Technologies, Institute for Technology of Nuclear and other Mineral Raw Materials, Belgrade 11000, Serbia; s.stankovic@itnms.ac.rs (S.S.); m.sokic@itnms.ac.rs (M.S.); b.markovic@itnms.ac.rs (B.M.)
[2] Innovation Centre of the Faculty of Technology and Metallurgy Ltd., Belgrade 11000, Serbia; divsic@tmf.bg.ac.rs
[3] IME Process Metallurgy and Metal Recycling, RWTH Aachen University, Intzestraße 3, 52056 Aachen, Germany; sstopic@metallurgie.rwth-aachen.de
[4] Faculty of Technology and Metallurgy, University of Belgrade, Belgrade 11000, Serbia; kamber@tmf.bg.ac.rs
* Correspondence: n.petronijevic@itnms.ac.rs; Tel.: +381-11-3691-722

Received: 25 November 2019; Accepted: 18 December 2019; Published: 20 December 2019

Abstract: Flotation tailings rich in carbonate minerals from the tailings deposit of the copper mine Majdanpek (Serbia) were applied for neutralization of the water taken from the extremely acidic Lake Robule (Bor, Serbia). Tests conducted in Erlenmeyer flasks showed that after neutralization of the lake water to pH 7, over 99% of aluminum (Al), iron (Fe), and copper (Cu) precipitated, as well as 92% of Zn and 98% of Pb. In order to remove residual Mn and Ag, the water was further treated with NaOH. After treatment with NaOH, all concentrations of the metals in the lake water samples were below discharge limits for munical wastewater according to the national legislation of the Republic of Serbia. The results of this work suggest that mining waste could be used for active neutralization of the acid mine drainage. The use of the mining waste instead of lime could reduce the costs of the active treatment of the acid mine drainage.

Keywords: acid mine drainage (AMD); flotation tailings; AMD neutralization; metals' precipitation; polluted site remediation; synergy of processes

---

1. Introduction

Environmental pollution by acid mine drainage (AMD) is a widespread problem in mining impacted areas [1,2]. One such area is located near the town of Bor, in the eastern part of Serbia. Mining activities in the region of Bor began in 1903. Approximately 7×10^8 tons of mining waste (overburden and flotation tailings) have been deposited in the close proximity of the town [3]. An extremely acidic water body named Lake Robule was formed at the foot of the overburden deposit, just a few kilometers from the center of the town. The length of the lake is approximately 400 m, and the width at its widest part is approximately 130 m. The water is characterized by a low pH (approximately 2–2.5) and very high concentrations of ferric iron, causing the deep red color of the lake. High concentrations of Cu, Zn, Al, Mn, and other metals were also detected in the lake water [4–6]. Approximately 10 L s^{-1} of acidic water flows from the lake to the Bor River through the drainage pipe [7]. One of the attempts to find an environmentally and economically feasible method to treat this accumulated AMD was conducted by Pavlović et al. [8]. They used a laboratory scale cascade line system with three reactors for selective precipitation of metals such as iron, copper, nickel, and arsenic from the synthetic solution that resembled acidic effluent from the open pit mine from the Bor area. The neutralizing agent was 1 M NaOH. Recently, Masuda et al. [9] applied hydrated lime ($Ca(OH)_2$) for neutralization of acidity and selective precipitation of metals from the water collected from Lake Robule using semi-industrial scale equipment for active AMD treatment with two chemical reactors. Stopic et al. [10] used red mud from Greece and Germany firstly for neutralization of AMD (pH value of 2.3) from South Africa and for precipitation of copper. Mwewa et al. [11] performed precipitation of poly-alumino-ferric sulfate coagulant for wastewater treatment using AMD solution from South Africa.

There are active and passive methods in the treatment of AMD [12,13]. The conventional method for active neutralization of the acid mine drainage is the application of $Ca(OH)_2$. This process is fast and efficient; hydrated lime increases the pH of the solution, resulting in precipitation of the dissolved metals in the form of metal hydroxides [14]. The main disadvantage of this technology is generation of the voluminous sludge. The price of the lime and need for the dehydration, solidification, and stabilization of the sludge after treatment increase the capital and operational costs of the AMD neutralization process [14–16]. More recently, passive methods have been attracting more attention due to reduced energy and maintenance costs [13]. With a view toward sustainable development, many studies have focused on finding alternative materials for AMD neutralization. Kaur et al. [17] investigated the application of the waste material from the alumina refining industry (Bayer liquor and precipitates formed by sea water neutralization of the Bayer liquor) to treat AMD from mine pit water. Kefeni et al. [18] reviewed technologies for AMD treatment, including alternative approaches for AMD neutralization by waste materials, such as the application of coal combustion by-products (fly ash, bottom ash, flue gas desulphurization materials), recycled concrete aggregates, and cryptocrystalline magnesite tailings from the gold extraction industry. Moodley et al. [19] also reviewed alternative methods for AMD remediation with a focus on industrial by-products, such as by-products from the paper mill and steel mill industries, meat industry, tire manufacturing, and phosphate waste rock.

The aim of the research is to investigate the ability of flotation tailings, a significant voluminous waste of the same industry that causes the formation of AMD in the Bor area, to neutralize and purify water collected from the acidic Lake Robule. Flotation tailings are a waste material generated during the production of mineral concentrate by froth flotation. Approximately 99 wt. % of the processed ore becomes flotation tailings [20]. Flotation tailings are consisted mostly of gangue minerals, but a considerable amount of carbonate minerals (calcite and dolomite) identified in particular flotation tailings samples collected from the dump of Copper Mine Majdanpek indicates a possible acid neutralization potential of the material. The motivation of the research is not only to investigate the ability of flotation tailings to neutralize and purify water collected from the acidic Lake Robule, but also to remove iron in order to prepare a solution for recovery of valuable elements such as copper, gold, aluminum, and silver in our future work. In order to predict the behavior of metals and especially their precipitation, the geochemical software will be tested and discussed with experimental results.

2. Materials and Methods

2.1. Sampling

Copper Mine Majdanpek is a part of the mining and smelting industry in Bor, Serbia. The mining operations started in 1963; and since then, 378 million tons of flotation tailings have been deposited in a large dump. Flotation tailings samples were dug out from cubes of a 0.04 m^3 volume (0.2 m × 0.2 m × 1 m) at 25 equidistant spatial locations (the distance between locations was 50 m), forming a square shaped network, and put in the separate bags. In the laboratory, the composite sample was prepared by the quartering method.

A total of 10 water samples was collected in March 2018 from the pipe that drains water from Lake Robule. The main physicochemical parameters (pH, Eh, temperature) were measured on the spot (Hanna Instruments HI98196, Hanna Instruments Deutschland GmbH, Vöhringen, Germany).

Containers (1 L, high density polyethylene) were rinsed with concentrated HNO_3 (Tehnohemija, Belgrade, Serbia) and deionized water before collecting samples. The filled containers were immediately transported to the laboratory the same day. Containers were stored at 4 °C in the refrigerator. A schematic geographic map of Serbia with the location of Copper Mine Majdanpek and the extremely acidic Lake Robule is given in Figure 1.

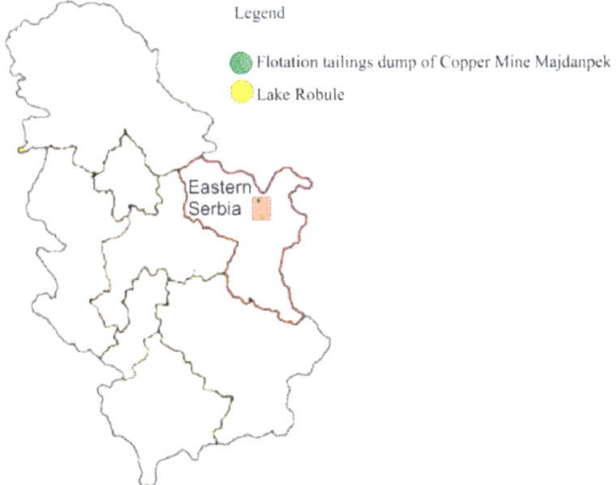

Figure 1. Schematic geographic map of Serbia with the location of Copper Mine Majdanpek and the extremely acidic Lake Robule.

2.2. Characterization of the Flotation Tailings

The chemical characterization of the flotation tailings was performed by dissolving a sample of the material in aqua regia and measuring the concentration of the selected elements by the Atomic Absorption Spectroscopy (AAS) method using Perkin Elmer Aanalyst 300 (PerkinElmer, Inc, Norwalk, CT, USA).

The concentration of the selected chemical elements in the water samples was measured by the Inductively Coupled Plasma Optical Emission Spectroscopy (ICP-OES) method (Spectro Genesis, Spectro Analytical Instruments, Kleve, Germany).

The sulfate concentration in the lake water sample was determined gravimetrically with $BaCl_2$. Carbonates and hydrogen carbonates were determined by titration with 0.1 M HCl using phenol phthalein and methyl orange (Tehnohemija, Belgrade, Serbia) as indicators.

Mineralogical characterization of the tailings was performed by the microscopy and XRD (X-Ray Diffraction, Zeiss-Jena, Oberkochen, Germany) methods. The polarizing microscope Carl-Zeiss, Model "JENAPOL-U" equipped with 10×, 20×, 50×, and 100× (oil immersion) objectives and a system for photomicrography ("Axiocam105 color" camera and "Carl Zeiss Axio Vision SE64 Rel. 4.9.1." software package with Multiphase module), was used for microscope investigations in reflected light. The XRD patterns were obtained using a Philips PW-1710 automated diffractometer with a Cu tube operated at 40 kV and 30 mA. The instrument was equipped with a diffracted beam curved graphite monochromator and an Xe filled proportional counter. The diffraction data were collected in the 2θ Bragg angle range of 4–65°, counting for 1 s. Semi-quantitative analysis of the data obtained by XRD was performed by "Powder Cell" computer software [21], (Federal Institute for Materials Research and Testing, Berlin, Germany).

2.3. Acid Neutralization Capacity

Acid Neutralization Capacity (ANC) represents the ability of a material to neutralize acid and remain stable under the external effluence of the environment [22]. In order to investigate and compare the ANC of flotation tailings before and after the treatment with the lake water, both samples of the tailings (fresh and treated) were subjected to the ANC test described by Stegemann and Cote [23]. The test procedure consisted of a series of leaching tests with increasing concentrations of nitric acid (HNO_3) and measuring the pH values of the leachates. Samples and solutions with different nitric acid contents, in the range from 0 (distilled water) to 2 H^+ eq/kg, with a solid to liquid ratio of 10, were rotated in polyethylene bottles ($V = 50$ mL) for 48 h prior to centrifugation and measurement of pH. Titration curves were obtained by plotting measured pH values versus acid content.

2.4. Determination of the Optimal Quantity of Flotation Tailings Required for Neutralization of the Lake Water Samples and Metal Precipitation

The neutralization test was performed with increasing quantities of the flotation tailings in order to identify the optimal solid to liquid ratio for neutralization of the lake water. The tests were conducted in eight 100 mL Erlenmeyer flasks containing 50 mL of water with increasing pulp density: 1, 3, 5, 10, 15, 20, 25, 30, and 40%. The Erlenmeyer flasks were put in an incubated orbital shaker (Heidolph Unimax 1010, Heidolph, Shwabach, Germany) at a temperature of 25 °C and a shaking speed of 250 rpm. After two hours of shaking, the samples were put to rest for seven days at room temperature.

After determination of the optimal quantity of the flotation tailings for the neutralization of the lake water, the precipitation of metal cations as a function of time and pH was determined. Nine 100 mL Erlenmeyer flasks were filled with 50 mL of the lake water, and 7.5 g of the flotation tailings were added to each flask. Flasks were placed in an orbital shaker under the same conditions as in the previous test. After agitation was stopped, the solution was left to stand for 30 min in order to let the particles settle out. Samples were collected after 5, 10, 15, and 30 min and 2, 4, 24, 72, and 168 h. The collected samples were filtered by using filter paper and analyzed by ICP-OES.

2.5. Treatment of the Lake Water Samples with Hydrated Lime and NaOH

After treatment of lake water samples with flotation tailings, the water was further treated with hydrated lime $Ca(OH)_2$ in order to increase its pH to 10, as is shown in Figure 2. A flask with 50 mL of the lake water previously treated with flotation tailings was put on a magnetic stirrer; the pH of the water was measured during the experiment by a laboratory pH meter (Hach Sension + MM340). Hydrated lime was carefully added until the pH of the solution reached a value of 10.

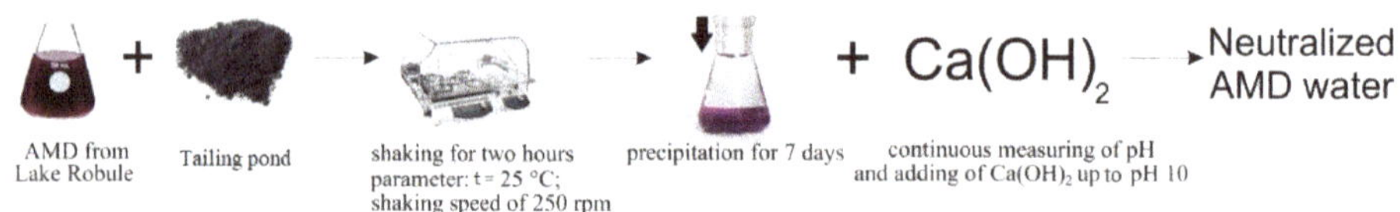

Figure 2. Schematic description of AMD water treatment with flotation tailings from the dump of Copper Mine Majdanpek.

In order to compare the results obtained by lake water neutralization with flotation tailings, the lake water was also neutralized with NaOH. Two Erlenmeyer flasks with 50 mL of the lake water were put on a magnetic stirrer and treated with NaOH and flotation tailings until the pH value of the water reached 7.

2.6. Simulation of Metal Precipitation Using PHREEQC Software

PHREEQC is a computer program designed to perform a variety of aqueous geochemical calculations [24], and in numerous research works [12,13,25,26], it was used to explain and support the experimental results addressed to the behavior of contaminants during a specific treatment. Here, the PHREEQC software (USGS, Reston, VA, USA) was applied for the modelling of the aqueous speciation of the water from Lake Robule and the saturation index calculations of the solid phases formed during the treatment of lake water by flotation tailings based on the pH and concentrations of elements in the solutions. An individual simulation was made for each defined time during the experiment according to the measured values of pH and the concentration of metals as input values for the calculations. Water volume, solid to liquid ratio, and temperature, as constant parameters during the experiment, were also input values for the calculation. Saturation Indices (SI) are defined as the logarithm of the ratio of the Ion Activity Product (IAP) to the solid solubility product (K_s), Equation (1).

$$SI = \log(IAP/K_s) \tag{1}$$

A positive SI indicates precipitation of the mineral, while a negative value of SI indicates mineral dissolution. Values of SI $\geq$ 0 were chosen to specify the controlling mineral for the constituent element's precipitation.

3. Results and Discussion

3.1. Physical and Chemical Properties of the Water Collected from Lake Robule

The physical and chemical properties of the water collected from Lake Robule are presented in Table 1. The acidic nature of the Lake Robule water was confirmed by a pH value of 2.47. The most dominant ions in the water samples were sulfate ions (7.5 g/L), Al^{2+} (1017.62 mg/L), and Fe ions (287 mg/L), almost completely in the form of Fe^{3+} (286.9 mg/L).

Table 1. Physico-chemical properties of the water collected from Lake Robule.

Characteristic	Unit	Value	Characteristic	Unit	Value
Temperature	°C	7	Fe^{3+}	mg/L	286.9
Color	-	Yes	Fe^{2+}	mg/L	0.013
Odor	-	None	Boron (B)	mg/L	0.0201
pH	-	2.47	Vanadium (V)	mg/L	<0.002
Eh	-	615.1	Cadmium (Cd)	mg/L	0.012
Iron (Fe)	mg/L	287	Selenium (Se)	mg/L	<0.001
Chromium (Cr)	mg/L	0.002	Silver (Ag)	mg/L	0.013
Copper (Cu)	mg/L	66.39	Carbonates (CO_3^{2-})	mg/L	1.2
Nickel (Ni)	mg/L	0.6	Hydrogen carbonates (HCO_3^-)	mg/L	2.44
Arsenic (As)	mg/L	<0.007	Antimony (Sb)	mg/L	<0.001
Zinc (Zn)	mg/L	17.6	Barium (Ba)	mg/L	<0.001
Lead (Pb)	mg/L	0.188	Beryllium (Be)	mg/L	0.013
Manganese (Mn)	mg/L	66	Cobalt (Co)	mg/L	<0.001
Aluminum (Al)	mg/L	1017.62	Sulfate (SO_4^{2-})	g/L	7.5

3.2. Characterization of the Flotation Tailings' Sample from Copper Mine Majdanpek

The chemical composition of the flotation tailings' composite sample collected from Copper Mine Majdanpek is presented in Table 2, while the results of the X-ray diffraction analysis are presented on Figure 3.

The most abundant minerals in the analyzed flotation tailings sample identified by X-ray diffraction analysis were quartz, pyrite, and carbonates (calcite and dolomite), followed by feldspar, clay minerals, mica, and illite, which were less abundant. The results of the semi-quantitative analysis of the data obtained by XRD were: quartz $\approx$ 50–55%, total carbonates $\approx$ 20–25%, kaolinite $\approx$ 5–10%, pyrite $\approx$ 5%, and illite $\approx$ 5%. High pyrite content supported the results of chemical analysis and also indicated the high potential of exposed tailings to generate AMD. The content of carbonate minerals (20–25%) indicated the acid neutralization capacity of the material. The concentrations of Mn (0.138%), Cu (0.072%), Zn (0.086%), and Pb (0.0079%) indicated the possibility of leaching of these metals from tailings in contact with acidic water from lake Robule.

Table 2. Chemical composition of the flotation tailings from Copper Mine Majdanpek.

Element	Cu (%)	Zn (%)	Fe (%)	S (%)	Ag (ppm)	Au (ppm)	Mn (%)	Al (%)	Cd (%)	Pb (%)
Content	0.072	0.086	10.7	7.01	1.667	0.405	0.138	5.07	0.0006	0.0079

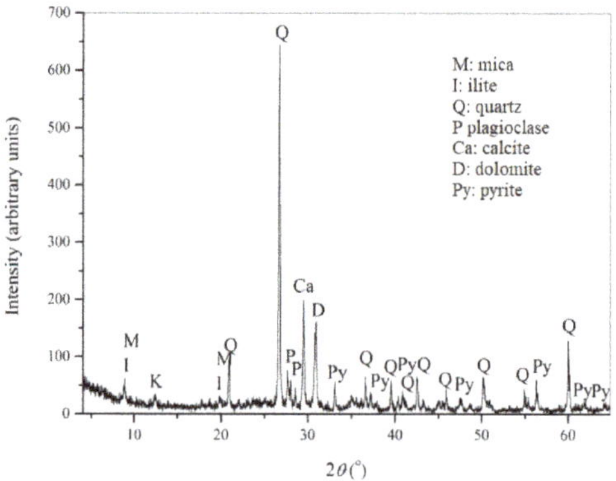

Figure 3. Diffractogram of the flotation tailings sample from Copper Mine Majdanpek.

3.3. Results of the Acid Neutralization Capacity Test

Remediation of the acid mine drainage is based on neutralization of the solution's acidity, which results in the precipitation of the metal cations, mostly in the form of the insoluble hydroxides and carbonates [8,14,27–30]. The acid neutralization capacity of mining waste, such as flotation tailings, depends on the relative reactivity of the contained minerals to interact with H^+ ions and undergo the neutralization process. According to Sverdrup [31], minerals can be divided into several groups based on their relative reactivity in the acid neutralization process: dissolving (calcite, dolomite, magnesite), fast weathering (anorthite, nepheline, forsterite, olivine, garnet, jadeite, leucite, spodumene, diopside, wollastonite), intermediate weathering (sorosilicates, pyroxenes, amphiboles, phyllosilicates), slow weathering (plagioclase, kaolinite), very slow weathering (K-feldspar), and inert (quartz). The most relevant minerals as potential neutralization agents contained in flotation tailings are carbonates, hydroxides and silicates [32,33]. The results of the ANC test of fresh and treated flotation tailings are shown in Figure 4. Common to both materials is a wide plateau, from 0.2 to 1.4 H^+ eq/kg, at pH values between 5 and 6 associated with the dissolution of carbonate minerals [34]. Carbonates, primarily calcite, are the main minerals involved in the acid neutralization reaction described by Equation (2) at pH > 6.3 and Equation (3) at pH < 6.3 [33].

$$CaCO_3 + H^+ \Leftrightarrow Ca^{2+} + HCO_3^- \tag{2}$$

$$CaCO_3 + 2H^+ \Leftrightarrow Ca^{2+} + H_2CO_3 \tag{3}$$

An interesting observation is that the treatment did not significantly reduce the buffering capacity of the material and that the amount of acid that the carbonates could neutralize was in the same range for both samples. Carbonate consumption for acid neutralization was reflected in the initial pH values after leaching with distilled water (0 H^+ eq/kg): 8.6 for fresh tailings and 7.5 for treated. Products of the treatment and their effect on the ANC were reflected in the position of the last plateau on the titration curves at a higher H^+ concentration for treated tailings (pH $\approx$ 4.5) than for fresh tailings (pH $\approx$ 3.0). This was associated with the formation of solid hydroxide minerals of metal cations with a valence of 3+ (Fe^{3+} and Al^{3+}) during the treatment. These metal hydroxides hydrolyzed with high ionic potential, thus representing important buffers controlling the pH at $\approx$ 4.3 ($Al(OH)_3$) and $\approx$ 3.5 ($Fe(OH)_3$) [22,33,34]. It can be concluded from the results of the ANC test that flotation tailings from Copper Mine Majdanpek had significant acid neutralization capacity and that these flotation tailings could be used multiple times for AMD neutralization. Furthermore, an important finding was that the buffering capacity of the tailings actually increased after treatment, due to the hydrolysis of the precipitated metal hydroxides.

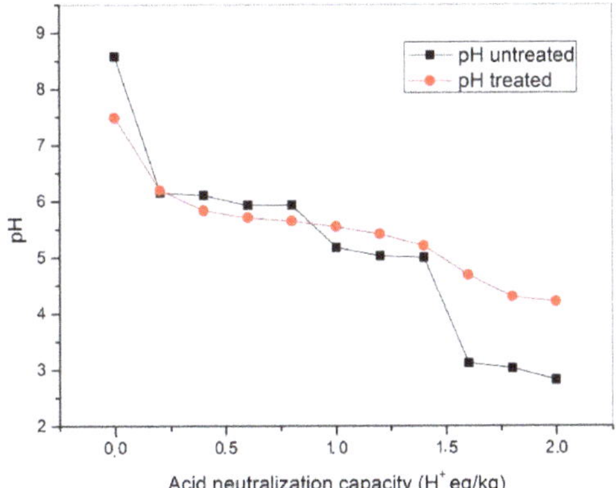

Figure 4. Results of the Acid Neutralization Capacity (ANC) test. Untreated: flotation tailings before neutralization of the lake water. Treated: flotation tailings after neutralization of the lake water.

3.4. Treatment of Water from Lake Robule by Using Flotation Tailings

3.4.1. Determination of the Optimal Quantity of Flotation Tailings Required for Neutralization of the Lake Water Samples

The next step was to find the optimal solid to liquid ratio (S:L) for neutralization of the water from Lake Robule. The results of the neutralization test with gradually increased amounts of flotation tailings are presented in Table 3.

Table 3. Determination of the optimal quantity of flotation tailings required for neutralization of the lake water samples to pH 7.

Pulp Density (%)	Mass (g)	Time (min)							
		15	30	60	120	240	1440	4320	10,080
1	0.5	2.53	2.93	3.06	3.08	3.1	3.8	4.23	4.24
3	1.5	2.89	3.23	3.97	4.12	4.21	4.34	4.71	5.3
5	2.5	2.67	3.85	4.25	4.33	4.43	4.77	5.56	6.46
10	5	2.85	4.44	4.81	5.19	5.43	5.92	6.67	6.76
15	7.5	2.88	4.76	5.47	5.81	5.92	6.23	6.8	7
20	10	3.14	5.03	5.71	6.02	6.16	6.32	6.86	7.06
25	12.5	3.16	5.29	5.86	6.1	6.35	6.51	6.86	7.04
30	15	4.11	5.44	5.93	6.15	6.26	6.51	6.88	7.07
40	20	4.44	5.62	6.01	6.28	6.41	6.57	6.95	7.2

The results of the experiment presented in Table 3 showed that following the increase in pulp density, the pH value dropped faster, but after seven days of settling, the difference in the final pH values between flasks with pulp densities of 15%, 20%, 25%, 30%, and 40% was not significant. Based on these results, a pulp density of 15% was identified as optimal for further experiments.

3.4.2. Experimental Results of the Treatment

Changes in the concentrations of the selected metals in solution during the treatment are presented in Table 4. Beside concentrations, the removal percentage of each element was calculated according to Equation (4).

$$Removal = \frac{final\ concentration}{starting\ concentration} \times 100\% \tag{4}$$

Table 4. Concentration of elements in solution during treatment time.

Time (min)	pH	Cu mg/L	Fe mg/L	Al mg/L	As mg/L	Cd mg/L	Mn mg/L	Ni mg/L	Zn mg/L	Pb mg/L
0	2.47	66.39	287	1017.62	<0.007	0.012	66	0.6	17.6	0.188
5	3.29	67.1	205	981	<0.007	0.021	67	0.6	19.4	0.185
10	3.54	67.1	151	916	<0.007	0.022	72.75	0.6	22	0.179
15	4.01	65	26	832	<0.007	0.022	73.75	0.6	25.5	0.153
30	4.27	63	5.18	471	<0.007	0.024	73.75	0.5	26	0.145
60	4.64	62	8.5	213	<0.007	0.022	83	0.5	28	0.131
120	4.78	58.9	0.268	21	<0.007	0.02	85.5	0.5	26.5	0.002
1440	6.60	3.98	0.16	0.288	<0.007	0.013	72	0.4	25	0.002
4320	6.82	2.49	0.09	0.1	<0.007	0.018	75.5	0.4	6.4	0.002
10080	7.01	0.49	0.059	0.1	<0.007	0.016	74	0.3	1.4	0.002
Precipitation (%)	-	>99%	>99%	>99%	-	+33%	+12%	50%	92%	98%

The + sign in the removal row means that the concentration of the certain metal increased after the neutralization experiment.

Experimental results showed that the concentration of all elements, except Mn and Cd, decreased during the treatment with the increasing value of pH. After 168 h of treatment, when the pH of the solution reached 7.01, the concentrations of all elements, except Mn, were below the discharge limits for municipal wastewaters prescribed by the national legislation of the Republic of Serbia [35], with over 99% of Fe, Al, and Cu precipitation, 98% removal of Pb, and 92% of Zn. Unlike all other metals, the concentrations of Mn and Cd in the solution increased during the treatment, indicating that these metals might be leached from the tailings.

3.5. Results of the PHREEQC Software Simulation of the Water Treatment

The results of the treatment simulation obtained by the PHREEQC program are given in Table 5. The results were based on the measured concentrations of the elements and pH values of the obtained solutions over time, as well as on the defined water volume (50 mL), solid to liquid ratio (15%), and temperature (18 °C) as constant values during the experiment. The results are presented in the form of calculated Saturation Indices (SI) of the corresponding mineral phases. The saturation indices of certain mineral phases, noted in the Table 5, increased and became positive at specific pH values. Such SI denote minerals whose constituent elements achieved thermodynamic equilibrium concentration in the solution during the neutralization reaction (SI = 0), after which they began to precipitate (SI > 0). The results of PHREEQC modeling partially complemented the experimental results and gave an additional explanation of the mechanism of metals' removal during the treatment.

Table 5. Results of the PHREEQC software simulation.

Element	Mineral	Time (min)	0	5	30	60	120	1440	Last
		pH	2.5	4.3	4.7	5.2	5.7	6.6	7.0
		Formula	-	-	-	-	-	-	-
Al	Al(OH)$_3$(a)	Al(OH)$_3$	−6.60	−4.12	−1.62	−1.59	−2.30	−0.21	−0.21
	Gibbsite	Al(OH)$_3$	−3.84	−1.37	1.14	1.16	0.46	2.54	2.54
Fe	Fe(OH)$_3$(a)	-	−0.80	0.72	1.61	1.62	1.85	2.12	2.09
	Goethite	FeOOH	4.83	6.35	7.24	7.25	7.49	7.75	7.73
	Hematite	Fe$_2$O$_3$	11.64	14.68	16.46	16.48	16.95	17.48	17.42
	Melanterite	FeSO$_4$·7H$_2$O	−10.03	−10.83	−10.37	−8.36	−8.15	−9.00	−10.00
	Siderite	FeCO$_3$	−12.25	−11.63	−8.44	−5.97	−5.56	−2.76	−3.75
Mn	Hausmannite	Mn$_3$O$_4$	−32.25	−29.71	−26.04	−23.62	−23.26	−8.25	−8.53
	Manganite	MnOOH	−10.87	−10.49	−9.73	−8.92	−8.88	−3.35	−3.52
	Pyrochroite	Mn(OH)$_2$	−13.57	−11.79	−9.65	−8.87	−8.58	−4.62	−4.57
	Pyrolusite	MnO$_2$·H$_2$O	−15.22	−16.24	−16.86	−16.01	−16.23	−9.13	−9.52
	Rhodochrosite	MnCO$_3$	−8.28	−6.51	−3.23	−2.45	−2.17	1.36	1.42

Table 5. *Cont.*

| Element | Mineral | Time (min) | 0 | 5 | 30 | 60 | 120 | 1440 | Last |
		pH	**2.5**	**4.3**	**4.7**	**5.2**	**5.7**	**6.6**	**7.0**
		Formula	-	-	-	-	-	-	-
Zn	Smithsonite	$ZnCO_3$	−10.13	−8.20	−4.99	−4.22	−3.94	−1.42	−0.37
	$Zn(OH)_2(e)$	$Zn(OH)_2$	−10.55	−8.61	−6.53	−5.75	−5.47	−1.48	−2.53
Pb	Anglesite	$PbSO_4$	−3.43	−2.85	−2.28	−2.24	−2.23	−2.93	−3.03
	Cerussite	$PbCO_3$	−8.81	−6.82	−3.50	−3.02	−2.80	0.15	0.05
	$Pb(OH)_2$	$Pb(OH)_2$	−9.42	−7.41	−5.23	−4.74	−4.52	−1.15	−1.25
Cd	$Cd(OH)_2$	$Cd(OH)_2$	−15.10	−13.12	−10.99	−10.22	−9.98	−6.11	−6.08
	$CdSO_4$	$CdSO_4$	−11.85	−11.29	−10.76	−10.45	−10.42	−10.63	−10.60
	Otavite	$CdCO_3$	−10.36	−8.39	−5.12	−4.36	−4.12	−0.68	−0.64

3.6. Mechanism of Acid Neutralization and Precipitation of Metals during the Treatment

Figures 5–7 present the changes in the concentrations of the metal cations as a function of the pH. After two hours of mixing, the pH of the solution increased to 4.78, and after seven days of settling, the pH of the solution further increased to 7. In samples collected from Lake Robule, over 99% of the total iron was in the ferric state. Deeper anoxic sections of the lake, which have not been disturbed nor sampled, could have iron in the ferrous (Fe^{2+}) state. During the first 5 min of the treatment, the pH value of the solution increased due to the carbonate dissolution and acid neutralization process and reached a value of 4.3, when the solution was supersaturated with Fe^{3+} ions that began to precipitate. The removal of iron was relatively rapid with a sharp decline in iron concentration in the pH range between 2 and 4. Approximately 99.8% of iron from the water samples was removed. Iron in the ferric state should readily precipitate as oxyhydroxide compounds ($FeO(OH)$), as shown in Equation (5), at pH values greater than 3.5 [17].

$$Fe^{3+}_{(aq)} + 2OH^{-}_{(aq)} \leftrightarrow FeO(OH)_{(s)} + H^{+}_{(aq)} \tag{5}$$

Equation (5) includes the dynamic information about the formation of goethite, but the analysis of the software included the equilibrium information, where the maximal possibility for precipitation was in the case of the hematite. Generally, as shown in Table 5, an increase of time and pH-values increased the possibility of the precipitation of iron hydroxide, goethite, and hematite with different values of calculated Saturation Indices (SI). In the case of the other elements such as zinc, cadmium, lead, and manganese, these SI-values were negative with small possibilities for precipitation.

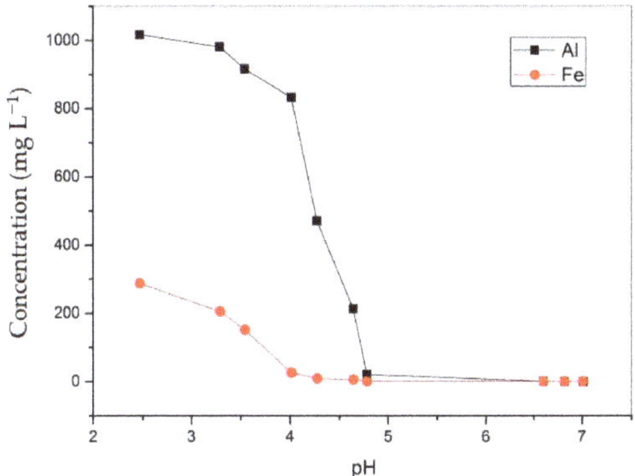

Figure 5. Changes in the concentrations of Al and Fe as a function of the pH.

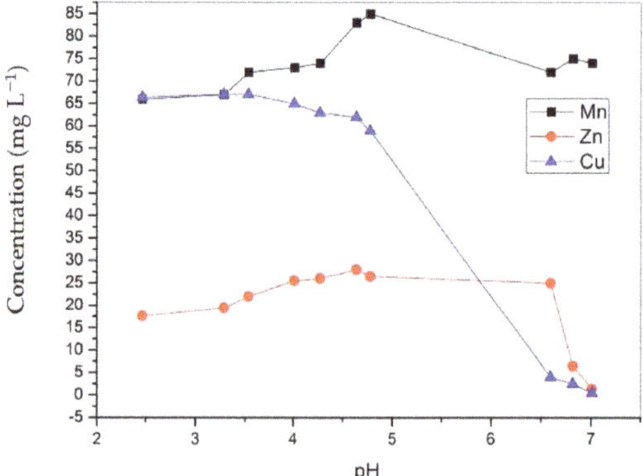

Figure 6. Changes in the concentrations of Mn, Zn, and Cu as a function of pH.

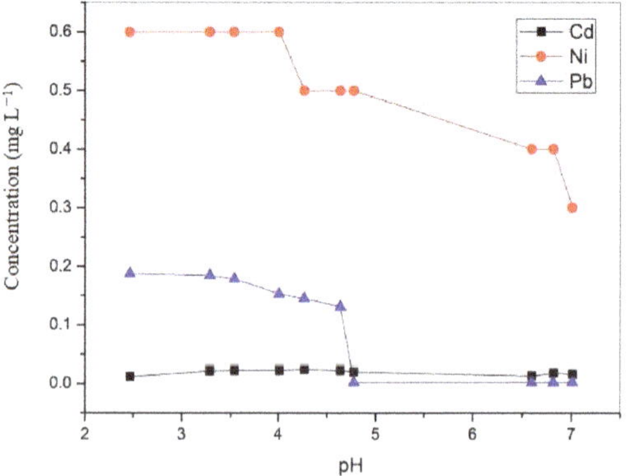

Figure 7. Changes in the concentrations of Cd, Ni, and Pb as a function of pH.

Park et al. [36] concluded that in synthetic multimetal solutions, ferric iron precipitated as an amorphous ferric oxyhydroxide and then was transformed by a slow dehydration reaction and internal atomic rearrangement to hematite. These two transformation processes compete with each other, and one of them is dominant depending on the reaction conditions [37].

$$\text{Fe}^{2+} \longrightarrow \text{Fe}^{3+} \longrightarrow \alpha\text{–tFeO(OH)} \longrightarrow \alpha\text{–tFe}_2\text{O}_3$$

$$\qquad\text{Oxidation} \qquad\qquad\quad \text{Precipitation} \qquad\qquad\quad \text{Dehydration}$$

After 30 min of the process, the pH value of the solution was 4.7, at which the Al^{3+} ions began to precipitate in the form of gibbsite ($Al(OH)_3$). This was consistent with the fact that in the presence of carbonate ions, the trivalent metals began to precipitate in the form of their hydroxides, Fe at pH ≥ 3.5 and Al at pH ≥ 4.5 [10]. Park et al. [36] also found that Al^{3+} from complex multimetal AMD solutions precipitated as amorphous $AlOHSO_4$. In the experiment presented in this paper, aluminum precipitated in the pH range from 3.5 to 5 (Figure 5).

The experimental results showed a significant decrease in Zn and Cu concentrations in the lake water samples during the treatment, although simulation results complemented only the precipitation of Zn in the form of smithsonite ($ZnCO_3$) and hydroxide ($Zn(OH)_2$) at pH values above 7, which was not achieved during the treatment. The majority of copper cations precipitated in the pH range between 5 and 7, and over 99% of copper was removed. Software simulation did not predict the precipitation of copper. Precipitation of copper from complex solutions depends on the interactions with other ions. Park et al. [36] investigated the precipitation of metals from quaternary mixtures

Fe/Al/Cu/Zn and Fe/Al/Cu/Ni and discovered that metals precipitated in the order Fe-Al-Cu-Zn or Ni. In these solutions, copper probably forms brochantite in the pH range from 5.0 to 6.6 and transforms to tenorite (CuO) above a pH of 6.6 [36,38]. Therefore, in this experiment, the forms of Cu precipitates were expected to be brochantite and tenorite. Theoretically, brochantite can be transformed to tenorite as follows [39]:

$$4Cu^{2+} + 6OH^- + SO4^{2-} \leftrightarrow Cu_4(SO_4)(OH)_6 \tag{6}$$

$$Cu_4(SO_4)(OH)_6 + 2OH^- \leftrightarrow 4CuO + SO4^{2-} + 2H_2O \tag{7}$$

$$Cu_4(SO_4)_{aq} + 2\,OH^- \leftrightarrow CuO + SO4^{2-} + H_2O \tag{8}$$

During the neutralization experiment, the concentration of zinc increased between pH 3.29 and 6.6 and then dropped to a minimal value of 1.4 mg L^{-1}, leading to the removal of 92% of Zn. The increase in zinc concentration was probably the consequence of the leaching of this metal from flotation tailings. Park et al. [36] concluded that in solutions that mimic AMD, zinc probably precipitates as hydrozincite $Zn_5(CO_3)_2(OH)_6$. The removal of Zn and Cu from multimetal solutions can be also attributed to the effect of co-precipitation or adsorption onto amorphous Fe and Al hydroxides [40]. At pH values between 6 and 7, the PHREEQC simulation predicted a precipitation of Pb in the form of carbonate mineral cerussite ($PbCO_3$) since $Pb(OH)_2$ precipitates at higher pH values between 9 and 10; as a result, 98% of lead precipitated at a pH value of approximately 4.8.

According to the software simulation, manganese should precipitate as $MnCO_3$, but no precipitation of manganese occurred during neutralization with flotation tailings. Instead, the concentration of manganese actually increased during the experiment probably due to leaching of Mn from the flotation tailings. The possible explanation is that the formation of $MnCO_3$ was not thermodynamically favorable under the experimental conditions and that also manganese hydroxide ($Mn(OH)_2$) precipitated at a pH above 10.

The concentration of cadmium also increased after neutralization because flotation tailings probably released some cadmium into the solution. The software simulation did not predict the precipitation of cadmium, which is an accordance with the experimental results. Fifty percent of nickel precipitated during the neutralization test. Park et al. [36] found that in the quaternary solution, Ni completely precipitated at pH 8.4. Precipitation of nickel started at pH 7, and the pH required for complete precipitation of Ni was not reached in this experiment. Nickel probably did not precipitate as $Ni(OH)_2$ because nickel hydroxide precipitates at approximately pH 10, but more likely as $NiCO_3$.

3.7. Comparison of the Lake Water Neutralization with Flotation Tailings and NaOH

In order to determine if other mechanisms of metal removal were included, such as adsorption of metal cations on the surface of the minerals that constituted flotation tailings, neutralization of the lake water to pH 7 was performed by application of flotation tailings and NaOH. The most significant differences were observed for the concentrations of Mn and Ag (Table 6). The concentrations of these metal cations were higher in water treated with flotation tailings than in water treated with NaOH. The most probable explanation is that these metals were leached from flotation tailings during neutralization. Changes in metal concentrations during neutralization with flotation tailings were the result of the increased pH of the solution; adsorption on the surface of the minerals had no significant effect on the metal removal efficiency.

Table 6. Comparison of the metals' precipitation efficiency after treatment with Flotation Tailings (FT) and NaOH.

Element	Concentration after FT (mg L^{-1})	Concentration after NaOH (mg L^{-1})
Copper (Cu)	0.54	0.48
Iron (Fe)	0.13	0.18
Aluminum (Al)	1.53	1.47
Arsenic (As)	<0.007	<0.007
Cadmium (Cd)	0.10	0.10
Manganese (Mn)	72.00	40.75
Nickel (Ni)	0.10	0.13
Zinc (Zn)	0.61	0.36
Lead (Pb)	0.57	0.66
Antimony (Sb)	0.43	0.41
Barium (Ba)	0.01	0.06
Beryllium (Be)	0.01	0.01
Boron (B)	0.14	0.15
Vanadium (V)	<0.002	<0.002
Cobalt (Co)	0.15	0.16
Chromium (Cr)	0.01	0.02
Selenium (Se)	0.02	0.03
Silver (Ag)	0.21	0.01

3.8. Post-Treatment of the Lake Water Samples with Hydrated Lime

In order to purify water from residual Mn and Ag, the sample of water that was treated with flotation tailings was further treated with hydrated lime, and the pH of the water was increased to 10. The concentrations of manganese and silver were reduced to 0.062 mg L^{-1} and 0.013 mg L^{-1} (Table 7). After this treatment, the concentrations of all metals in the water were below the discharge limits for municipal wastewaters according to national legislation of the Republic of Serbia [35].

Table 7. Results of the treatment of the lake water sample with Ca(OH)$_2$ after neutralization with flotation tailings. Results were compared with national discharge limits for municipal wastewater of the Republic of Serbia [35].

Element	Concentration (mg L^{-1})	National Discharge Limits for Municipal Waste Waters (mg L^{-1})
Copper (Cu)	0.023	2
Iron (Fe)	0.137	200
Aluminum (Al)	0.074	/
Arsenic (As)	0.007	0.2
Cadmium (Cd)	<0.001	0.1
Manganese (Mn)	0.062	5
Nickel (Ni)	0.002	1
Zinc (Zn)	0.012	2
Lead (Pb)	<0.002	0.2
SO$_4{}^{2-}$	4340	/
Antimony (Sb)	<0.017	/
Barium (Ba)	0.025	/
Beryllium (Be)	0.007	/
Boron (B)	0.099	/
Vanadium (V)	<0.002	/
Cobalt (Co)	<0.001	/
Chromium (Cr)	<0.001	1
Selenium (Se)	<0.001	/
Silver (Ag)	0.013	/

/: not defined.

3.9. Characterization of the Solid Residue after the Treatment of Water from Lake Robule with Flotation Tailings

Results of the XRD analysis of the flotation tailings residue after the neutralization experiment with 15% pulp density are presented in Figure 8. The content of carbonate minerals (calcite and dolomite) only slightly decreased after water treatment. The results of the semi-quantitative analysis of the XRD data were: quartz 55–60%, total carbonates 15–20%, mica/illite ≈ 10%, pyrite ≈ 5%, and kaolinite ≈ 5%. These results confirmed that the same sample of the flotation tailings could be used multiple times for neutralization of the water from Lake Robule.

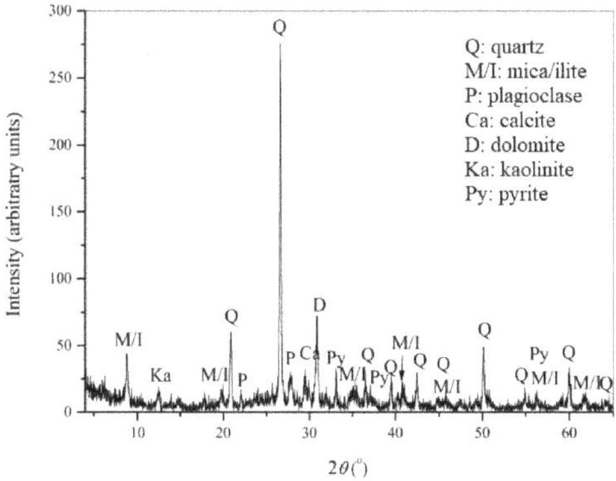

Figure 8. Diffractogram of the flotation tailings residue after neutralization of the lake water sample.

4. Conclusions

- Flotation tailings' samples collected from the flotation tailings dump of Copper Mine Majdanpek were rich in carbonate minerals (calcite and dolomite) and possessed significant acid neutralization capacity.
- Laboratory experiments confirmed that the flotation tailings could be applied in order to neutralize water from the extremely acidic Lake Robule located near the town of Bor (Serbia).
- After neutralization with flotation tailings, over 99% of Al, Fe, and Cu precipitated, 98% of Pb, and 92% of Zn. The concentrations of Cd and Mn increased due to leaching of these metals from flotation tailings.
- In order to remove Mn below discharge limits, application of hydrated lime was required in order to increase the pH up to 10.
- The flotation tailings could be applied in the active treatment of AMD in combination with hydrated lime. The results of this research were a step further in the implementation of the principles of the sustainable development including minimization, treatment, and reuse of waste within the same industry.
- Our future activities shall be performed in order to propose a continuous neutralization process for the application of the flotation tailings as an alternative material for acid mine drainage remediation.
- Furthermore, this methodology will be studied in scale-up conditions, in order to obtain more realistic data on the consumption of flotation tailing amount and percent of acid mine water

Author Contributions: Conceptualization, investigation, and experimental design, N.P. and S.S.; methodology and editing, B.M.; formal analysis and editing, D.R.; writing, review and editing, S.R.S.; supervision, conceptualization, and editing, M.S. and Z.K. All authors have read and agreed to the published version of the manuscript.

Funding: This work was realized in the frame of the projects TR 34023 and TR 34033, supported by the Ministry of Education, Science and Technological Development of the Republic of Serbia.

Conflicts of Interest: The authors declare no conflict of interest.

References

1. Dimitrijević, M.D. *Pyrite Oxidation and Acid Mine Drainage*; University of Belgrade: Belgrade, Serbia, 2013; pp. 75–95. (In Serbian)

2. Park, I.; Tabelin, C.B.; Jeon, S.; Li, X.; Seno, K.; Ito, M.; Hiroyoshi, N. A review of recent strategies for acid mine drainage prevention and mine tailings recycling. *Chemosphere* **2019**, *219*, 588–606. [CrossRef] [PubMed]

3. Bogdanović, G.; Trumić, M.; Trumić, M.; Antić, D.V. Mining waste management: Genesis and possibility of processing. *Recycl. Sustain. Dev.* **2011**, *4*, 37–43. (In Serbian)

4. Korać, M.; Kamberović, Ž. Characterization of wastewater streams from Bor site. *Metall. Mater. Eng.* **2007**, *13*, 41–51.

5. Stevanović, Z.; Obradović, L.; Marković, R.; Jonović, R.; Avramović, L.; Bugarin, M.; Stevanović, J. Mine waste water management in the Bor municipality in order to protect the Bor River water. In *Waste Water—Treatment Technologies and Recent Analytical Developments*; Einschlag, F.S.G., Carlos, L., Eds.; InTech: Rijeka, Croatia, 2013. [CrossRef]

6. Stanković, S.; Morić, I.; Pavić, A.; Vasiljević, B.; Johnson, D.B.; Cvetković, V. Investigation of the microbial diversity of an extremely acidic metal-rich water body (Lake Robule, Bor, Serbia). *J. Serb. Chem. Soc.* **2014**, *79*, 729–741. [CrossRef]

7. Beškoski, V.P.; Papić, P.; Dragišić, V.; Matić, V.; Vrvić, M.M. Long term studies on the impact of thionic bacteria on the global pollution of waters with toxic ions. *Adv. Mater. Res.* **2009**, *71*, 105–108. [CrossRef]

8. Pavlović, J.; Stopić, S.; Friedrich, B.; Kamberović, Z. Selective removal of heavy metals from metal-bearing wastewater in a cascade line reactors. *Environ. Sci. Pollut. Res. Int.* **2007**, *14*, 518–522. [CrossRef]

9. Masuda, N.; Marković, R.; Bozić, D.; Bessho, M.; Inoue, T.; Hoshino, K.; Ishiyama, D.; Stevanović, Z. Experimental results of metal recovery by a two-step neutralization process from AMD from a copper mine in Serbia. In Proceedings of the IMWA 2019, Conference "Mine Water: Technological and Ecological Challenges", Perm, Russia, 15–19 July 2019; Wolkersdorfer, C., Khayrulina, E., Polyakova, S., Bogush, A., Eds.; Perm State University: Perm, Russia, 2019; pp. 232–237.

10. Stopic, S.; Dertmann, C.; Xakalashe, B.; Lucas, H.; Alkan, G.; Aygmurlu, B.; Friedrich, B. A near zero waste valorization vision for bauxite residue through experimental results. In Proceedings of the XXI YuCorr International Conference, Tara Mountain, Serbia, 17–20 September 2019; Pavlović, M., Pavlović, M., Eds.; Serbian Society of Corrosion and Materials Protection (UISKoZaM): Belgrade, Serbia, 2019; pp. 120–125.

11. Mwewa, B.; Stopic, S.; Ndlovu, S.; Simate, G.; Xakalashe, B.; Friedrich, B. Synthesis of poly-alumino-ferric sulphate coagulant from acid mine drainage by precipitation. *Metals* **2019**, *9*, 1166. [CrossRef]

12. Masindi, V.; Ndiritu, J.G.; Maree, J.P. Fractional and step-wise recovery of chemical species from acid mine drainage using calcined cryptocrystalline magnesite nano-sheets: An experimental and geochemical modelling approach. *J. Environ. Chem. Eng.* **2018**, *6*, 1634–1650. [CrossRef]

13. Pérez-López, R.; Castillo, J.; Quispe, D.; Nieto, J.M. Neutralization of acid mine drainage using the final product from CO_2 emissions capture with alkaline paper mill waste. *J. Hazard. Mater.* **2010**, *177*, 762–772. [CrossRef]

14. Johnson, D.B.; Hallberg, K.B. Acid mine drainage remediation options: A review. *Sci. Total. Environ.* **2005**, *338*, 3–14. [CrossRef]

15. Aube, B.; Zinck, J. Lime treatment of acid mine drainage in Canada. In Proceedings of the Brazil-Canada Seminar on Mine Rehabilization: Techological innovations, Florianopolis, Brazil, 1–3 December 2003; Barbosa, J.P., Ed.; CETEM/MCT: Rio de Janiero, Brazil, 2003; pp. 26–30.

16. Ivšić-Bajčeta, D.; Kamberović, Ž.; Korać, M.; Gavrilovski, M. A solidification/stabilization process for wastewater treatment of sludge from primary copper smelter. *J. Serb. Chem. Soc.* **2013**, *78*, 725–739. [CrossRef]

17. Kaur, G.; Couperthwaite, S.J.; Hatton-Johnes, B.W.; Millar, G.J. Alternative neutralisation materials for acid mine drainage treatment. *J. Water Process. Eng.* **2018**, *22*, 46–58. [CrossRef]

18. Kefeni, K.K.; Msagati, T.A.M.; Mamba, B.B. Acid mine drainage: Prevention, treatment options, and resource recovery: A review. *J. Clean. Prod.* **2017**, *151*, 475–493. [CrossRef]

19. Moodley, I.; Sheridan, C.M.; Kappelmeyer, U.; Akcil, A. Environmentally sustainable acid mine drainage remediation: Research developments with a focus on waste/by-products. *Miner. Eng.* **2018**, *126*, 207–220. [CrossRef]

20. Dold, B. Sustainability in metal mining: From exploration, over processing to mine waste management. *Rev. Environ. Sci. Bio/Technol.* **2008**, *7*, 275–285. [CrossRef]

21. Kraus, W.; Noize, G. POWDER CELL—A program for the representation and manipulation of crystal structures and calculation of the resulting X-ray powder patterns. *J. Appl. Cryst.* **1996**, *29*, 301–303. [CrossRef]

22. Wahlström, M.; Laine-Ylijoki, J.; Kaartinen, T.; Hjelmar, O.; Bendz, D. *Acid Neutralization Capacity of Waste—Specification of Requirement Stated in Landfill Regulations*; Nordic Council of Ministers: Copenhagen, Denmark, 2009; p. 51.

23. Stegemann, J.A.; Cote, P.L. Summary of an investigation of test methods for solidified waste evaluation. *Waste Manag.* **1990**, *10*, 41–52. [CrossRef]

24. Parkhurst, D.L.; Appel, C.A.J. *Description of Input and Examples for PHREEQC Version 3—A Computer Program for Speciation, Batch-Reaction, One-Dimensional Transport, and Inverse Geochemical Calculations*; CreateSpace Independent Publishing Platform: Scotts Valley, CA, USA, 2014; p. 497.

25. Štulović, M.; Radovanović, D.; Kamberović, Ž.; Korać, M.; Anđić, Z. Assessment of leaching characteristics of solidified products containing secondary alkaline lead slag. *Int. J. Environ. Res. Publi. Health* **2019**, *16*, 2005. [CrossRef]

26. Madzivire, G.; Gitari, W.M.; Vadapalli, V.R.K.; Ojumu, T.V.; Petrik, L.F. Fate of sulphate removed during the treatment of circumneutral mine water and acid mine drainage with coal fly ash: Modelling and experimental approach. *Miner. Eng.* **2011**, *24*, 1467–1477. [CrossRef]

27. Rodriguez, J.; Schweda, M.; Stopić, S.; Friedrich, B. Techno-economical comparison between chemical precipitation and electrocoagulation for heavy metal removal in industrial wastewater. *Met. Berl.* **2007**, *61*, 208–214.

28. Dimitrijević, M.D. Acid mine drainage. *Bakar* **2012**, *37*, 33–44. (In Serbian)

29. Dimitrijević, M.D.; Alagić, S.Č. Passive treatment of acid mine drainage. *Bakar* **2012**, *37*, 57–68. (In Serbian)

30. Dimitrijević, M.D.; Nujkić, M.M.; Milić, S.M. The lime Treatment of acid mine drainage. *Bakar* **2012**, *37*, 45–56. (In Serbian)

31. Sverdrup, H.U. *The Kinetics of Base Cation Release Due to Chemical Weathering*; Lund University Press: Lund, Sweden, 1990; p. 245.

32. Lawrence, R.W.; Scheske, M. A method to calculate the neutralization potential of mining waste. *Environ. Geol.* **1997**, *32*, 100–106. [CrossRef]

33. Dold, B. Acid rock drainage prediction: A critical review. *J. Geochem. Explor.* **2017**, *172*, 120–132. [CrossRef]

34. Komonweeraket, K.; Cetin, B.; Benson, C.H.; Aydilek, A.H.; Edil, T.B. Leaching characteristics of toxic constituents from coal fly ash mixed soils under the influence of pH. *Waste Manag.* **2015**, *38*, 174–184. [CrossRef]

35. Masuda, N.; Marković, R.; Bessho, M.; Božić, D.; Obradović, L.; Marinković, V.; Ishiyama, D.; Stevanović, Z. A new approach to recover dissolved metals in AMD by two-step pH control on the neutralization method. In Proceedings of the 13th International Mine Water Association Congress "Mine Water and Circular Economy" (IMWA 2017), Lappeenranta, Finland, 25–30 June 2017; Wolkersdrofer, C., Sartz, L., Silanpaa, M., Hakkinen, A., Eds.; Lappeenranta University of Technology: Lappeenranta, Finland, 2017; pp. 1111–1118.

36. Park, S.M.; Yoo, J.C.; Ji, S.W.; Yang, J.S.; Baek, K. Selective recovery of dissolved Fe, Al, Cu and Zn in acid mine drainage based on modeling to predict precipitation pH. *Environ. Sci. Pollut. Res. Int.* **2015**, *22*, 3013–3022. [CrossRef]

37. Cudennec, Y.; Lecerf, A. The transformation of ferrihydrite into goethite or hematite revisited. *J. Solid State Chem.* **2006**, *179*, 716–722. [CrossRef]

38. Giannopoulou, I.; Panias, D. Differential precipitation of copper and nickel from acidic polymetallic aqueous solutions. *Hydrometallurgy* **2008**, *90*, 137–146. [CrossRef]

39. Park, S.M.; Yoo, J.C.; Ji, S.W.; Yang, J.S.; Baek, K. Selective recovery of Cu, Zn, and Ni from acid mine drainage. *Environ. Geochem. Health* **2013**, *35*, 735–743. [CrossRef]

40. Patterson, J.W.; Allen, H.E.; Scala, J.J. Carbonate precipitation for heavy metals pollutants. *J. Water Pollut. Control. Fed.* **1977**, *49*, 2397–2410.

 *metals*

Article

Structural and Electrochemical Properties of Nesting and Core/Shell Pt/TiO$_2$ Spherical Particles Synthesized by Ultrasonic Spray Pyrolysis

Milica G. Košević [1], Milana M. Zarić [1,2], Srećko R. Stopić [3], Jasmina S. Stevanović [1,2], Thomas E. Weirich [4], Bernd G. Friedrich [3] and Vladimir V. Panić [1,2,5,*]

[1] Institute of Chemistry, Technology and Metallurgy, University of Belgrade, National institute, Njegoševa 12, 11000 Belgrade, Serbia; milica.kosevic@ihtm.bg.ac.rs (M.G.K.); milana.zaric@ihtm.bg.ac.rs (M.M.Z.); j.stevanovic@ihtm.bg.ac.rs (J.S.S.)
[2] Centre of Excellence in Environmental Chemistry and Engineering—ICTM, University of Belgrade, 11000 Belgrade, Serbia
[3] IME Process Metallurgy and Metal Recycling, RWTH Aachen University, 52056 Aachen, Germany; SStopic@metallurgie.rwth-aachen.de (S.R.S.); bfriedrich@ime-aachen.de (B.G.F.)
[4] GFE Central Facility for Electron Microscopy and IFK Institute of Crystallography, RWTH Aachen University, 52074 Aachen, Germany; weirich@gfe.rwth-aachen.de
[5] Chemical-Technological Department, State University of Novi Pazar, 36300 Novi Pazar, Serbia
* Correspondence: panic@ihtm.bg.ac.rs; Tel.: +381-11-3640231

Received: 15 October 2019; Accepted: 17 December 2019; Published: 20 December 2019

Abstract: Pt/TiO$_2$ composites were synthesized by single-step ultrasonic spray pyrolysis (USP) at different temperatures. In an in-situ method, Pt and TiO$_2$ particles were generated from tetra-*n*-butyl orthotitanate and chloroplatinic acid, and hydrothermally-prepared TiO$_2$ colloidal dispersion served as Pt support in an ex-situ USP approach. USP-synthesized Pt/TiO$_2$ composites were generated in the form of a solid mixture, morphologically organized in nesting huge hollow and small solid spheres, or TiO$_2$ core/Pt shell regular spheroids by in-situ or ex-situ method, respectively. This paper exclusively reports on characteristic mechanisms of the formation of novel two-component solid composites, which are intrinsic from the USP approach and controlled precursor composition. The generation of the two morphological components within the in-situ approach, the hollow spheres and all-solid spheres, was indicated to be caused by characteristic sol-gel/solid phase transition of TiO$_2$. Both the walls of the hollow spheres and the cores of all-solid ones consist of TiO$_2$ matrix populated by 10 nm-sized Pt. On the other hand, spherical, uniformly-sized, Pt particles of a few nanometers in size created a shell uniformly deposited onto TiO$_2$ spheres of *ca*. 150 nm size. Activities of the prepared samples in an oxygen reduction reaction and combined oxygen reduction and hydrogen evolution reactions were electrochemically tested. The ex-situ synthesized Pt/TiO$_2$ was more active for oxygen reduction and combined oxygen reduction and hydrogen reactions in comparison to the in-situ Pt/TiO$_2$ samples, due to better availability of Pt within a core/shell structure for the reactions.

Keywords: electrocatalysis; supported Pt nanoparticles; Pt/TiO$_2$ synthesis; Titanium oxide colloid

---

1. Introduction

Increasingly, a popular topic nowadays is the investigation of fuel cells (FC) as alternative power devices. Many studies have explored possibilities for improving their performance and durability [1–3]. Some of the most investigated electrode materials [4–7] in an FC are platinum-based nanoparticles supported on carbons, due to the high activity of Pt in the kind of reactions FC functioning are based on. Pt can be synthesized in various sizes and shapes, such as spherical [8], cubical [8,9], nanodendritical [10,11], etc. Commonly, the loading of Pt in the cathode of FCs is rather high, at 80–90 mass % [12]. Bearing in mind the low abundance and high cost of platinum, a lot of research aims to reduce the platinum amount in FC catalysts [12–16]. Carbon's advantages, if compared to other supporting materials, are good conductivity and high surface area [17]. However, the usage of carbon as a Pt support has disadvantages, due its low stability and tendency to corrode [18–20]. Namely, the major issue in using carbon-supported platinum as a catalyst is its degradation due to the low stability of carbon during operating and startup/shutdown modes [21]. Carbon degradation, i.e., corrosion, leads to the agglomeration of the supported catalyst particles, which leads to a decrease in catalyst activity. In addition, Pt particle agglomeration and Pt dissolution/re-deposition also contribute to catalyst degradation.

There have been attempts to replace carbon with other more stable and possibly interactive supporting materials [22–25]. TiO_2 is seen as a promising candidate [26–28], due to its contribution to stabilization and electroactivity enhancement when used as a Pt support [29,30].

Different methods and conditions of Pt/TiO_2 synthesis have been performed in order to improve the characteristics of synthetized material. Since particle size plays an important role in synthesis and catalysis, we compared different types of techniques for synthesis of TiO_2 particles and investigated the morphology and structure of obtained material. Various sizes of Pt synthesized within Pt/TiO_2 composites are reported in the literature, such as 3.6–6.1 nm in 20 mass % Pt deposited on nano-stick shaped TiO_2 [31], 7 nm Pt on TiO_2 films [32], and 3 nm-sized Pt in low loading Pt/TiO_2 composites (0.1–0.5 mass %) [33]. Pt sized less than 50 nm was deposited on TiO_2 nanotubes [34] and 10–20 nm-sized Pt was deposited on alkali-treated nanoporous TiO_2 material [35]. In our previous work, Pt/TiO_2 composites were synthesized by an ex-situ sol-gel procedure [36].

The goal of this research was to improve the synthesis of Pt/TiO_2 electrocatalysts in order to obtain electrocatalytic material of ordered structure. An additional aim was to decrease Pt loading and simultaneously gain good electrocatalytic efficiency for synthesized Pt. Different types of nanoparticle synthesis by ultrasonic spray pyrolysis (USP) have been reported [37–42]. In this work, a novel hydrothermally-based USP process for the controlled synthesis of a catalytically-improved catalyst/support ordered structure is reported. The core/shell hierarchy includes active material, such as Pt, deposited in the form of thin layer (e.g., shell) on a core (in our case, TiO_2). The core/shell structure enables better utilization of Pt, because it consists of a reactive form of Pt over a less noble core. It is expected that this structure set has an important benefit in decreased Pt loading, because it is uniformly dispersed in locally higher amounts as a thin shell, but is fully operative within a composite shell/support assembly [43]. When a monolayer of Pt is formed over a core, the Pt loading can be significantly reduced [13]. This can contribute to cost reduction in comparison to some commercially available catalysts consisting of 40 mass % (Pt/C HISPEC produced by Johnson Matthey, London, UK, and Pt supported on carbon black by Merck, Kenilworth, NJ, USA) and 50 mass % Pt (Tanaka Kikinzoku Intl., Hiratsuka, Japan). There is a wide range of Pt loading reported in the literature, such as below 5 mass % [33], 5–10 [14,16,44], 20 and 30% [15,45], 30–40 [15,46] 45–50% [14,47].

Although TiO_2 by itself is considered inert, it can be interactive due to its synergetic hypo–hyper-d-electronic interactive effect [48]. Furthermore, controlling the synthesis of TiO_2 support can favor the core/shell arrangement of the particles, and therefore improve the electrocatalytic properties of Pt/TiO_2.

USP in-situ synthesis and ex-situ methods for the synthesis of Pt/TiO_2 composites are reported herein. In the in-situ method, TiO_2 and Pt are simultaneously synthesized using orthotitanate and H_2PtCl_6 solutions as precursors. In the ex-situ synthesis, TiO_2 was introduced as colloidal dispersion prepared prior to USP synthesis, while Pt was synthesized during the USP synthesis from a H_2PtCl_6 solution. The USP synthesis enabled the generation of multi-component fully-ordered core/shell composites. The ex-situ synthesis process is expected to give better control of the structure of TiO_2 support using the USP methodology and colloidal TiO_2, previously synthesized by forced hydrolysis procedure. The morphology, structure, and electrochemical behavior of different set of synthesized particles have been tested. Cyclic voltammetry and linear sweep voltammetry have been performed to test the electrochemical properties of synthesized Pt/TiO$_2$. The activity of the synthesized material for oxygen reduction reaction (ORR) and combined ORR and hydrogen evolution reaction (HER) have been checked.

2. Materials and Methods

2.1. Material Synthesis and Preparation

2.1.1. Synthesis of the Pt/TiO$_2$ Powders

Two sets of experiments were performed in order to synthesize the Pt/TiO$_2$ composites. In the ex-situ synthesis, TiO$_2$ colloidal dispersion, which is hydrothermally prepared from TiCl$_3$ solution (Sigma-Aldrich) [36], and 2.0 g·dm^{-3} chloroplatinic acid were used as precursors. The TiO$_2$ colloid was prepared by adding TiCl$_3$ solution dropwise into a boiling HCl solution. During the 90 min of boiling of the mixture under reflux, TiO$_2$ particles were generated. The concentration of solid TiO$_2$ in the obtained colloidal dispersion was adjusted to 0.4 g·L^{-1} using ultrafiltration upon dilution with water to pH 4.0.

In the in-situ synthesis, the mixture of 0.1 mol·dm^{-3} tetra-n-butyl orthotitanate (Merck) stabilized by hydrochloric acid, and 2.0 g·dm^{-3} chloroplatinic acid (Alfa Aesar) served as the precursor solution.

Both sets of syntheses, i.e., in-situ and ex-situ, were performed at three different temperatures typical for the USP procedure and equipment: 500, 650, and 800 °C [49,50]. In addition, nominal Pt loadings of 5 and 20 mass % were projected at every synthesis temperature. The precursors for TiO$_2$ and Pt were mixed to give Ti:Pt mole ratios of 50:1 and 10:1 for 5 and 20 mass % Pt, respectively. The mixture was introduced into the ultrasonic spray pyrolysis equipment, as seen in Figure 1 (the details are described in a previously published paper [51]). Only the samples which showed electrochemical activity were subjected to structural characterization, as described in Section 2.2.1.

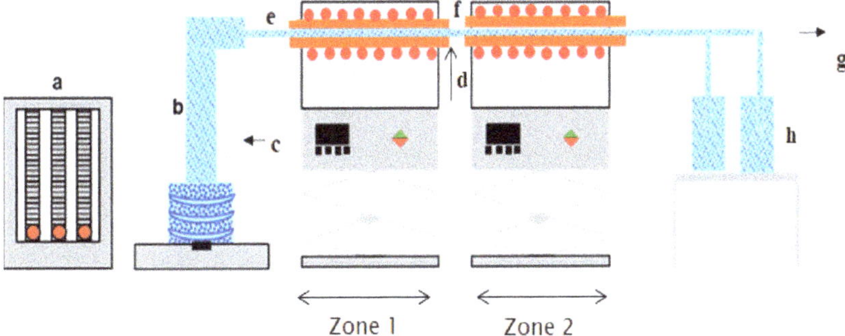

Figure 1. Experimental set-up used for ultrasonic spray syntheses (**a**) gas flow regulation of N$_2$ and H$_2$, (**b**) ultrasonic generator with precursor solution (**c**) gas inlet 1, N$_2$ (**d**) gas inlet 2, H$_2$, (**e**) furnace 1, (**f**) furnace 2, (**g**) gas outlet and (**h**) collection bottles

During the USP synthesis of spray pyrolysis, precursors were driven in the first zone by N$_2$ with the flow of 1.0 L·min^{-1} through a two-tube furnace upon ultrasonic atomization (Gapusol 9001, RBI atomizer, Meylan, France, operating at 2.5 MHz). In the second zone, H$_2$ is introduced with the flow of 1.0 L min^{-1} for the Pt reduction to take part. This results in an N$_2$ flow with 1.0 L·min^{-1} in the first zone, and a mixture of N$_2$ and H$_2$ flow with 2.0 L·min^{-1} in the second zone. The tube furnaces were pre-heated to the targeted temperature. Synthesized particles were collected in collector bottles filled with ethanol.

2.1.2. Preparation of Pt/TiO$_2$ Coating on Glassy Carbon from Synthesized Powders

Powders collected in the bottles filled with ethanol where centrifuged at 10,000 rpm for 3 min and separated from the ethanol. Obtained precipitates were dried at 90 °C for 24 h in air. In order to check if there were any residues in powders, from either the ethanol or precursors, which could have an impact on the electrochemical response, a part of the obtained powders was washed with water and thermally treated in N$_2$ for 3 h. All of synthesized powders were suspended in water and ultrasonically homogenized during 1 h (40 kHz, 70 W), in order to obtain suspensions of 3 mg·cm^{-3}. Suspensions were deposited onto a glassy carbon (GC) disc in the form of a 0.31 mg·cm^{-2} thin layer. The applied suspension was left to dry at room temperature. The deposited layer was stable and showed good adhesion to GC, hence there was no need to use any binder. The prepared electrode served as a working electrode.

2.2. Material Characterization

2.2.1. Composition, Morphology, and Structural Characterization

The morphology and elemental composition of the synthesized powders were analyzed by scanning transmission electron microscopy, STEM Tecnai F20 (FEI Company, Eindhoven, The Netherlands) and a system (EDAX Inc., Mahwah, NJ, USA) equipped with energy dispersive spectroscopy (EDX) operated at 200 kV for the analysis of characteristic X-ray emissions. Particle size was analyzed from the recorded STEM images by Image J 1.40 software, (University of Wisconsin, Madison, WI, USA), and the average values with standard deviations are reported.

2.2.2. Electrochemical Characterization

The cyclic voltammetry (CV) and linear sweep voltammetry (LSV) measurements were performed in 100 mL of 1 mol·dm^{-3} H$_2$SO$_4$ (pH close to zero), at room temperature, in a three-compartment electrode cell that consisted of a platinum plate as a counter electrode, saturated calomel electrode (SCE) or Ag/AgCl as a reference electrode (all potentials in this paper are given in SCE scale) and working electrode. The electrolyte was saturated with oxygen gas for LSV measurements. Cyclic voltammograms and polarization curves were obtained at scan rates of 50 and 1 mV·s^{-1}, respectively. All measurements were performed at ambient temperature using Bio-Logic SAS, potentiostat/galvanostat, model SP-200, (BioLogic Science Instruments, Seyssinet-Pariset, France).

3. Results and Discussion

3.1. STEM and EDX Characterization of Pt/TiO$_2$ Samples

3.1.1. USP Pt/TiO$_2$ Samples Synthesized In-Situ

A typical STEM annular dark field (ADF) image of the Pt/TiO$_2$ composite particles from the in-situ synthesis with nominal 20 mass % Pt loading, synthesized at 800 °C, is shown in Figure 2a. The STEM image proves the regular spherical nature of the generated nanoparticles, which range between about 100 and 500 nm in size. The average particle diameter in this sample is 260 ± 80 nm. The typical geometrical motif in this sample is one larger particle, associated with a few smaller particles. Moreover, some of the larger particles are characterized by a circular darker zone in the particle center. Since the contrast in STEM ADF images at a particular position is governed by the atomic number and the sample thickness, the interpretation is straightforward and thus suggests the presence of hollow cores.

The assumption of hollow cores is also supported by local EDX measurements performed at the bright and dark areas (Table 1). The hollow particles have sizes of about 330 ± 90 nm and are thus larger than the average particles. Nevertheless, the results from the semiquantitative EDX in Table 1 show that the detected Pt loadings are lower than the expected nominal values (the detection uncertainty for Pt is about ± 3%). For 20% nominal Pt loading, it was 17.1 mass % (14.5% decrease), while for the

5% nominal Pt loading it was much smaller, i.e., 2.4 mass % (52% decrease). This indicates that the in-situ USP procedure requires additional optimization of the synthesis parameters, since mass and stoichiometric controls appear not to be ensured. Consequently, different types of Pt losses during the synthesis procedure can take place, causing lower Pt loading in synthesized materials. In addition, the EDX data reveal that the outer surface of the sphere contains slightly more Pt than at the inner surface of hollow sphere.

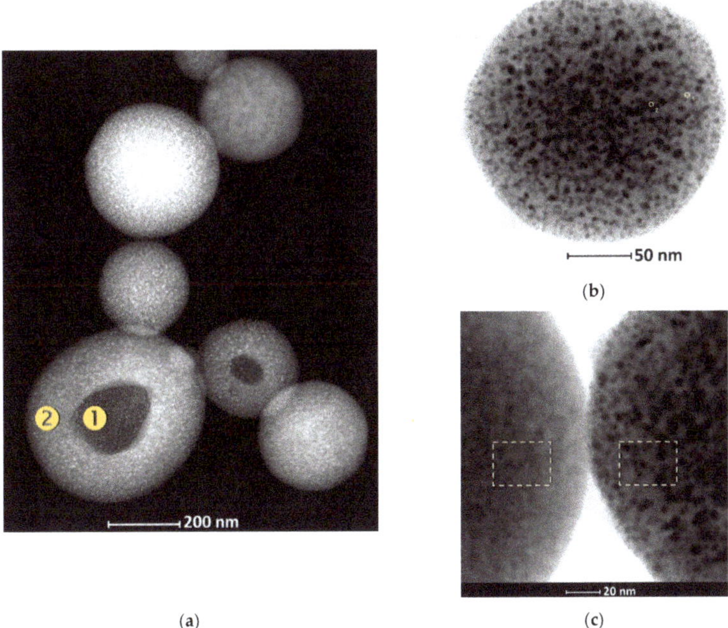

Figure 2. (**a**) Typical annular dark field (ADF) scanning transmission electron microscopy (STEM) image of an agglomerate of the Pt/TiO$_2$ composite particles that have been synthesized by in-situ ultrasonic spray pyrolysis (USP). Image contrast and EDX (energy dispersive spectroscopy) analysis at the indicated numbered positions prove that a part of the particles has a core-shell structure with hollow core. (**b**) The ADF STEM shows a typical spherical in-situ Pt/TiO$_2$ particle (at 650 °C) at larger magnification. It should be noted that the speckled appearance of the particle indicates that the nanoparticle itself is composed of a large number of nanosized crystallites. (**c**) The STEM bright-field image shows part of two adjacent spherical particles with nominal 5 mass % (left) and 20 mass % (right) Pt loadings (the boxed areas mark the positions that have been analyzed by EDX spectroscopy).

These findings allow us to propose the following mechanism for the USP generation of a Pt/TiO$_2$ composite, as schematically presented in Figure 3. Fine droplets of aerosol are initially formed from the precursor solution in the nebulizer, which are subsequently transferred to the high-temperature zones of the USP furnace. If the droplets pass through a gel phase upon sudden heating, the solid shell around the gelled sphere could be formed during continued heating. It should be noted that it was observed that the precursor solution turns into sol after several weeks, which indicates the possibility of a sol-gel transition. Owing to thermal tensions during the solidification of a gel, it appears that the shell shrinks more rapidly than the gel is being solidified, which causes the shell to crack. The remaining highly tense gel phase flings off the shell, leaving a circular hole in the spherical wall, i.e., a new gelled sphere is delivered by its parent, which turned into a solid hollow sphere. If the shrinkage in a newly-born gelled sphere would prevail in the solidification of the interior, the delivering mechanism could be repeated.

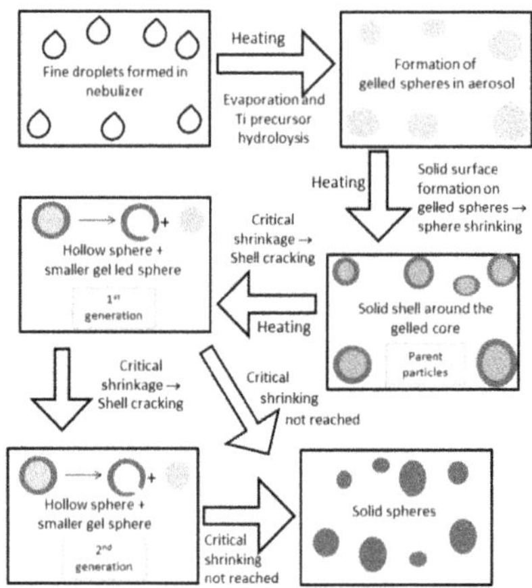

Figure 3. Schematic illustration of the in-situ USP mechanism for the formation of hollow and solid spheres of TiO$_2$/Pt.

The second-generation hollow spheres with holes of smaller diameter will then be formed, as seen in Figure 2a right beside the largest hollow sphere (first generation). The gelled sphere, being small enough, will turn solid throughout the volume without a breaking of the shell, which produces the solid spheres. These spheres are seen in Figure 2a with no holes on their surfaces, and they have sizes below the average (131 ± 42 nm). This mechanism can be visualized as the "opening of the Russian nesting doll", which produces the particulate composition of the novel USP-synthesized Pt/TiO$_2$ composite, colloquially defined as "Russian nesting doll"-like Pt/TiO$_2$ spherical particles.

The separation of the initial gelled sphere into the parental hollow sphere and the newly-born smaller gelled sphere keeps the former with the Pt content distributed throughout the shell (Table 1). TiO$_2$ tends to solidify faster than the Pt, thus causing every subsequent generation of the gelled, and afterwards solidified, sphere to be of lower Pt content with respect to its parental one. While the Pt content at the external shell side indicatively decreases, the relative decrease of Pt content for parental–newborn sphere pair (Table 1) also decreases. Here we consider the nominal loading as "parental" to that at the outer side of the first generation of the spheres (bright zone), while the latter plays the parent role for the loadings at the outer side of the second generation (the dark zone, Table 1). It follows that every next generation of the spheres in the mechanism of "the opening of the Russian nesting dolls" should get richer in Pt in the core of the gelled phase. Hence, the rest of Pt will be "trapped" within the core of the last generation of completely solidified sphere.

Table 1. Pt loadings (mass %) in the particle walls found by semiquantitative EDX analysis for the in-situ synthesized Pt/TiO$_2$ particles and the relative decrease with respect to nominal (20 mass % Pt) or Pt loading at the outer side of parental sphere. The composition at the inner side is considered as that of the outer side of a newly-born sphere.

Zone	Nominal Pt Loading, Mass %	Found Pt Loading Mass %	RelativeDecrease, %
Bright (outer side of the sphere wall, Figure 2a, area 2)	20	17.1	14.5
Dark (inner side of the sphere wall, Figure 2a, area 1)		15.4	9.94

STEM images of spherical Pt/TiO_2 particles of near-average diameters are shown in Figure 2b,c. Figure 2b illustrates the granular appearance of the Pt (20 mass %)/TiO_2 sphere, with the surface populated by particles within the 3.4–7.8 nm size range, of an average diameter (±SD) of 5.2 ± 0.9 nm. However, the structure of the surface of the sphere walls depends on Pt content. The STEM appearances of an area of the two spheres of similar size, synthesized with nominal Pt loadings of 5 and 20 mass %, are compared in Figure 2c. The surface of the sphere with lower Pt loading (on the left in Figure 2c) appears smoother and more compact than the sphere with the higher Pt loading (on the right in Figure 2c). Higher Pt loading causes the sphere surface to show black spots-like particles that are 5.2 ± 0.9 nm in size. In order to identify the composition of the morphological elements at the sphere surface, EDX analysis has been performed, and the data related to area and point markers in Figure 2 are given in Tables 2 and 3.

Table 2. Pt loadings (mass%) on the outer side of particle walls found by EDX analysis for the in-situ synthesized Pt/TiO_2 particles with nominal Pt loadings of 5 and 20 mass %. EDX sampling was performed according to the markers given in Figure 2b,c.

EDX Sampling	Nominal Pt Loading, 20 mass%	Sphere Diameter, nm
Figure 2b, area 1	22.3	194
Figure 2b, point 1/point 2	3.2/25.8	
Figure 2c, area right	15.2	198

Table 3. Comparing the Pt loading of 5 and 20 mass % on the outer side of particle walls found by EDX analysis for the in-situ synthesized Pt/TiO_2 particles with nominal Pt loadings. EDX sampling was performed according to the areas given in Figure 2c.

EDX Sampling	Mass %	Nominal Pt Loading, mass%	Sphere Diameter, nm
Figure 2c, area right	20	15.2	198
Figure 2c, area left	5	3.0	278

The surface composition of the smaller spheres was found to be higher (area 1 in Figure 2b) with respect to the biggest hollow sphere of the first generation (Figure 2a and Table 1). The sphere from Figure 2b is 194 nm in size, which is well below the average size—the criterion adopted for full solidification of the descendant gelled spheres in the mechanism of USP particle generation. Differently-shaded elements on the sphere surface were found to differ considerably in composition. The dark spots on a relatively uniform gray background (Figure 2c, sphere on the right) have considerably higher Pt content with respect to gray zones rich in TiO_2 (Table 3, area right and left). Considerably different contents indicate that black spots could be Pt laying on the continuous spherical TiO_2 matrix generated by the sol-gel transition of the oxide solid phase. Table 2 shows that 194-nm-sized sphere has 22.3 mass % Pt on the scanned area size of 1900 nm^2, while on a sphere of 198 nm size, 15.2 mass % Pt was found on scanned area of 420 nm^2. Since 5-nm-sized Pt particles lay on the TiO_2 matrix, relative occupation of the scanned surface by Pt particles should increase with scanned surface area, which makes the EDX data on Pt content higher for a larger scanned area.

It follows from the preceding considerations about the mechanism of the in-situ USP generation of the Pt/TiO_2 composite that, although being USP-intrinsic and intriguing, its structure and composition are not very promising for electrocatalytic applications. Despite that the average Pt loading in the synthesized spheres is fairly close to the nominal values (Tables 1–3), there is some deficiency of Pt at the surface of most of the bigger spheres. Moreover, residual Pt is likely distributed within smaller spheres, which prevents its addition to the catalytic activity of the composite. The reactants in aqueous medium easily approach the interior of the hollow sphere if the internal surface of the sphere wall contributes to the reaction. However, this should not be the case for hollow Pt/TiO_2 spheres.

3.1.2. USP Pt/TiO$_2$ Samples Synthesized Ex-Situ from Colloidal TiO$_2$

Figure 4 shows STEM ADF images of Pt (20 mass %)/TiO$_2$ synthesized by USP at 500 and 800 °C from previously hydrothermally-prepared colloidal TiO$_2$ and chloroplatinic acid. The spheres appear smaller than those from the in-situ procedure, with average diameters (independent of USP temperature) of 137 ± 56 nm and 131 ± 42 nm at 500 and 800 °C, respectively. Although it is not obvious from Figure 4, the size distribution of the ex-situ spheres is more uniform (SD around 50 nm) with respect to in-situ spheres (*SD* around 80 nm). Statistical counting returned data showing extremely uniform size distribution of the particles: the surface of the 500 °C spheres are populated by particles with the sizes in the 1.1–1.3 nm range. Larger particles of the sizes in a wider range of 1.9–2.7 nm are formed at 800 °C. Besides the much smaller particles that are formed by the ex-situ approach, another crucial difference to the in-situ procedure is the looser and thus less compact structure of the spheres. This is expressed by their less well-defined outer contours.

Although the structure of the core of the spheres is not clear in Figure 4, in both point and line types of data collection, marked with red color in the inset of Figure 4a, EDX analysis showed the presence of Ti. Figure 5 shows the EDX spectra recorded at point 1 in both insets in Figure 4a,b. The corresponding Pt loadings are presented in Table 4.

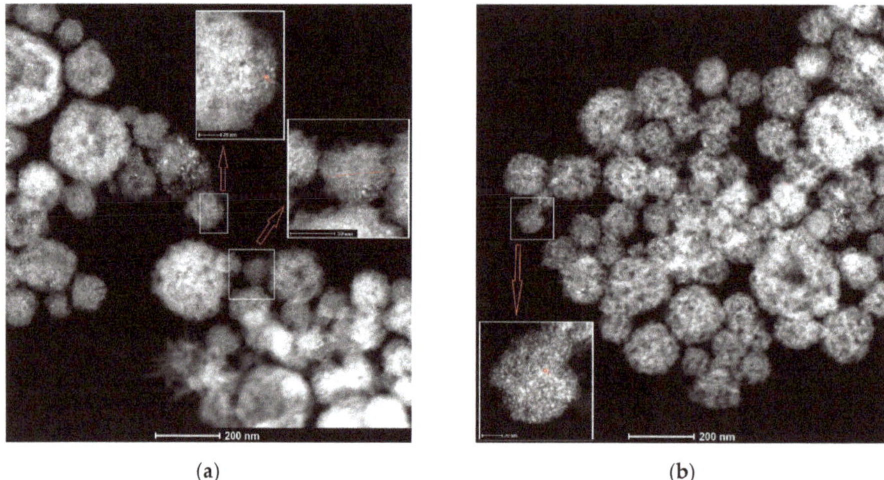

(a)   (b)

Figure 4. STEM ADF images of ex-situ Pt/TiO$_2$ composites obtained from the mixture of TiO$_2$ colloid and H$_2$PtCl$_6$ solution with projected mass of 20% Pt/TiO$_2$ at (**a**) 500 °C and (**b**) 800 °C.

Table 4. Pt loadings (mass %) found by EDX analysis for the ex-situ synthesized Pt/TiO$_2$ with nominal Pt loading of 20 mass % at different USP temperatures. EDX sampling is performed according to the markers given in Figure 4.

Sampling Site	USP Temperature, °C	Pt Loading of 20 mass %
Figure 4a, point 1	500	48.0
Figure 4a, line 1	500	27.6
Figure 4b, point 1	800	43.2

In-point EDX spectra from Figure 5 show clear reflections of Ti, O, and Pt for samples synthesized ex-situ at 500 and 800 °C. Peaks over 8 eV come from Cu (the substrate used for EDX measurements) and Pt. The intensities of Ti and O reflections are suppressed for the 500 °C sample relatively to the 800 °C sample, while the intensities of main Pt peaks are almost the same. Although the coordinates of the EDX sampling points with respect to the sphere center differ (in the case of the inset in Figure 4a, it is closer to the edge of sphere), the suppressing might indicate that reflections related to TiO_2 come from internal parts of the sphere. On the other hand, the similarity of the Pt peak intensities suggests that Pt is placed on the sphere surface in almost pure form, and it is formed of nm-sized spherical particles, as seen in Figure 4. These considerations appear to be reasonable, since the Pt loadings in the 500 °C and the 800 °C samples are similar, with that in the former being slightly larger. Another indication about ordered positions of the two components is found in the in-line EDX analysis, as shown in the inset in Figure 4a. The average Pt loading down the line across the sphere is considerably lower than in-point loading. Since TiO_2 is located in the core, the surface EDX analysis should register the increase in Pt loading at the line ends (sphere edges) with respect to central sphere zones. The size and shape of hydrothermally-prepared TiO_2 particles in colloidal dispersion were found to be quite similar [36] to the spheres from Figure 4. For all samples synthesized ex-situ, the EDX analysis found considerable excess of Pt with respect to the nominal loading. Hence, the hierarchical ordering of the Pt nm-sized particles which coat the TiO_2 spherical core is reached. Also, the size of particles obtained by both methods, in-situ and ex-situ synthesis, was below 10 nm, which is the optimal size for the improved electrocatalytical activity of Pt as reported in literature [31,32].

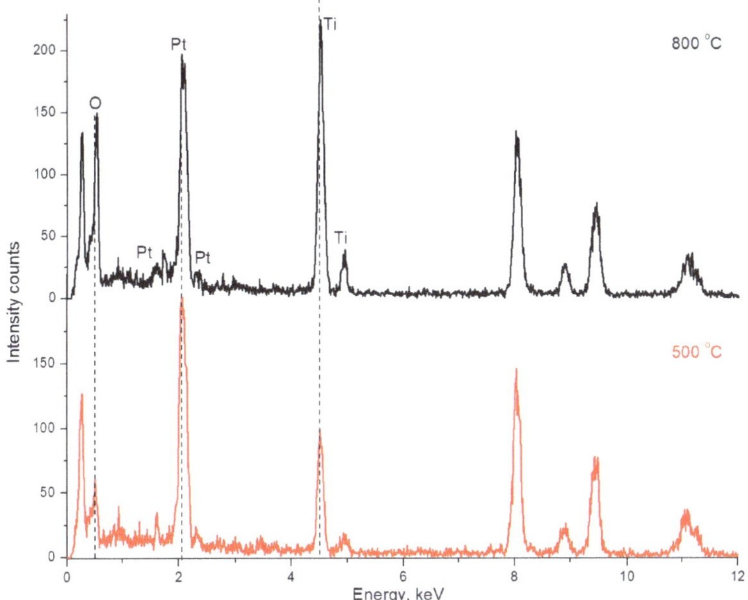

Figure 5. EDX Spectra of USP ex-situ prepared Pt/TiO$_2$ at 800 °C and 500 °C.

3.2. Electrochemical Measurements

Characteristic steady-state cyclic voltammograms of synthesized Pt/TiO$_2$, compared with those obtained for commercial Pt black (Johnson Matthey), are shown in Figure 6. The samples synthesized in-situ with nominal Pt loading of 5 mass % were not considered, since the loading is too low for measuring electrochemical characteristics.

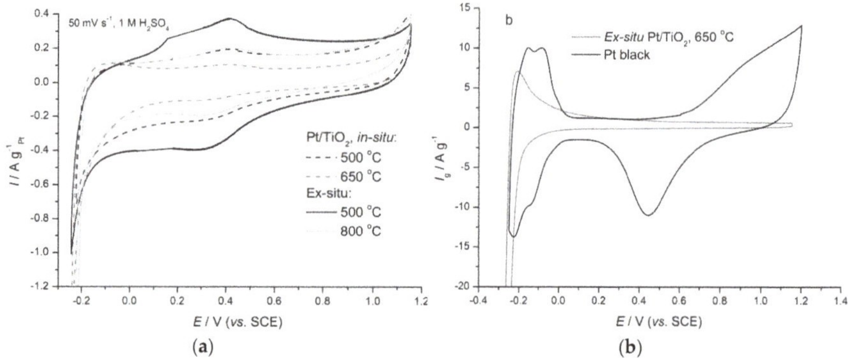

Figure 6. Comparison of cyclic voltammograms of Pt/TiO$_2$ composites and Pt black (**a**) and ex-situ at 650 °C (**b**) in de-aerated 1.0 M H$_2$SO$_4$, sweep rate 50 mV·s^{-1}.

All registered CVs of USP synthesized Pt/TiO$_2$ are almost featureless in the potential region above 0 V, without clearly developed Pt oxide formation and reduction, as in the case of Pt black.

With the exception of the sample synthesized ex-situ at 650 °C (Figure 6b), values of CV currents of all the other Pt/TiO$_2$ samples are below 5 µA. Additionally, hydrogen reduction/oxidation peaks of considerable currents [52] at around –0.2 V are clearly registered only for the 650 °C sample with 20 mass % Pt. None of the samples show hydrogen adsorption/desorption region typical for Pt, as seen for Pt black (Figure 6b). Compared to the literature data on a 40 mass % Pt/TiO$_2$ electrocatalyst [53], the value of hydrogen reduction/oxidation currents obtained for this 650 °C is not significantly larger, i.e., 7 A·g^{-1} in comparison with 1.25 A·g^{-1}, although the Pt loading is twice larger.

The electrochemical properties of the ex-situ synthesized Pt/TiO$_2$ show considerable dependence on USP temperature. It appears that the suitable hierarchical structure is achieved at 650 °C, with electrochemically advantageous Pt size and distribution over the TiO$_2$ core, along the trend discussed in relation to the STEM images from Figure 4.

Although a full CV fingerprint of Pt was not registered for the synthesized Pt/TiO$_2$ composites, the samples show electrochemical activity for an oxygen reduction reaction and combined oxygen reduction and hydrogen evolution reactions. Initial quasi-steady-state polarization curves for competing oxygen reduction (OR)/hydrogen evolution (HE) at low and moderate currents are shown in Figure 7.

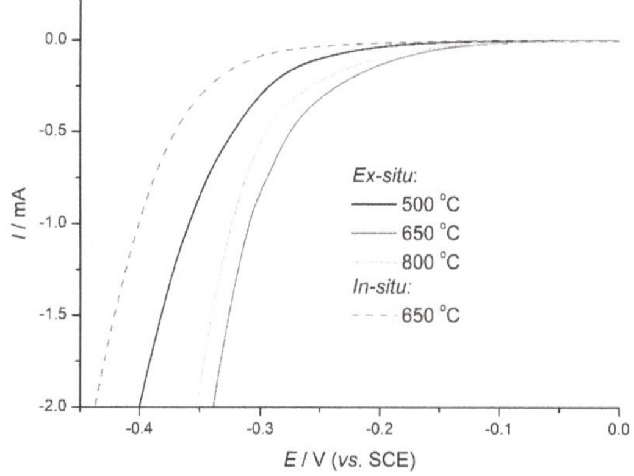

Figure 7. Initial quasi-steady-state polarization curves for combined oxygen reduction and hydrogen evolution reactions at low and moderate currents of Pt/TiO$_2$ composites. Electrolyte: O$_2$ purged 1 M H$_2$SO$_4$; 1 mV·s^{-1}.

As already observed in CV responses (Figure 6), where the highest currents in HE region were registered for the 650 °C sample, this sample shows best polarization characteristics, as seen in Figure 7. In general, the ex-situ synthesized samples have considerably higher OR/HE activity compared to the activity of samples synthesized in-situ. The differences in activity are to be assigned to different component distributions and surface compositions of the spheres as constituents of in-situ and ex-situ synthesized Pt/TiO_2. Pt loading at the surface of ex-situ synthesized spheres is higher and exceeds nominal loading, since small Pt particles decorate uniformly and cover regularly the surface of TiO_2 spherical core. Pt particles are smaller than those prepared in-situ. This hierarchy is beneficial for full availability of high-surface-area Pt for the OR/HE process. The in-situ prepared Pt/TiO_2 has lower surface Pt loading, with most of the active components constituting the walls of huge hollow spheres or the inactive cores of smaller spheres. These morphological elements have randomly mixed components and larger Pt particles which are hardly accessible as reaction sites (those facing the interior of the hollow spheres or being buried inside the solid spheres). Hence, the differences seen in Figure 6a,b are due to larger number of active sites readily accessible for the reaction in the case of the ex-situ synthesized Pt/TiO_2 composite.

Tafel representations of quasi-steady-state polarization curves for the Pt/TiO_2 composites from Figure 7 are shown in Figure 8.

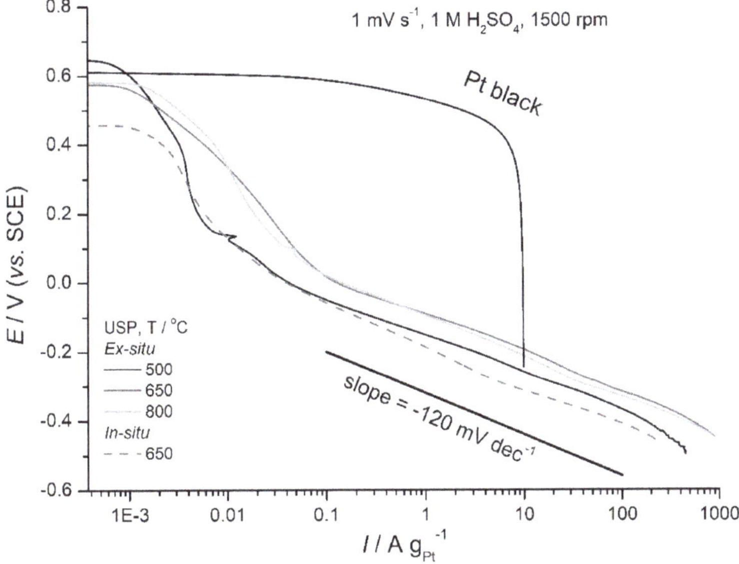

Figure 8. Tafel representation of quasi-steady-state polarization curves for oxygen reduction reaction (ORR) (above -0.2 V) and combined ORR/hydrogen evolution reaction (HER) (below −0.2 V) of Pt/TiO_2 composites compared to Pt black. Electrolyte: O_2-purged 1 M H_2SO_4; 1 mV·s⁻¹.

In the low overpotential region, 0.6–0.0 V, all Pt/TiO_2 composites show poor activity, exhibiting high values of the slopes. At 0 V, these values shifted to around 120 mV·dec⁻¹, characteristic for the ORR, but also for HER at high overpotential [54–56]. If compared to Pt polarization, the ORR on Pt/TiO_2 has high initial overpotential, and hence, has considerably shifted cathodically. The curves take the slope close to −120 mV/dec around 0.0 V vs. SCE, while the curve for Pt black reaches the ORR limiting current already at ca. 0.4 V. Since the potential of 0.0 V is positive to the thermodynamic onset of HER (−0.243 V vs. SCE), it follows that Pt/TiO_2 shows measurable ORR activity at the potentials as negative at 0.0 V. At the potentials negative to −0.2 V, the thermodynamic condition for the onset of

HER is fulfilled and polarization curves for Pt/TiO_2 represent joint ORR/HER activity. No changes in the slope upon the HER onset are detected, which indicates that both ORR and HER could proceed on Pt/TiO_2 with high surface coverage and surface recombination/desorption as the rate-determining steps. Since Pt oxide formation/reduction is suppressed on Pt/TiO_2 (Figure 6b), the polarization data indicate that Pt states in mixture with or on a TiO_2 surface have poor activity towards the adsorption of oxygen-containing species and moieties, due to Pt interaction with TiO_2 [48]. However, the interaction appears to stabilize the adsorbed hydrogen, causing the high surface coverage. As indicated in Figure 7, and seen also in Figure 8, ex-situ synthesized Pt/TiO_2 samples are considerably more active for ORR/HER than in-situ synthesized samples.

4. Conclusions

Spherical Pt/TiO_2 composite materials of novel structure and characteristic component distribution were synthesized at different temperatures using ultrasonic spray pyrolysis (USP). In the ex-situ synthesis, TiO_2 colloid and chloroplatinic acid were used as precursors, while the in-situ approach was based on tetra-n-butyl orthotitanate and chloroplatinic acid in an hydrochloric acid solution as precursors. While in-situ USP synthesis generates the mixture of huge hollow spheres and 100-nm-sized solid spheres with deficient Pt loading due to intrinsic aerosol-gel/solid phase transition, the regular TiO_2 spherical cores, with the surface uniformly covered by the shell of nm-sized Pt particles, are synthesized via the ex-situ approach. Accordingly, the ex-situ synthesized samples are more active for electrochemical oxygen reduction and combined oxygen reduction and hydrogen evolution reactions than the in-situ samples. The prepared samples were electrochemically checked using cyclic (CV) and linear sweep voltammetry (LSV) and collected responses were compared to those obtained for Pt black. The benefit of the USP prepared samples compared to Pt black is considerably increased CV cathodic currents in the hydrogen reduction region. The highest CV currents were registered for the ex-situ sample synthesized at 650 °C as a moderate USP temperature, with nominal 20 mass % Pt. This sample also showed the highest activity for oxygen reduction/combined oxygen reduction and hydrogen evolution reactions. Higher affinity of Pt in Pt/TiO_2 toward hydrogen than for the adsorption of oxygen-containing species is clear.

Author Contributions: Conceptualization, J.S.S. and B.G.F.; methodology, V.V.P. and S.R.S.; formal analysis, T.E.W.; investigation, M.G.K. and M.M.Z.; data curation, M.G.K.; writing—original draft preparation, M.G.K. and M.M.Z.; writing—review and editing, V.V.P. and T.E.W.; supervision, V.V.P., B.G.F. and J.S.S.; project administration, V.V.P. and S.R.S. All authors have read and agreed to the published version of the manuscript.

Funding: This research was funded by Ministry of Education, Science and Technological Development of the Republic of Serbia, grant number 172060 and by the funds of the bilateral research project, ID 451-03-01413/2016-09/7, supported by DAAD, Germany.

Acknowledgments: This work was supported by the Ministry of Education, Science and Technological Development of the Republic of Serbia (Grant No. 172060). We would like to thank the German Academic Exchange Service DAAD for the financial support at the project No. 57334757 under title "Novel designs of synthesis for tailoring the ordered structures of multicomponent metal oxides as uniform coatings of activated titanium anodes" between 01.01.2017 to 31.12.2018. Especially, the authors would like to thank Cleopatra Herwartz, GFE Central Facility for Electron Microscopy, RWTH Aachen University for recording STEM images and EDS spectra of the samples.

Conflicts of Interest: The authors declare no conflict of interest.

References

1. Berber, M.R.; Hafez, I.H.; Fujigaya, T.; Nakashima, N. A highly durable fuel cell electrocatalyst based on double-polymer-coated carbon nanotubes. *Sci. Rep.* **2015**, *5*, 16711. [CrossRef]
2. Wang, J. System integration, durability and reliability of fuel cells: Challenges and solutions. *Appl. Energy* **2017**, *189*, 460–479. [CrossRef]
3. Bae, S.J.; Kim, S.-J.; Lee, J.-H.; Song, I.; Kim, N.-I.; Seo, Y.; Kim, K.B.; Lee, N.; Park, J.-Y. Degradation pattern prediction of a polymer electrolyte membrane fuel cell stack with series reliability structure via durability data of single cells. *Appl. Energy* **2014**, *131*, 48–55. [CrossRef]

4. Wang, X.X.; Tan, Z.H.; Zeng, M.; Wang, J.N. Carbon nanocages: A new support material for Pt catalyst with remarkably high durability. *Sci. Rep.* **2014**, *4*, 1–11. [CrossRef] [PubMed]

5. Sharma, R.; Gyergyek, S.; Li, Q.; Andersen, S.M. Evolution of the degradation mechanisms with the number of stress cycles during an accelerated stress test of carbon supported platinum nanoparticles. *J. Electroanal. Chem.* **2019**, *838*, 82–88. [CrossRef]

6. Teran-Salgado, E.; Bahena-Uribe, D.; Márquez-Aguilar, P.A.; Reyes-Rodriguez, J.L.; Cruz-Silva, R.; Solorza-Feria, O. Platinum nanoparticles supported on electrochemically oxidized and exfoliated graphite for the oxygen reduction reaction. *Electrochim. Acta* **2019**, *298*, 172–185. [CrossRef]

7. Kim, Y.; Lee, D.; Kwon, Y.; Kim, T.-W.; Kim, K.; Kim, H.J. Enhanced electrochemical oxygen reduction reaction performance with Pt nanocluster catalysts supported on microporous graphene-like 3D carbon. *J. Electroanal. Chem.* **2019**, *838*, 89–93. [CrossRef]

8. Teng, X.; Yang, H. Synthesis and electrocatalytic property of cubic and spherical nanoparticles of cobalt platinum alloys. *Front. Chem. Eng. China* **2010**, *4*, 45–51. [CrossRef]

9. Krstajić Pajić, M.N.; Stevanović, S.I.; Radmilović, V.V.; Gavrilović-Wohlmutherc, A.; Radmilović, V.R.; Gojković, S.L.; Jovanović, V.M. Shape evolution of carbon supported Pt nanopar-ticles: From synthesis to application. *Appl. Catal. B Environ.* **2016**, *196*, 174–184. [CrossRef]

10. Mourdikoudis, S.; Chirea, M.; Altantzis, T.; Pastoriza-Santos, I.; Pe'rez-Juste, J.; Silva, F.; Bals, S.; Liz-Marzan, L.M. Dimethylformamide-mediated synthesis of water-soluble platinum nanodendrites for ethanol oxidation electrocatalysist. *Nanoscale* **2013**, *5*, 4776–4784. [CrossRef]

11. Gao, Q.; Gao, M.R.; Liu, J.W.; Chen, M.Y.; Cui, C.H.; Lia, H.H.; Yu, S.H. One-pot synthesis of branched palladium nanodendrites with superior electrocatalytic performance. *Nanoscale* **2013**, *5*, 3202–3207. [CrossRef] [PubMed]

12. Sui, S.; Wang, X.; Zhou, X.; Su, Y.; Riffat, S.; Liu, C.-J. A comprehensive review of Pt electrocatalysts for the oxygen reduction reaction: Nanostructure, activity, mechanism and carbon support in PEM fuel cells. *J. Mater. Chem. A* **2017**, *5*, 1808–1825. [CrossRef]

13. Wang, R.; Wang, H.; Luo, F.; Liao, S. Core–Shell-Structured Low-Platinum Electrocatalysts for Fuel Cell Applications. *Electrochem. Energy Rev.* **2018**, *1*, 324–387. [CrossRef]

14. Jayawickrama, S.M.; Han, Z.; Kido, S.; Nakashima, N.; Fujigaya, T. Enhanced platinum utilization efficiency of polymer-coated carbon black as an electrocatalyst in polymer electrolyte membrane fuel cells. *Electrochim. Acta* **2019**, *312*, 349–357. [CrossRef]

15. Devrim, Y.; Arıca, E.D. Multi-walled carbon nanotubes decorated by platinum catalyst for high temperature PEM fuel cell. *Int. J. Hydrog. Energy* **2019**, *44*, 18951–18966. [CrossRef]

16. Berghian-Grosan, C.; Radu, T.; Biris, A.R.; Dan, M.; Voica, C.; Watanabe, F.; Biris, A.S.; Vulcu, A. Platinum nanoparticles coated by graphene layers: A low-metal loading catalyst for methanol oxidation in alkaline media. *J. Energy Chem.* **2020**, *40*, 81–88. [CrossRef]

17. Alcaide, F.; Álvarez, G.; Miguel, O.; Lázaro, M.J.; Moliner, R.; López-Cudero, A.; Solla-Gullón, J.; Herrero, E.; Aldaz, A. Pt supported on carbon nanofibers as electrocatalyst for low temperature polymer electrolyte membrane fuel cells. *Electrochem. Commun.* **2009**, *11*, 1081–1084. [CrossRef]

18. Fraser, A.; Zhang, Z.; Merle, G.; Gbureck, U.; Ye, S.; Gostick, J.; Barralet, J. Composite Carbon Nanotube Microsphere Coatings for Use as Electrode Supports. *Adv. Funct. Mater.* **2018**, *28*, 1803713. [CrossRef]

19. Lafforgue, C.; Zadick, A.; Dubau, L.; Maillard, F.; Chatenet, M. Selected Review of the Degradation of Pt and Pd-based Carbon-supported Electrocatalysts for Alkaline Fuel Cells: Towards Mechanisms of Degradation. *Fuel Cells* **2018**, *18*, 229–238. [CrossRef]

20. Kreitmeier, S.; Wokaun, A.; Büchi, F.N. Local Catalyst Support Degradation during Polymer Electrolyte Fuel Cell Start-Up and Shutdown. *J. Electrochem. Soc.* **2012**, *159*, F787–F793. [CrossRef]

21. Zhang, Y.; Chen, S.; Wang, Y.; Ding, W.; Wu, R.; Li, L.; Qi, X.; Wei, Z. Study of the degradation mechanisms of carbon-supported platinum fuel cells catalyst via different accelerated stress test. *J. Power Sources* **2015**, *273*, 62–69. [CrossRef]

22. Wu, Z.; Dang, D.; Tian, X. Designing Robust Support for Pt Alloy Nanoframes with Durable Oxygen Reduction Reaction Activity. *ACS Appl. Mater. Interfaces* **2019**, *11*, 9117–9124. [CrossRef] [PubMed]

23. Liu, F.; Wu, Z.; Dang, D.; Wang, G.; Tian, X.; Yang, X. Three dimensional titanium molybdenum nitride nanowire assemblies as highly efficient and durable platinum support for methanol oxidation reaction. *Electrochim. Acta* **2019**, *295*, 50–57. [CrossRef]

24. Huynh, T.T.; Pham, H.Q.; Nguyen, A.V.; Bach, L.G.; Ho, V.T.T. Advanced Nanoelectrocatalyst of Pt Nanoparticles Supported on Robust $Ti_{0.7}Ir_{0.3}O_2$ as a Promising Catalyst for Fuel Cells. *Ind. Eng. Chem. Res.* **2019**, *58*, 675–684. [CrossRef]

25. Liu, F.; Dang, D.; Tian, X. Platinum-decorated three dimensional titanium copper nitride architectures with durable methanol oxidation reaction activity. *J. Hydrog. Energy* **2019**, *44*, 8415–8424. [CrossRef]

26. Kim, S.-W.; Han, T.H.; Kim, J.; Gwon, H.; Moon, H.-S.; Kang, S.-W.; Kim, S.O.; Kang, K. Fabrication and Electrochemical Characterization of TiO_2 Three-Dimensional Nanonetwork Based on Peptide Assembly. *ACS Nano* **2009**, *3*, 1085–1090. [CrossRef]

27. Diebold, U. The surface science of titanium dioxide. *Surf. Sci. Rep.* **2003**, *48*, 53–229. [CrossRef]

28. Fovet, Y.; Gal, J.Y.; Toumelin-Chemla, F. Influence of pH and fluoride concentration on titanium passivating layer: Stability of titanium dioxide. *Talanta* **2001**, *53*, 1053–1063. [CrossRef]

29. Lewera, A.; Timperman, L.; Roguska, A.; Alonso-Vante, N. Metal-support interactions between nanosized Pt and metal oxides (WO_3 and TiO_2) studied using X-ray photoelectron spectroscopy. *J. Phys. Chem. C* **2011**, *115*, 20153–20159. [CrossRef]

30. Dulub, O.; Hebenstreit, W.; Diebold, U. Imaging cluster surfaces with atomic resolution: The strong metal-support interaction state of Pt supported on TiO_2 (110). *Phys. Rev. Lett.* **2000**, *84*, 3646–3649. [CrossRef]

31. Kim, D.-S.; Zeid, A.; Kim, Y.-T. Additive treatment effect of TiO2 as supports for Pt-based electrocatalysts on oxygen reduction reaction activity. *Electrochim. Acta* **2010**, *55*, 3628–3633. [CrossRef]

32. Wang, Y.; Mohamedi, M. Hierarchically organized nanostructured TiO2/Pt on microfibrous carbon paper substrate for ethanol fuel cell reaction. *Int. J. Hydrog. Energy* **2017**, *42*, 22796–22804. [CrossRef]

33. Amer, M.S.; Ghanem, M.A.; Al-Mayouf, A.M.; Prabhakarn, A.N.H. Low-loading of oxidized platinum nanoparticles into mesoporous titanium dioxide for effective and durable hydrogen evolution in acidic media. *Arab. J. Chem.* **2018**. [CrossRef]

34. Shaddad, M.N.; Al-Mayouf, A.M.; Ghanem, M.A.; AlHoshan, M.S.; Singh, J.P.; Al-Suhybani, A.A. Chemical Deposition and Electrocatalytic Activity of Platinum Nanoparticles Supported on TiO_2 Nanotubes. *Int. J. Electrochem. Sci.* **2013**, *8*, 2468–2478.

35. Guo, P.; Xu, W.; Zhu, S.; Yang, X.; Inoue, A. Preparation and electrocatalytic performance of the Pt supported on the alkali-treated nanoporous TiO_2 material. *Ionics* **2015**, *21*, 2863–2869. [CrossRef]

36. Košević, M.; Šekularac, G.; Živković, L.; Panić, V.; Nikolić, B. TiO_2 From Colloidal Dispersion as Support in Pt/TiO_2 Nanocomposite for Electrochemical Applications. *Croat. Chem. Acta* **2017**, *90*, 251–258. [CrossRef]

37. Fugare, B.Y.; Lokhande, B.J. Study on structural, morphological, electrochemical and corrosion properties of mesoporous RuO_2 thin films prepared by ultrasonic spray pyrolysis for supercapacitor electrode application. *Mater. Sci. Semicond. Process.* **2017**, *71*, 121–127. [CrossRef]

38. Muñoz-Fernandez, L.; Alkan, G.; Milošević, O.; Rabanal, M.E.; Friedrich, B. Synthesis and characterisation of spherical core-shell Ag/ZnO nanocomposites using single and two—Steps ultrasonic spray pyrolysis (USP). *Catal. Today* **2019**, *321–322*, 26–33. [CrossRef]

39. Alkan, G.; Diaz, F.; Matula, G.; Stopic, S.; Friedrich, B. Scaling up of nanopowder collection in the process of ultrasonic spray pyrolysis. *World Metall.-ERZMETALL* **2017**, *70*, 97–101.

40. Mata, V.; Maldonado, A.; Olvera, M.L. Deposition of ZnO thin films by ultrasonic spray pyrolysis technique. Effect of the milling speed and time and its application in photocatalysis. *Mater. Sci. Semicond. Process.* **2018**, *75*, 288–295. [CrossRef]

41. Liang, F.; Chen, S.; Xie, W.; Zou, C. The decoration of Nb-doped TiO2 microspheres by reduced graphene oxide for enhanced CO gas sensing. *J. Phys. Chem. Solids* **2018**, *114*, 195–200. [CrossRef]

42. Alkan, G.; Rudolf, R.; Bogovic, J.; Jenko, D.; Friedrich, B. Structure and Formation Model of Ag/TiO_2 and Au/TiO_2 Nanoparticles Synthesized through Ultrasonic Spray Pyrolysis. *Metals* **2017**, *7*, 389. [CrossRef]

43. Zhang, J.; Mo, Y.; Vukmirovic, M.B.; Klie, R.; Sasaki, K.; Adzic, R.R. Platinum Monolayer Electrocatalysts for O_2 Reduction: Pt Monolayer on Pd(111) and on Carbon-Supported Pd Nanoparticles. *J. Phys. Chem. B* **2004**, *108*, 10955–10964. [CrossRef]

44. Esmaeilifar, A.; Rowshanzamir, S.; Eikani, M.H.; Ghazanfari, E. Synthesis methods of low-Pt-loading electrocatalysts for proton exchange membrane fuel cell systems. *Energy* **2010**, *35*, 3941–3957. [CrossRef]

45. Arici, E.; Kaplan, B.Y.; Mert, A.M.; Gursel, S.A.; Kinayyigit, S. An effective electrocatalyst based on platinum nanoparticles supported with graphene nanoplatelets and carbon black hybrid for PEM fuel cell. *Int. J. Hydrog. Energy* **2019**, *44*, 14175–14183. [CrossRef]

46. Yang, Z.; Chen, M.; Xia, M.; Wang, M.; Wang, X. An effective and durable interface structure design for oxygen reduction and methanol oxidation electrocatalyst. *Appl. Surf. Sci.* **2019**, *487*, 655–663. [CrossRef]

47. Huang, W.; Wang, H.; Zhou, J.; Wang, J.; Duchesne, P.N.; Muir, D.; Zhang, P.; Han, N.; Zhao, F.; Zeng, M.; et al. Highly active and durable methanol oxidation electrocatalyst based on the synergy of platinum–nickel hydroxide–graphene. *Nat. Commun.* **2015**, *6*, 1–8. [CrossRef]

48. Jakšić, M.M. Advances in electrocatalysis for hydrogen evolution in the light of the Brewer-Engel valence-bond theory. *J. Mol. Catal.* **1986**, *38*, 161–202. [CrossRef]

49. Zheng, H.; Wang, C.; Zhang, X.; Li, Y.; Ma, H.; Liu, Y. Control over energy level match in Keggin polyoxometallate-TiO$_2$ microspheres for multielectron photocatalytic reactions. *Appl. Catal. B* **2018**, *234*, 79–89. [CrossRef]

50. Yang, J.; Wang, G.; Wang, D.; Liu, C.; Zhang, Z. A self-cleaning coating material of TiO$_2$ porous microspheres/cement composite with high-efficient photocatalytic depollution performance. *Mater. Lett.* **2017**, *200*, 1–5. [CrossRef]

51. Stopic, S.; Friedrich, B.; Schroeder, M.; Weirich, T.E. Synthesis of TiO$_2$ core/RuO$_2$ shell particles using multistep ultrasonic spray pyrolysis. *Mater. Res. Bull.* **2013**, *48*, 3633–3635. [CrossRef]

52. Kundu, M.K.; Bhowmik, T.; Mishra, R.; Barman, S. Pt Nanostructures/N-Doped Carbon hybrid, an Efficient Catalyst for Hydrogen Evolution/Oxidation Reactions: Enhancing its Base Media Activity through Bi-functionality of the Catalyst. *ChemSusChem* **2018**, *11*, 2388–2401. [CrossRef] [PubMed]

53. Dhanasekaran, P.; Vinod Selvaganesh, S.; Sarathi, L.; Santoshkumar, D.B. Rutile TiO2 Supported Pt as Stable Electrocatalyst for Improved Oxygen Reduction Reaction and Durability in Polymer Electrolyte Fuel Cell. *Electrocatalysis* **2016**, *7*, 495–506. [CrossRef]

54. Chiwata, M.; Kakinuma, K.; Wakisaka, M.; Uchida, M.; Deki, S.; Watanabe, M.; Uchida, H. Oxygen Reduction Reaction Activity and Durability of Pt Catalysts Supported on Titanium Carbide. *Catalysts* **2015**, *5*, 966–980. [CrossRef]

55. Antoine, O.; Bultel, Y.; Durand, R. Oxygen reduction reaction kinetics and mechanism on platinum nanoparticles inside Nafion®. *J. Electroanal. Chem* **2001**, *499*, 85–94. [CrossRef]

56. Creţu, R.; Kellenberger, A.; Vaszilcsin, N. Enhancement of hydrogen evolution reaction on platinum cathode by proton carriers. *Int. J. Hydrog. Energy* **2013**, *38*, 11685–11694. [CrossRef]

Article

Synthesis of Poly-Alumino-Ferric Sulphate Coagulant from Acid Mine Drainage by Precipitation

Brian Mwewa [1,2,*], Srećko Stopić [3], Sehliselo Ndlovu [1,2], Geoffrey S. Simate [1], Buhle Xakalashe [3] and Bernd Friedrich [3]

1   School of Chemical and Metallurgical Engineering, University of the Witwatersrand, Private Bag 3, Johannesburg 2050, South Africa; Sehliselo.Ndlovu@wits.ac.za (S.N.); Geoffrey.Simate@wits.ac.za (G.S.S.)
2   DST/NRF SARChI: Hydrometallurgy and Sustainable Development, School of Chemical and Metallurgical Engineering, University of the Witwatersrand, Private Bag 3, Johannesburg 2050, South Africa
3   IME Process Metallurgy and Metal Recycling, RWTH Aachen University, Intzestraße 3, 52056 Aachen, Germany; sstopic@ime-aachen.de (S.S.); BXakalashe@metallurgie.rwth-aachen.de (B.X.); bfriedrich@ime-aachen.de (B.F.)
*   Correspondence: sirbhimself@yahoo.co.uk; Tel.: +27-117177516

Received: 3 October 2019; Accepted: 24 October 2019; Published: 29 October 2019

Abstract: The wastes generated from both operational and abandoned coal and metal mining are an environmental concern. These wastes, including acid mine drainage (AMD), are treated to abate the devastating effects they have on the environment before disposal. However, AMD contains valuable resources that can be recovered to subsidize treatment costs. Two of the major constituents of coal AMD are iron and aluminium, which can be recovered and engineered to function as coagulants. This work examines the potential of producing a poly-alumino-ferric sulphate (AMD-PAFS) coagulant from coal acidic drainage solutions. The co-precipitation of iron and aluminium is conducted at pH values of 5.0, 6.0 and 7.0 using sodium hydroxide in order to evaluate the recovery of iron and aluminium as hydroxide precipitates while minimizing the co-precipitation of the other heavy metals. The precipitation at pH 5.0 yields iron and aluminium recovery of 99.9 and 94.7%, respectively. An increase in the pH from 5.0 to 7.0 increases the recovery of aluminium to 99.1%, while the recovery of iron remains the same. The precipitate formed at pH 5.0 is used to produce a coagulant consisting of 89.5% and 10.0% iron and aluminium, respectively. The production of the coagulant is carried out by dissolving the precipitate in 5.0% (w/w) sulphuric acid. Subsequently, the treatment of the brewery wastewater shows that the AMD-PAFS coagulant is as efficient as the conventional poly ferric sulphate (PFS) coagulant. The turbidity removal is 91.9 and 87.8%, while the chemical oxygen demand (COD) removal is 56.0 and 64.0% for AMD-PAFS and PFS coagulants, respectively. The developed process, which can easily be incorporated into existing AMD treatment plants, not only reduces the sludge disposal problems but also creates revenue from waste.

Keywords: acid mine drainage; precipitation; iron; aluminium; coagulation; water treatment

---

1. Introduction

Acid mine drainage (AMD) is one of the largest environmental threats facing the world today. It is rated second only to global warming and stratospheric ozone depletion in terms of its ecological effects [1]. Environmentalists have termed AMD the single most significant threat to South Africa's environment. AMD is caused by the oxidation of sulfur, present in the mineral pyrite (Fe_2S). When exposed to water and air, either during mining operations, once the mine has been abandoned or as a result of natural weathering, the pyrite is oxidized, which leads to the generation of high acidity and ferrous iron-impacted waters [2,3]. There are a series of reactions and side reactions involved during the formation of AMD, with the overall reaction given by Equation (1). The presence of AMD has the potential to devastate streams, rivers and aquatic life [4–8]. Such devastating scenarios necessitate the treatment of AMD to abate the effects it has on the environment.

$$4FeS_2 + 14H_2O + 15O_2 \rightarrow 4Fe(OH)_3 + 16H^+ + 8SO_4^{2-} \tag{1}$$

For many decades, the most widely applied method for the treatment of AMD is an active treatment process involving chemical-neutralization reagents [9,10]. This technology entails the addition of lime to acidic waters to raise the pH and precipitate the dissolving of metals. This process produces a hydroxide sludge, which typically contains 2–5% solids [11]. The voluminous sludge is difficult to dispose of because of the scarcity of land. In addition, the process produces metastable phases whose long-term stability has not been established. Therefore, post-precipitate stabilization before final solids disposal is required. However, the metals in AMD can be recovered with the objective of obtaining valuable products while meeting the effluent discharge limitation [12]. This is one of the potential ways to extend the use of natural resources. This paradigm shift has led to a number of studies being conducted to investigate the recovery of valuable products from AMD, including iron oxide pigments for production of paint [13–15]; ferric oxide nanoparticles [16]; inorganic pigments [17]; metals like Fe, Al, Zn, and Cu [18–21]; and acid and water [22–25], and the use of AMD neutralization sludge in brick and cement production, and as an artificial soil additive [26–28].

There is also huge potential to recover alternative coagulants from AMD for water treatment. The coagulants that are widely used to remove a broad range of impurities from effluent, including colloidal particles and dissolved organic substances, are metal salts such as aluminium sulphate $Al_2(SO_4)_3.5H_2O$, aluminium chloride $AlCl_3$, polyaluminium chloride $AlCl_3$, ferric sulphate $Fe_2(SO_4)_3.5H_2O$ and ferric chloride $FeCl_3$ [29]. The actual coagulant species involved in the coagulation process are formed after the coagulant chemicals are added to water. The addition of these cationic species to water results in colloidal destabilization as they specifically interact with and neutralize the negatively charged colloidal particles. For example, when aluminium sulphate/chloride is dissolved in water, the Al ion Al^{3+}, immediately coordinates with six water molecules, $Al(H_2O)_6^{3+}$ [29,30]. The hydrolysis reactions (e.g., Equation (2)) proceed with the formation of numerous mononuclear species, e.g., $Al(OH)^{2+}$, $Al(OH)_2^+$, $Al(OH)_3$ (molecule) and $Al(OH)_4^-$, followed by the formation of three polynuclear species including but not limited to $Al_2(OH)_2^{4+}$, $Al_3(OH)_4^{5+}$ and $Al_{13}O_4(OH)_{24}^{7+}$, as well as a solid precipitate $[Al(OH)_3]$. The hydrolysis of Fe is very similar in many respects to that of Al. Flynn Jr. [31] studied the hydrolysis of ferric iron and reported five mononuclear species Fe^{3+}, $Fe(OH)^{2+}$, $Fe(OH)_2^+$, $Fe(OH)_3$ molecule and $Fe(OH)_4^-$, and dimeric species $Fe_2(OH)_2^{4+}$ and $Fe_3(OH)_4^{5+}$.

$$2Al^{3+} + 2H_2O \Leftrightarrow Al_2(OH)_2^{4+} + 2H^+ \tag{2}$$

The high concentration of Fe and Al in AMD, as high as 5000 mg/L for Fe and 500 mg/L for Al, has led to studies that have focused on developing an understanding of its potential reuse as a coagulant in wastewater treatment. For example, a novel application of AMD for coagulation/flocculation of microalga biomass was developed by Salama et al. [32]. A coagulation efficiency of 89% and 93% was obtained for *S. obliquus* and *C. vulgaris*, respectively, with a 10% dose of AMD as a coagulant. Lopes et al. [33] used mine water directly as a coagulant for the treatment of sewage wastewater. AMD was effective in the removal of suspended solids, organic matter, phosphorus and bacteria of the coliform group. Another process for the direct use of AMD as a coagulant in municipal wastewater treatment was tested by Rao et al. [34] and compared with $FeCl_3$. The AMD was found to be as effective as $FeCl_3$. However, the treated water contained high residual heavy metals from AMD. This precluded its general use in water treatment without pretreatment to remove heavy metals. This led to other studies being conducted to recover ferric sulphate coagulant by reacting the ferric hydroxide precipitate formed from AMD at pH 3.5–3.6 with sulphuric acid [34,35]. The use of dodecylamine surfactant to avoid co-precipitation of other metals, thereby improving the purity of the precipitate, was also tested.

The recovered coagulant was effective in municipal wastewater treatment and compared favourably with conventional coagulants.

This study is motivated by the work done by Jiang and Graham [36], who produced a poly-alumino-iron sulphate (PAFS) coagulant using chemical grade aluminium sulphate $Al_2(SO_4)_3.5H_2O$ and ferric sulphate $Fe_2(SO_4)_3.5H_2O$ salts as the two primary raw materials. The coagulant was evaluated for the removal of colour and dissolved organic carbon from drinking water and showed similar or better performance to conventional coagulants. In addition, the PAFS achieved the lowest residual metal-ion (Fe and Al) concentration when compared to ferric sulphate and aluminium sulphate. The high Fe and Al concentration in AMD means similar coagulants can be recovered from such mine-impacted waters. The specific objective of this study is to evaluate the recovery of an AMD-derived poly-alumino-ferric sulphate (AMD-PAFS) coagulant from coal AMD using chemical precipitation between pH 5.0 and 7.0. The efficiency of the AMD-PAFS is compared with conventional PFS coagulant in the treatment of brewery wastewater to remove turbidity, COD total dissolved solids (TDS). The effect of the coagulants on the electric conductivity (EC) of the wastewater is also evaluated.

2. Materials and Methods

2.1. Materials

The AMD sample was collected from Mpumalanga, South Africa. The sample was stored in a sealed polyethylene container. Before an experimental run, the solid debris and all the precipitated iron were removed by filtration using a grade 4 Whatman filter paper. Analytical-grade sodium hydroxide and sulphuric acid were used to prepare solutions for pH adjustment. All the solutions were prepared using deionised water. The conventional coagulant poly ferric sulphate (PFS) used for comparative tests was supplied by Merck, South Africa. The AMD-PAFS coagulant and PFC coagulant were tested on the brewery wastewater obtained from South African (SA) Breweries.

2.2. Experimental Procedure

2.2.1. Iron and Aluminium Co-Precipitation

All precipitation experiments were conducted in a 2 L reactor, shown in Figure 1. The agitator was fitted with a two-radial-blade impeller, and a speed of 300 rpm was used for all the experimental runs. In order to maximize mixing, the reactor was fitted with four equally spaced baffles. The reactor closure had ports for electrodes to measure pH and temperature. The experimental procedure involved oxidization of Fe (II) to Fe (III) by aeration for a period of 24 h. The oxidation of Fe (II) to Fe (III) is essential to the precipitation of Fe at low pH. Fe (III) precipitates at the pH range of 3–4, while Fe (II) does not precipitate at a pH < 6 [37]. The Fe (II) concentration was monitored by wet chemistry using potassium dichromate titration method [38]. Table 1 presents the summary of the experimental conditions. All experiments were performed in triplicates. The experimental procedure involved maintaining the temperature of the reactor contents at ambient temperature (25 °C) using the infrared heater. The agitation was then increased to the required speed, and the pH was adjusted by automatically injecting either 4.0 M sodium hydroxide or 0.1 M sulphuric acid using a Glass Chem reactor system, which has an automatic titrator. The accuracy of the pH control was 0.1 pH units. After attaining the required pH, the experiment was allowed to proceed for a period of one hour. The precipitate was separated from the effluent by vacuum filtration, followed by washing with deionised water to remove the entrained effluent solution. The precipitate was then left in the oven for 24 h at 80 °C to dry. 6 g of the dried precipitate was dissolved in 50 mL of 5.0% (w/w) sulphuric acid to obtain a clear solution, which was then used as a coagulant.

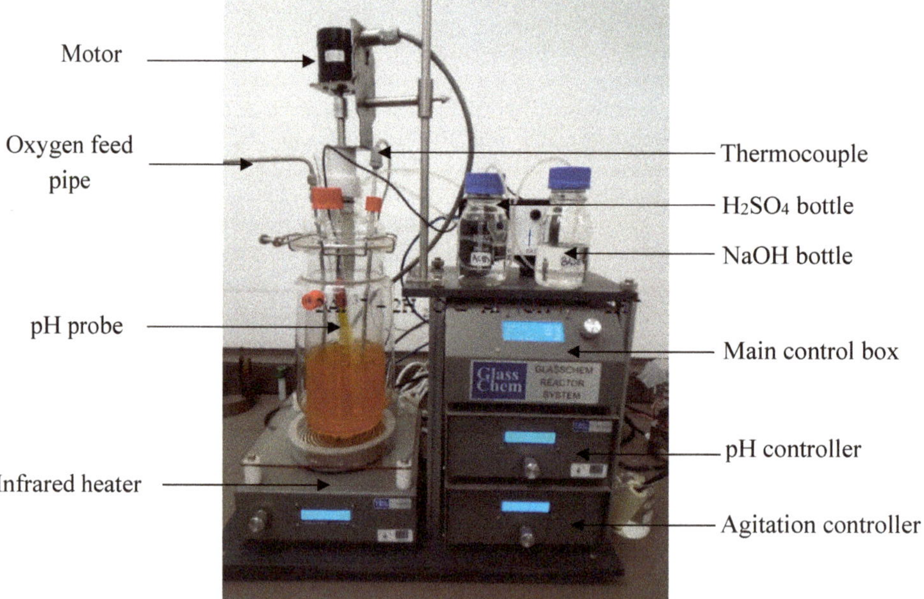

Figure 1. Picture of the experimental setup for the acid mine drainage (AMD) precipitation experiments using NaOH and H_2SO_4 at 25 °C.

Table 1. Factors and values selected for co-precipitation of iron and aluminium.

Factors	Values
Reaction temperature, °C	25
Fe oxidation time, hours	24
Type of oxidant	O_2
Agitation, rpm	300
Precipitation pH	5.0, 6.0, 7.0
Precipitation time, hours	1
NaOH concentration, molar	4
H_2SO_4 concentration, molar	0.1

2.2.2. Metal Analysis

All solution and precipitate samples were analyzed for Fe, Al, Ca, Mn, Mg, Cu, Zn, Ni and Co using inductive coupled plasma mass spectroscopy (ICP-MS 7700X), from Agilent Chemetrix. The concentration of the sulphate was determined using ion chromatography. Metal recovery (R) was calculated according to Equation (3), as follows:

$$R = \frac{C_0 - C_1}{C_0} \tag{3}$$

where C_0 is the concentration of a particular metal species in raw AMD (mg/L) and C_1 is the concentration of a metal species in the effluent (mg/L) after precipitation. Tabak et al. [39] defined the precipitate purity as the ratio of a desired precipitated metal species to the sum of all the metal species that have been precipitated. Based on this definition, the precipitate purity (P) was calculated according to Equation (4), as follows:

$$p = \frac{C_i}{\sum_i^n C_j} \times 100\% \tag{4}$$

where C_i is the concentration of the individual or sum of the species of interest (%), n is the total number of metal species and C_j is the concentration of all the metal species precipitated (%). In this case, C_i was regarded as the total concentration of iron and aluminium in the precipitate.

2.3. Water Treatment by Coagulation

A six-beaker jar tester apparatus was used with each beaker containing 500 mL of brewery wastewater samples. The same concentration of the AMD-PAFS and conventional PFS was added to the water and pH adjusted to 7.0. The water samples were agitated for 3 min at a paddle speed of 200 rpm, followed by 10 min of slow mixing at a speed of 20 rpm and sedimentation of 30 min. Supernatant samples were withdrawn at 5 cm below the surface of the water samples. The performance evaluation was based on pH, EC, turbidity, COD and TDS measurement. The pH, EC and TDS were measured using the Hanna HI 9812-5 pH/EC/TDS/temperature portable meter (Hanna Instruments, Johannesburg, South Africa). The meter was calibrated with standard solutions of pH 4.0 and 7.0 before use. The supernatant was measured for turbidity and COD using a Merck Pharo 300 spectroquant, (Merck, Johannesburg, South Africa). The unit of measurement for turbidity was the Formazin attenuation units (FAU). The analysis methods followed the "Standard Method for Examination of Water and Wastewater" [40].

3. Results and Discussion

3.1. Precipitation

The general characteristic of the raw AMD is presented in Table 2. The characteristics of AMD are typical of the South African coal AMD solutions [41,42]. As can be seen from the table, this included high concentration of Fe, Al, Ca, Mg and Mn with minor concentrations of Ni, Zn, Cu and Co. The total Fe composition in the raw AMD was 80% as Fe (II) and 20% as Fe (III). The table also shows the SA standard for wastewater discharge into a water resource as well as the characteristics of the effluents obtained in this study at different precipitating pH values. When the pH was raised to 5.0, the Fe and Al concentrations were 2.7 and 14.0 mg/L in the effluent, respectively. This translated to 99.9 and 96.5% Fe and Al removal, respectively, calculated using Equation (3). These recoveries, which are averages of the triplicate results, are depicted in Figure 2 with the error bars related to the standard deviation. An increase in pH to 7.0 resulted in Al concentration of 3.7 mg/L in the effluent and the recovery being 99.1%, but the iron recovery remained at 99.9%. Other workers have also found similar results [16,19,21]. For example, during the synthesis of magnetic nanoparticles from AMD, Wei et al. [16] reduced Fe from 169 mg/L at pH 2.6 to 0.09 mg/L at pH 6.7 and Al from 71 mg/L to 0.2 mg/L under the same pH conditions. This represented 99.9% and 99.7% Fe and Al recovery, respectively. Figure 3 presents the effect of pH on the solubility of the other major heavy metals. The results show that the precipitation of Ca is almost negligible in the tested pH range. However, Mn and Mg effluent concentration were reduced from 93.9 mg/L to 83.6 mg/L and 474.0 mg/L to 457 mg/L, respectively. This represented 10.9% and 3.6% co-precipitation of Mn and Mg, respectively. Other minor elements, including Zn, Cu and Co, did not precipitate.The precipitate purity was calculated using Equation (4) and gave 99.0, 99.0 and 98.0% for pH 5.0, 6.0 and 7.0, respectively. The results obtained in the study are comparable with results obtained by other researchers. Michalková et al. [18] obtained less than 0.05% of Zn, Co, Cu and Ni in the AMD precipitated using sodium hydroxide at pH 6.9. In a study by Wei et al. [16], the precipitation of Ca, Mg, Mn and Ni during neutralization at pH 6.7 was 6.02, 5.57, 16.67 and 37.27%, respectively.

Table 2. Summary of the chemical composition of raw AMD and effluents after precipitation.

Parameter	pH	Fe (Total)	Al	Ca	Mg	Mn	Zn	Ni	Co	Cu
		Metal Concentration, mg/L								
Raw AMD	2.1	4290.0	396.0	503.0	474.0	93.9	14.3	1.7	1.4	0.5
SA effluent standard [a]	5.5–9.5	0.3	NA	NA	NA	0.1	0.1	NA	NA	0.01
Effluent after precipitation	5.0	2.7	14.0	500.0	454.0	89.3	14.1	1.6	1.4	0.5
SD	NA	0.03	1.50	2.01	9.53	2.10	2.13	NA	NA	NA
	6.0	2.1	9.0	501	461.0	86.7	13.3	1.6	1.4	0.5
SD		0.04	0.92	1.15	3.60	3.10	1.67	NA	NA	NA
	7.0	2.1	3.7	500	457.0	83.6	13.9	1.6	1.4	0.5
SD		0.03	0.67	2.02	4.58	1.87	3.87	NA	NA	NA

[a] data from [43] (NA = not applicable).

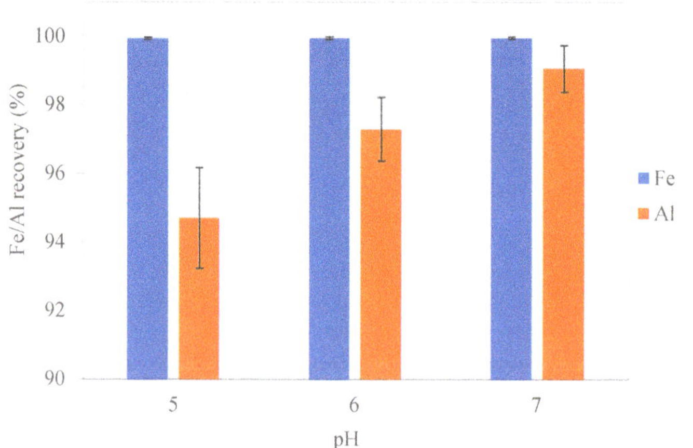

Figure 2. Effect of pH on Fe and Al recovery through precipitation of the AMD using NaOH and H$_2$SO$_4$ 25 °C.

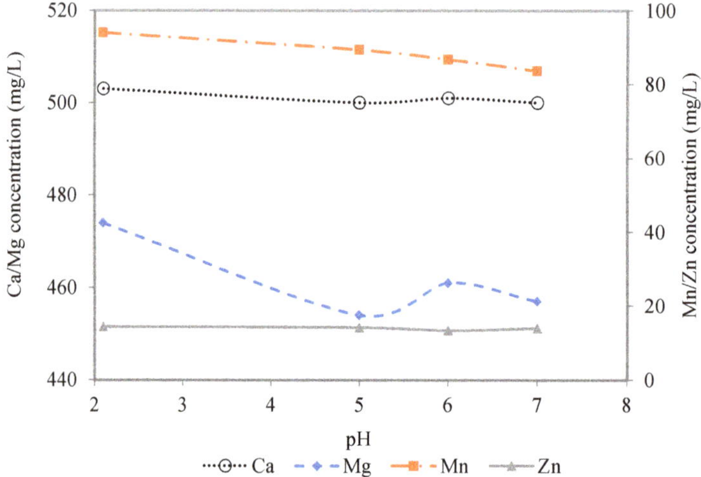

Figure 3. Solubility of the major heavy metals as a function of pH in the raw and treated AMD solutions at 25 °C.

3.2. Coagulation

3.2.1. Coagulant Production and Testing

The precipitate obtained at pH 5.0 was dissolved in 5.0% (w/w) sulphuric acid to produce a coagulant. Table 3 summarizes the characteristics of the AMD-PAFS as well as the commercial PFS and polyaluminium sulphate (PAS) coagulants. The coagulant was composed of 89.5% Fe and 10.0% Al with trace amounts of Mn, Mg and Ca. The total mass concentration of Fe and Al in the AMD-PAFS coagulant was 96,644 mg/L, which compares well with the commercial PFS with the Fe mass concentrations 115,000 mg/L. The AMD-PAFS was compared with the commercial FPS in the treatment of brewery wastewater. Table 4 shows the characteristics of the brewery wastewater used in the study.

Table 3. Chemical composition of the poly-alumino-ferric sulphate coagulant produced by precipitation at pH and conventional poly ferric sulphate and poly aluminium sulphate.

Parameter	AMD-PAFS	PFS [b]	PAS [b]
Fe (Total), mg/L	88,768.84	115,000	112.5
Al, mg/L	9876.54	4419	47,668
Mg, mg/L	444.43	160.6	0.38
Ca, mg/L	6.1	56.8	8.4
Mn, mg/L	102.23	1585	1.3
Zn, mg/L	ND	22.4	3.80
Ni, mg/L	ND	ND	ND
Co, mg/L	ND	ND	ND
Cu, mg/L	ND	11.5	<0.0004
SO_4^{2-}, mg/L	122,000	130,800	53,000

[b] data from [11] (ND = not detected).

Table 4. Characteristics of the brewery wastewater that was treated with the AMD-PAFS and PFS coagulant.

Parameter	Value
pH	7.1 ± 0.2
COD, mg/L	3160 ± 90
TDS, mg/L	1810 ± 30
Turbidity, FAU	99 ± 7
Electric conductivity, µS/cm	3510 ± 133

3.2.2. Effect of the Coagulant on Turbidity and Chemical Oxygen Demand Removal

Turbidity, which is the cloudiness of the water, has long been the targeted substance during the coagulation and flocculation processes and is largely used as an indicator for the efficiency of the coagulation process [44]. It is the principal physical characteristic of water and expresses the optical property that causes light to be scattered and absorbed by particles and molecules rather than transmitted in a straight line through the water sample. The turbidity removal efficiency was determined by adding different doses of the coagulants from 10 mg/L to 150 mg/L. As shown in Figure 4, the percentage removal of the turbidity of the brewery water samples increased from 18.1% at 10 mg/L to 91.92% at 150 mg/L AMD-PAFS. This compared favourable with results obtained from the use of PFS, where 22.3% and 87.8% turbidity removal at 10 and 150 mg/L PFS were obtained, respectively. The increment in the removal of turbidity was due to the increment of the activity site of the coagulants. The AMD-PAFS coagulant not only compared favourably with PFS coagulants but also with other synthetic coagulants. For example, a poly-aluminium-silicate-chloride coagulant (PSiFAC) was synthesized and tested in the treatment of simulated surface water [45]. A 99% turbidity removal was obtained at a PSiFAC concentration of 100 mg/L. The relatively high performance of the PSiFAC can be attributed to the presence of the silicate species. The silicate species increases the bridge effect and thereby slows down the formation of $Fe(OH)_3$ precipitate, which results in enhanced coagulation [46].

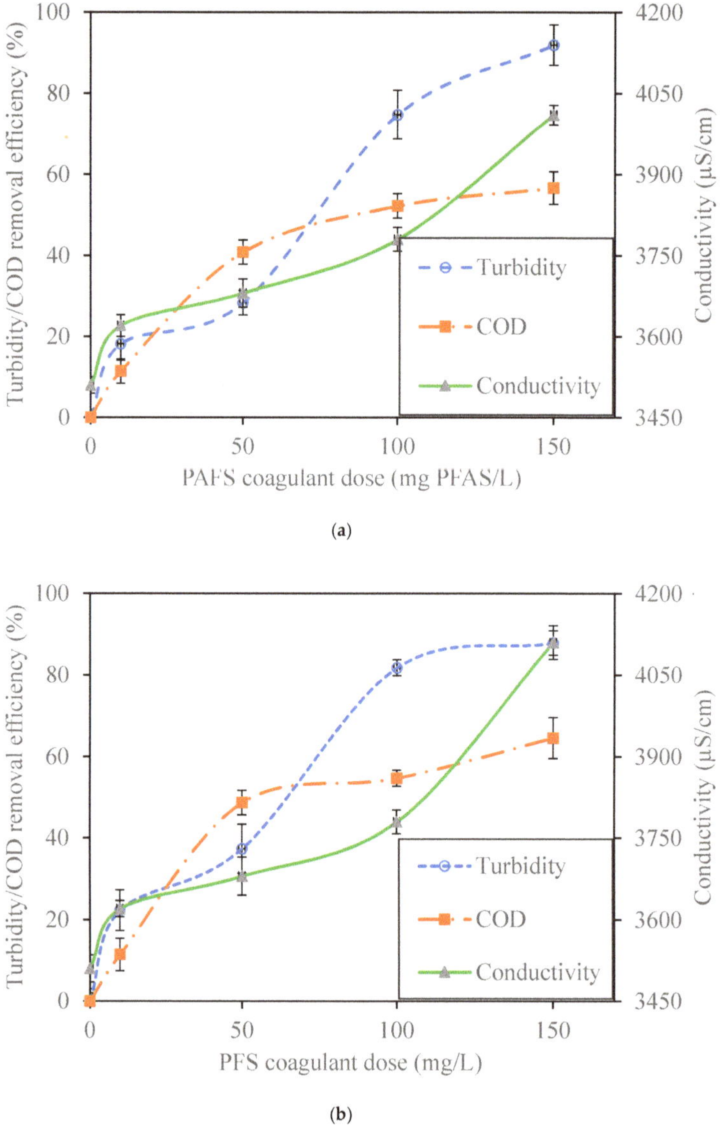

Figure 4. Performance evaluation of (**a**) poly-alumino-ferric sulphate (AMD-PAFS) coagulant and (**b**) poly ferric sulphate coagulant (initial pH 7.0, temperature = 24.6 °C) during the treatment of brewery wastewater.

The COD is the amount of oxygen required to break down an inorganic pollutant in water or wastewater. Contrary to turbidity removal, the COD removal at 150 mg/L PFS was 64%, which was higher than the COD removal of 56% obtained at the same concentration of AMD-PAFS. This result is consistent with the previous studies such as the study by Xing and Sun [47], who obtained 72.4% COD removal from antibiotic fermentation wastewater, which had an initial COD concentration of 3279 mg/L by using 200 mg/L PFS coagulant. However, one important observation from this study was the formation of the emulsion at 150 mg/L AMD-PAFS, which could be an indication of excess coagulant.

3.2.3. Effect of the Coagulants on Electric Conductivity

The EC is the measure of the dissolved ionic components in water and hence the electric characteristics. The EC gives an indication of the amount of total dissolved substitution in water [48]. As shown in Figure 4, the electric conductivity of the brewery wastewater increased as the dose of the coagulants increased. The conductivity of the original water sample was 3510 µS/cm, but it was increased to 4010 and 4110 µS/cm for AMD-PAFS and PFS coagulants, respectively. The sporadic rise in EC observed in all the samples tested could be due to the presence of the dissolved ions of the wastewater coupled with the dissolved ions of the coagulants and the pH regulator (NaOH). Similar observations have been made by other researchers [49,50].

3.2.4. Effect of the Coagulant Dose on Total Dissolved Solids

TDS is one of the key parameters that can be used for water quality analysis. It is related to the quantity of material in water that can pass a filter size of 2 µm. The TDS increases the conductivities of water due to the presence of dissolved impurities [51]. The TDS in water influence the quality of drinking water such as taste, alkalinity, hardness and corrosion properties. As shown in Figure 5, the TDS of the untreated brewery wastewater was 1810 mg/L. The TDS increases only slightly with an increase in coagulant dose for both the AMD-PAFS and PFS coagulants. In general, the increase in TDS is due to an increase in the number of solute particles or ions as a result of coagulant addition. The principal anions contributing to the TDS value include the carbonate, bicarbonate, chloride, sulphate and nitrates, and cations such as calcium, magnesium, potassium and sodium [52]. The sulphate components of the tested coagulants contributed to the increase in the TDS of the treated water when the coagulant dose was increased.

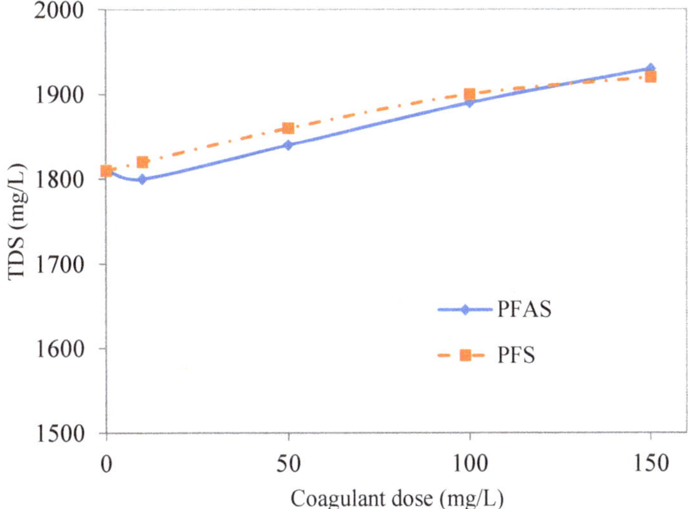

Figure 5. Total dissolved solids as a function of coagulant dose during treatment of brewery wastewater.

4. Conclusion

The recovery of Fe and Al from coal generated AMD at pH 5.0 was 99.9% Fe and 94.7% Al. With an increased pH of up to 7.0, the overall Al recovery increased to 99.1%. Although Al precipitation was 99.1% at pH 7.0, the precipitate formed at pH 5.0 was chosen for coagulant production due to the reduced chances of co-precipitation of other impurities should they exist in substantially higher concentrations. Dissolution of precipitate in 5.0% (w/w) sulphuric acid produced a coagulant containing 89.5% Fe and 10.0% Al. The coagulant produced had comparable characteristics to the PFS commercial coagulant. The subsequent brewery wastewater treatment tests showed that the AMD-derived coagulant was as effective as the conventional coagulants in the removal of COD and turbidity. This process can be easily integrated in existing AMD treatment plants, which would provide revenue and thereby subsidize the treatment costs. Furthermore, the issues associated with disposal of the voluminous sludge could be avoided, as the coagulant recovery would reduce the sludge volume by 95.0%.

Author Contributions: Conceptualization, B.M.; Methodology, B.M., S.N., G.S.S., B.X.; Validation, B.M.; Formal analysis, B.M., G.S.S., S.S., S.N., B.X.; Investigation, B.M., and B.X.; Resources, S.N., B.F.; Data curation, B.M.; Writing—Original draft preparation, B.M.; Writing—Review and editing, B.M., B.X., G.S.S., S.S., and S.N.; Visualization, B.M., B.X., and G.S.S.; Supervision, S.N., G.S.S., S.S., and B.F.; Project administration, S.N., and B.F.; Funding acquisition, S.N., and B.F.

Funding: This research was funded by the National Research Foundation (NRF) and the Department of Science and Technology (DST) of South Africa through the Germany-South Africa collaborative Project "AddWater" ref# 105,879 and the SARChI chair in Hydrometallurgy and Sustainable Development (SARCI150223114415 Grant# 98350) and the international office of the BMBF in Germany under the AddWater Project (No. 01DG17024).

Acknowledgments: The National Research Foundation (NRF) and the Department of Science and Technology (DST) of South Africa are gratefully acknowledged for their financial contribution to this work through the Germany/South Africa Collaborative Project "AddWater" ref# 105,879 and the SARChI chair in Hydrometallurgy and Sustainable Development (SARCI150223114415 Grant# 98350). The international office of the BMBF in Germany for the financial support under the AddWater Project (No. 01DG17024) is also acknowledged.

Conflicts of Interest: The authors declare no conflict of interest. The sponsors had no role in the design of the study; in the collection, analysis or interpretation of data; in the writing of the manuscript, and in the decision to publish the results.

References

1. Manders, P.; Godfrey, L.; Hobbs, P. *Briefing Note: Acid Mine Drainage in South Africa*; CSIR: Pretoria, South Africa, 2009.

2. Jacobs, J.A.; Lehr, J.H.; Testa, S.M. *Acid Mine Drainage, Rock Drainage, and Acid Sulfate Soils: Causes, Assessment, Prediction, Prevention, and Remediation*; John Wiley & Sons: New York, NY, USA, 2014.

3. Stumm, W.; Morgan, J.J. *Aquatic Chemistry: Chemical Equilibria and Rates in Natural Waters*; John Wiley & Sons: New York, NY, USA, 2012.

4. Fripp, J.; Ziemkiewicz, D.P.F.; Charkavorki, H. *Acid Mine Drainage Treatment*; ERDC TN-EMRRP-SR-14; Ecosystem Management and Restoration Research Program: Baltimore District, MD, USA, 2000; Volume 7.

5. Garland, R. Acid mine drainage-can it affect human health? *Quest* **2011**, *7*, 46–47.

6. Monachese, M.; Burton, J.P.; Reid, G. Bioremediation and Tolerance of Humans to Heavy Metals through Microbial Processes: A Potential Role for Probiotics? *Appl. Environ. Microbiol.* **2012**, *78*, 6397–6404. [CrossRef] [PubMed]

7. Van Dyk, L.D.; Simate, G.S.; Seepe, L.; Sibanda, V.; Shemi, A.; Ndlovu, S. The removal of Co^{2+}, V^{3+} and Cr^{3+} from waste effluents using cassava waste. *S. Afr. J. Chem. Eng.* **2013**, *18*, 51–69.

8. Singh, R.; Gautam, N.; Mishra, A.; Gupta, R. Heavy metals and living systems: An overview. *Indian J. Pharmacol.* **2011**, *43*, 246–253. [CrossRef] [PubMed]

9. Coulton, R.; Bullen, C.; Hallett, C. The design and optimisation of active mine water treatment plants. *Land Contam. Reclam.* **2003**, *11*, 273–279. [CrossRef]

10. Johnson, D.B.; Hallberg, K.B. Acid mine drainage remediation options: A review. *Sci. Total Environ.* **2005**, *338*, 3–14. [CrossRef]

11. Menezes, J.; Silva, R.; Arce, I.; Schneider, I.; Silva, R.D.A. Production of a poly-alumino-iron sulphate coagulant by chemical precipitation of a coal mining acid drainage. *Miner. Eng.* **2010**, *23*, 249–251. [CrossRef]

12. Simate, G.S.; Ndlovu, S. Acid mine drainage: Challenges and opportunities. *J. Environ. Chem. Eng.* **2014**, *2*, 1785–1803. [CrossRef]

13. Ryan, M.; Kney, A.; Carley, T. A study of selective precipitation techniques used to recover refined iron oxide pigments for the production of paint from a synthetic acid mine drainage solution. *Appl. Geochem.* **2017**, *79*, 27–35. [CrossRef]

14. Hedin, R.S. Recovery of marketable iron oxide from mine drainage in the USA. *Land Contam. Reclam.* **2003**, *11*, 93–97. [CrossRef]

15. Macingova, E.; Luptakova, A. Recovery of iron from acid mine drainage in the form of oxides. *Inżynieria Miner.* **2014**, *15*, 193–198.

16. Wei, X.; Viadero, R.C. Synthesis of magnetite nanoparticles with ferric iron recovered from acid mine drainage: Implications for environmental engineering. *Colloids Surf. A Physicochem. Eng. Asp.* **2007**, *294*, 280–286. [CrossRef]

17. Marcello, R.; Galato, S.; Peterson, M.; Riella, H.; Bernardin, A.; Bernardin, A. Inorganic pigments made from the recycling of coal mine drainage treatment sludge. *J. Environ. Manag.* **2008**, *88*, 1280–1284. [CrossRef] [PubMed]

18. Michalková, E.; Schwarz, M.; Pulišová, P.; Máša, B.; Sudovský, P. Metals Recovery from Acid Mine Drainage and Possibilities for their Utilization. *Pol. J. Environ. Stud.* **2013**, *22*, 1111–1118.

19. Park, S.M.; Yoo, J.C.; Ji, S.W.; Yang, J.S.; Baek, K. Selective recovery of dissolved Fe, Al, Cu, and Zn in acid mine drainage based on modeling to predict precipitation pH. *Environ. Sci. Pollut. Res.* **2015**, *22*, 3013–3022. [CrossRef] [PubMed]

20. Sahinkaya, E.; Gungor, M.; Bayrakdar, A.; Yucesoy, Z.; Uyanik, S. Separate recovery of copper and zinc from acid mine drainage using biogenic sulfide. *J. Hazard. Mater.* **2009**, *171*, 901–906. [CrossRef] [PubMed]

21. Wei, X.; Viadero, R.C.; Buzby, K.M. Recovery of Iron and Aluminum from Acid Mine Drainage by Selective Precipitation. *Environ. Eng. Sci.* **2005**, *22*, 745–755. [CrossRef]

22. Nleya, Y. Removal of Toxic Metals and Recovery of Acid from Acid Mine Drainage Using Acid Retardation and Adsorption Processes. Ph.D.; Thesis, University of Witwatersrand, Johannesburg, South Africa, 2016. Available online: http://wiredspace.wits.ac.za/handle/10539/21051 (accessed on 2 October 2019).

23. Buzzi, D.C.; Viegas, L.S.; Rodrigues, M.A.S.; Bernardes, A.M.; Tenório, J.A.S. Water recovery from acid mine drainage by electrodialysis. *Miner. Eng.* **2013**, *40*, 82–89. [CrossRef]

24. Martí-Calatayud, M.; Buzzi, D.; Gabaldón, M.G.; Ortega, E.; Bernardes, A.; Tenório, J.; Pérez-Herranz, V.; Bernardes, A. Sulfuric acid recovery from acid mine drainage by means of electrodialysis. *Desalination* **2014**, *343*, 120–127. [CrossRef]

25. Masindi, V. Recovery of drinking water and valuable minerals from acid mine drainage using an integration of magnesite, lime, soda ash, CO_2 and reverse osmosis treatment processes. *J. Environ. Chem. Eng.* **2017**, *5*, 3136–3142. [CrossRef]

26. Silsbee, M. *The Use of Sludge Generated by the Neutralization of Acid Mine Drainage in the Cement Industry*; EPD06021; United States Environmental Protection Agency: Washington, DC, USA, 2006.

27. Evenson, C.; Nairn, R. Enhancing phosphorus sorption capacity with treatment wetland iron oxyhydroxides. In Proceedings of the 17th National Meeting of the American Society for Surface Mining and Reclamation, Tampa, FL, USA, 11–15 June 2000.

28. Dudeney, A.W.; Neville, K.J.; Tarasova, I.; Heath, A.O.; Smith, S.R. Utilisation of Ochreous Sludge as a Soil Amendment. In Proceedings of the Symposium: Mine Water, Newcastle, UK, 1 July 2004; pp. 19–23.

29. Saunders, F.M.; Roeder, L.M. *Coagulant Recovery: A Critical Assessment*; The Water Research Foundation: Denver, CO, USA, 1991.

30. Baes, C.; Mesmer, R. *The Hydrolysis of Cations*; Wiley: New York, NY, USA, 1976.

31. Flynn, C.M. Hydrolysis of inorganic iron (III) salts. *Chem. Rev.* **1984**, *84*, 31–41. [CrossRef]

32. Salama, E.-S.; Kim, J.R.; Ji, M.-K.; Cho, D.-W.; Abou-Shanab, R.A.; Kabra, A.N.; Jeon, B.-H. Application of acid mine drainage for coagulation/flocculation of microalgal biomass. *Bioresour. Technol.* **2015**, *186*, 232–237. [CrossRef] [PubMed]

33. Lopes, F.A.; Menezes, J.C.; Schneider, I.A. *Acid Mine Drainage as Source of Iron for the Treatment of Sewage by Coagulation and Fenton's Reaction*; IMWA: Aachen, Germany, 2011.

34. Rao, S.R.; Gehr, R.; Riendeau, M.; Lu, D.; Finch, J.A. Acid mine drainage as a coagulant. *Min. Eng.* **1992**, *5*, 1011–1020. [CrossRef]

35. Rao, S.R.; Leroux, M.; Finch, J. Resource Recovery from Acid Mine Drainage. In *Metals Removal from Acidic Drainage-Chemical Methods*; MEND Project 3-21.2a; Noranda Technology Center: Pointe Claire, QC, Canada, 1996.

36. Jiang, J.-Q.; Graham, N.J.D. Development of Optimal Poly-Alumino—Iron Sulphate Coagulant. *J. Environ. Eng.* **2003**, *129*, 699–708. [CrossRef]

37. Snoeyink, V.L.; Jenkins, D. *Water Chemistry*; John Wiley: New York, NY, USA, 1980.

38. Mendham, J. *Vogels Textbook of Quantitative Chemical Analysis*; Pearson Education: New Delhi, India, 2006.

39. Tabak, H.H.; Scharp, R.; Burckle, J.; Kawahara, F.K.; Govind, R. Advances in biotreatment of acid mine drainage and biorecovery of metals: 1. Metal precipitation for recovery and recycle. *Biodegradation* **2003**, *14*, 423–436. [CrossRef] [PubMed]

40. Clescerl, L.; Greenberg, A.; Eaton, A. *Standard Methods for Examining Water and Wastewater*; APHA: Washington, DC, USA; AWWA: Washington, DC, USA; WEF: Washington, DC, USA, 1999.

41. Gitari, W.M.; Petrik, L.F.; Akinyemi, S.A. Treatment of Acid Mine Drainage with Coal Fly Ash: Exploring the Solution Chemistry and Product Water Quality. In *Coal Fly Ash Beneficiation—Treatment of Acid Mine Drainage with Coal Fly Ash*; Akinyemi, S.A., Gitari, M.W., Eds.; InTech: Limpopo, South Africa, 2018. [CrossRef]

42. Petrik, L.F.; Fatoba, O.O.; Missengue, R. *Treatment of Mine Water Using a Combination of Coal Fly Ash and Flocculants in a Jet Loop Reactor System*; WRC Report No. 2129/1/18; Water Research Commission: Pretoria, South Africa, 2017.

43. National Gazette. National Gazette No. 36820, 6 September 2013, Vol. 579. Available online: https://www.greengazette.co.za/documents/national-gazette-36820-of-06-september-2013-vol-579_20130906-GGN-36820.pdf (accessed on 20 September 2019).

44. Cheng, W.P.; Chi, F.H.; Yu, R.F.; Lee, Y.C. Using Chitosan as a Coagulant in Recovery of Organic Matters from the Mash and Lauter Wastewater of Brewery. *J. Polym. Environ.* **2005**, *13*, 383–388. [CrossRef]

45. Tolkou, A.; Zouboulis, A. Synthesis and coagulation performance of composite poly-aluminum-ferric-silicate-chloride coagulants in water and wastewater. *Desalin. Water Treat.* **2015**, *53*, 3309–3318. [CrossRef]

46. Wang, D.; Tang, H. Modified inorganic polymer flocculant-PFSi: Its preparation, characterization and coagulation behavior. *Water Res.* **2001**, *35*, 3418–3428. [CrossRef]

47. Xing, Z.-P.; Sun, D.-Z. Treatment of antibiotic fermentation wastewater by combined polyferric sulfate coagulation, Fenton and sedimentation process. *J. Hazard. Mater.* **2009**, *168*, 1264–1268. [CrossRef]

48. Oyem, H.; Oyem, I.; Ezeweali, D. Temperature, pH, Electrical Conductivity, Total Dissolved Solids and Chemical Oxygen Demand of Groundwater in Boji-BojiAgbor/Owa Area and Immediate Suburbs. *Res. J. Environ. Sci.* **2014**, *8*, 444–450. [CrossRef]

49. Aniyikaiye, T.E.; Oluseyi, T.; Odiyo, J.O.; Edokpayi, J.N. Physico-Chemical Analysis of Wastewater Discharge from Selected Paint Industries in Lagos, Nigeria. *Int. J. Environ. Res. Public Health* **2019**, *16*, 1235. [CrossRef]

50. Miranda, R.; Latour, I.; Hörsken, A.; Jarabo, R.; Blanco, A. Enhanced Silica Removal by Polyamine- and Polyacrylamide-Polyaluminum Hybrid Coagulants. *Chem. Eng. Technol.* **2015**, *38*, 2045–2053. [CrossRef]

51. Kazi, T.; Virupakshi, A. Treatment of tannery wastewater using natural coagulants. *Int. J. Environ. Res. Public Health* **2013**, *2*, 4061–4068.

52. Beyene, H.D.; Hailegebrial, T.D.; Dirersa, W.B. Investigation of Coagulation Activity of Cactus Powder in Water Treatment. *J. Appl. Chem.* **2016**, *2016*, 1–9.

 *metals*

Article

Chemical Stability of Zirconolite for Proliferation Resistance under Conditions Typically Required for the Leaching of Highly Refractory Uranium Minerals

Aleksandar N. Nikoloski [1,*], Rorie Gilligan [1], Jonathan Squire [2] and Ewan R. Maddrell [3]

[1] College of Science, Health, Engineering and Education, Murdoch University, Perth 6150, Australia; Rorie.Gilligan@murdoch.edu.au

[2] Sellafield Ltd., Seascale, Cumbria CA20 1PG, UK; jon.squire@sellafieldsites.com

[3] National Nuclear Laboratory, Workington, Cumbria CA14 3YQ, UK; ewan.r.maddrell@nnl.co.uk

* Correspondence: A.Nikoloski@murdoch.edu.au; Tel.: +61-8-9360-2835; Fax: +61-8-9360-6343

Received: 11 July 2019; Accepted: 12 September 2019; Published: 1 October 2019

Abstract: In this study, synthetic zirconolite samples with a target composition $Ca_{0.75}Ce_{0.25}ZrTi_2O_7$, prepared using two different methods, were used to study the stability of zirconolite for nuclear waste immobilisation. Particular focus was on plutonium, with cerium used as a substitute. The testing of destabilisation was conducted under conditions previously applied to other highly refractory uranium minerals that have been considered for safe storage of nuclear waste, brannerite and betafite. Acid (HCl, H_2SO_4) leaching for up to 5 h and alkaline ($NaHCO_3$, Na_2CO_3) leaching for up to 24 h was done to enable comparison with brannerite leached under the same conditions. Ferric ion was added as an oxidant. Under these conditions, the synthetic zirconolite dissolved much slower than brannerite and betafite. While the most intense conditions were observed previously to result in near complete dissolution of brannerite in under 5 h, zirconolite was not observed to undergo significant attack over this timescale. Fine zirconolite dissolved faster than the coarse material, indicating that dissolution rate is related to surface area. This data and the long term stability of zirconolite indicate that it is a good material for long-term sequestration of radioisotopes. Besides its long term durability in the disposal environment, a wasteform for fissile material immobilisation must demonstrate proliferation resistance such that the fissile elements cannot be retrieved by leaching of the wasteform. This study, in conjunction with the previous studies on brannerite and betafite leaching, strongly indicates that the addition of depleted uranium to the wasteform, to avert long term criticality events, is detrimental to proliferation resistance. Given the demonstrated durability of zirconolite, long term criticality risks in the disposal environment seem a remote possibility, which supports its selection, above brannerite or betafite, as the optimal wasteform for the disposition of nuclear waste, including of surplus plutonium.

Keywords: uranium; zirconolite; brannerite; betafite; leaching; kinetics

---

1. Introduction

Zirconolite, $CaZrTi_2O_7$ is one of several titanate phases present in synthetic titanate ceramics developed for the immobilisation of actinides and fission products in spent nuclear fuel. Other phases include pyrochlore, brannerite and zircon [1].

These minerals frequently contain uranium and/or thorium. In zirconolite, uranium undergoes extensive substitution onto the calcium site [2]. In brannerite, uranium is an essential element while in pyrochlore and zirconolite, actinides and light rare earth elements (REEs) can also take the place of calcium. Synthetic forms of these minerals can substitute other actinides as well, such as plutonium, americium and curium, in the calcium site, being likely too large for the other sites. The high chemical durability of these materials suggests that they could be ideal for the sequestration of surplus plutonium and other actinides present in spent nuclear fuel. There are reports of zirconolites that have been observed to show evidences of post-crystallization corrosion [3].

Zirconolite has been identified in weathered gravel in Sri Lanka [4], in Western Australian dolerite intrusions [5], in lunar granite [6] and other varied geological settings and/or minero-genetic conditions [7]. Zirconolites up to 650 million years old have been identified in which $^{206}Pb/^{238}U$, $^{207}Pb/^{235}U$ and $^{206}Pb/^{207}Pb$ isotope ratios give consistent ages, indicating that no uranium has been lost from the zirconolite despite the host rock having undergone extensive weathering over the 650 million years since formation [2]. It is also worth noting the existence of 2 billion year old zirconolite from Phalaborwa in South Africa as referred in some of the references of available papers [2]. This makes zirconolite potentially ideal for sequestering the radioactive elements present in spent nuclear fuel over the millennia required for them to decay into less harmful substances.

Cerium is often used as a substitute for plutonium in studies of nuclear waste ceramics. Cerium and plutonium have very close ionic radii (Ce^{4+} = 97 pm and Pu^{4+} = 96 pm) when coordinated by eight other atoms [8], as in the Ca site in zirconolite [1]. However, cerium is far safer to work with than plutonium.

Lumpkin [1] compared several mineral phases for waste immobilisation. The advantages of zirconolite over others include its high aqueous durability and chemical flexibility, though it is less tolerant to radiation dose than some other phases. The relative aqueous stability of several phases from pH 2–12 is as follows: zirconolite > pyrochlore > brannerite >> perovskite [1].

Brannerite is known to dissolve quickly in sulphuric and hydrochloric acids under oxidising conditions [9,10], while betafite (pyrochlore) will dissolve at a lower rate under similar conditions [11]. By comparing the leaching of zirconolite with the leaching of brannerite and betafite under these conditions, the stability of zirconolite as a host for actinides can be evaluated and demonstrated.

2. Materials and Methods

2.1. Sample Preparation

Synthetic zirconolite samples were prepared using two different methods. The zirconolite target composition was $Ca_{0.75}Ce_{0.25}ZrTi_2O_7$, with Ce as a substitute for Pu. Assuming that the feed mixtures are homogeneous, the composition of the feed mixtures for both methods should be CaO 11.5 wt. %, Ce_2O_3 11.2 wt. %, ZrO_2 33.7 wt. %, and TiO_2 43.6 wt. %. The oxidation state of Ti was expected to be mixed 4/3+ to maintain charge balance. The preparation methods are outlined below.

i. Alkoxide route

Required quantities of zirconium n-propoxide and titanium isopropoxide were hydrolysed with a solution containing the necessary amounts of calcium and cerium nitrate. The slurry was then stir-dried in a stainless steel beaker on a hot plate. Once dried the product was calcined at 750 °C in air for 8 h.

ii. Oxide route

ZrO_2 and TiO_2 as approximately 1 μm particle size powders, CeO_2 as a 5 μm particle size powder and calcium nitrate were combined to form a slurry. This was stir dried and calcined as for the alkoxide route.

The powders produced by both routes were then planetary milled as a slurry for 20 min with propan-2-ol as a carrier fluid, dried and sieved. The powders were then blended with 2.2 wt% Ti metal in a Turbula mixer and packed into stainless steel hot isostatic pressing (HIP) cans 3.5 cm diameter by 5 cm high. The Ti metal acted as an in-can reducing agent to convert Ce^{4+} to Ce^{3+} and Ti^{4+} to Ti^{3+} to ensure correct charge balance in the zirconolite. The HIP cans were then sealed and evacuated, and then hot isostatically pressed at 1320 °C and 100 MPa for 2 h.

2.2. Sample Characterisation

Mineralogical analysis by X-ray diffraction (XRD) was performed with a GBC Enhanced Multi-material Analyser (EMMA) (GBC Scientific Equipment, Braeside, Victoria, Australia) at Murdoch University. Samples were placed directly onto X-ray absorbing silicon discs within circular metal sample holders. Samples were introduced under a drop of ethanol and the ethanol was allowed to evaporate prior to the analysis.

The X-ray tube was operated at a voltage of 35.0 kV and current of 28.0 mA. Diffraction patterns were collected over a range of $20° \leq 2\theta \leq 70°$ using a 1° diverging slit, a 0.2° receiving slit and a 1° scattering slit. A step size of 0.02° was used, with a speed of 1°/min (1.2 s per step) with five passes. Cu Kα X-rays were used. A Kα_2 strip was performed on the diffraction patterns, with a Kα_2/Kα_1 ratio of 0.51. Initial scans showed no peaks of interest below 20°.

Scanning electron microscopy (SEM) observations were performed with a JEOL JCM-6000 bench top SEM with an energy dispersive X-ray spectroscopy (EDX) analyser (JEOL Ltd., Tokyo, Japan). An accelerating voltage of 15 kV was used to produce the SEM images of the samples. Both secondary electron (SE) and backscattered electron (BSE) modes were utilised. Particles were mounted on carbon discs. The cross-sections of the particles were prepared by embedding in epoxy resin and subsequent polishing with silicon carbide. A 15 kV accelerating voltage was used for the semi-quantitative EDX analyses, the highest possible with the instrument used in this study. All EDX analyses were run for 60 s. All images associated with EDX analyses were taken in BSE. For line-scan analyses, the counting time was set to 15 s per step. X-ray elemental maps were produced with a resolution of 384×512 pixels and a counting time of 10×0.2 ms per pixel. The standard colour scheme for the element maps adhered to throughout this report is red for calcium, green for zirconium and blue for titanium. Cerium was not included on the element maps due to the overlap of the Ce Lα peak at 4.83 keV with the Ti Kβ peak at 4.93 keV.

All aqueous samples were analysed for calcium, titanium, zirconium and cerium with a Thermo-Fisher iCAP-Q ICP-MS instrument (Thermo-Fisher Scientific, Bremen, Germany) at Murdoch University. The purity of zirconolite samples was verified by digestions and ICP-MS analysis performed at a commercial minerals laboratory. The coarse zirconolite was assayed twice.

2.3. Leaching Study

Similar conditions for the leaching study were used to those previously reported for brannerite leaching [9,10,12]. Acid (HCl, H$_2$SO$_4$) leaching experiments were run for five hours and alkaline leaching tests were run for 24 h to enable comparison with brannerite leached under the same conditions. Coarse zirconolite was used in the majority of the experiments. The highest temperature experiment for each lixiviant was repeated with fine zirconolite. The conditions used in the leaching experiments are listed in Table 1.

Table 1. Leaching conditions used in this study.

Lixiviant	Temperature (°C)	Lixiviant Concentration (mol/L)	Size Range (µm)	Duration (h)
HCl	50	0.25	125–250	5
HCl	85	0.25	125–250	5
HCl	50	1.00	125–250	5
HCl	85	0.25	63–125	5
H$_2$SO$_4$	30	0.25	125–250	5
H$_2$SO$_4$	50	0.25	125–250	5
H$_2$SO$_4$	70	0.25	125–250	5
H$_2$SO$_4$	85	0.25	125–250	5
H$_2$SO$_4$	50	1.00	125–250	5
H$_2$SO$_4$	85	0.25	63–125	5
NaHCO$_3$, Na$_2$CO$_3$	70	0.67, 0.33	125–250	24
NaHCO$_3$, Na$_2$CO$_3$	70	0.67, 0.33	63–125	24

As with the brannerite leaching experiments, Fe^{3+} was added as an oxidant. Iron was added as 0.05 mol/L $FeCl_3$ in the chloride leaching experiments, 0.05 mol/L $Fe(SO_4)_{1.5}$ in the sulphate leaching experiments and 0.025 mol/L $K_3Fe(CN)_6$ in the carbonate leaching experiments.

3. Results and Discussion

3.1. Feed Characterisation

The two feed samples produced using the different methods had different size distributions—63–125 µm for the alkoxide route sample and 125–250 µm for the oxide route sample. Wet screening was used to narrow down the size range of each sample. These are labelled 'fine' sample and 'coarse' sample, respectively.

3.1.1. Feed Assays

Chemical analyses of the synthetic zirconolite by ICP-MS presented in Table 2; Table 3 show that the synthetic zirconolite from both methods was of high purity. Hafnium was the main non-formula element identified. Hafnium is often found with zirconium, and separating the two presents a significant technical challenge.

Table 2. Major elements (>0.1 wt%) in the zirconolite feed samples.

Element	Coarse Zirconolite	Fine Zirconolite
Zr	25.88%	25.54%
Ti	25.75%	25.68%
Ce	8.20%	8.12%
Ca	7.83%	7.88%
Hf	0.63%	0.63%
Si	0.19%	0.25%

Table 3. Minor elements (>100 ppm) in the zirconolite feed samples in ppm.

Element	Coarse Zirconolite	Fine Zirconolite
Na	800	900
Ag	707	849
Nb	650	620
Fe	475	645
Mg	355	420
S	350	400
Al	280	430
P	245	240
Ga	240	275
K	200	300

These zirconolite specimens had average formulas of $Ca^{2+}_{0.71}Ce^{3+}_{0.21}(Zr_{1.03}Hf_{0.01})Ti_{1.95}O_7$ for the coarse zirconolite and $Ca^{2+}_{0.72}Ce^{3+}_{0.21}(Zr_{1.02}Hf_{0.01})Ti_{1.95}O_7$ for the fine zirconolite. Si has been excluded based on EDX results, showing that it was present in a separate minor SiO_2 phase. Both were slightly Ti deficient compared to the ideal zirconolite composition, but had a higher amount of Ti than typical natural samples. Titanium is commonly replaced by Fe^{3+}/Nb^{5+} in natural zirconolite. Tantalum may also be present in this site in small amounts [13] Cerium, REEs and actinides replace calcium in the zirconolite crystal structure [2,4,14–16].

3.1.2. XRD

XRD also showed feed samples produced by both methods to be effectively pure zirconolite, a solid solution $(Ca_{0.75}Ce_{0.25})ZrTi_2O_7$. The XRD data showed the presence of zirconolite and perovskite as major phases. Zirconolite exists in three polytypes [17]; and reference diffraction patterns for them have been superimposed on the measured diffraction pattern (Figure 1). Titanium dioxide (rutile) was detected in small amounts. The other polymorphs of titanium dioxide, anatase and brookite, were not detected. Perovskite was detected as a minor phase.

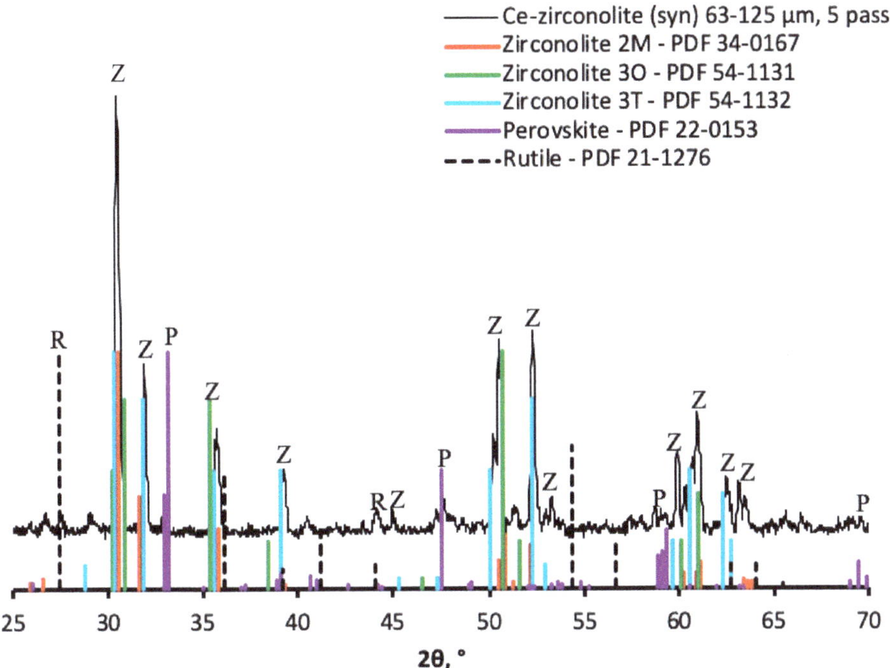

Figure 1. X-ray diffraction (XRD) pattern of the zirconolite material with relevant PDF references.

3.1.3. SEM, EDX

The examination by SEM showed the feed samples contained very small inclusions of a second phase, and possibly some unreacted ZrO_2. Backscattered electron images (Figure 2) showed that the inclusions have a lower average atomic mass. EDX analyses (Figures 3 and 4) indicated that this material was titanium dioxide, though it is not possible to tell from the EDX analyses which polymorph of titanium dioxide was present. The XRD results indicate that it was most likely rutile, possibly a relic of the Ti metal added for redox control during the HIP process. Neither silicon nor hafnium were detected in EDX analyses of zirconolite. When silicon was detected, it occurred as a separate phase (SiO_2), while hafnium at 0.6% of the mass was below the detection limit for EDX analyses.

EDX analyses of the zirconium-free regions showed that they contained cerium along with calcium and titanium (Figures 3 and 4). This material was probably the same perovskite phase identified by XRD. The brightness of this phase in the BSE images suggests that was not pure Ca perovskite however. Pure Ca perovskite has a low average atomic number (Z_{avg}) (16.5) close to that of rutile (16.4) [18] while the calculated Z_{avg} of zirconolite exceeds 22. While there were some subtle variations in the BSE brightness of the zirconolite/perovskite regions, these variations did not clearly correlate with variations in composition as determined by EDX spectra or elemental maps.

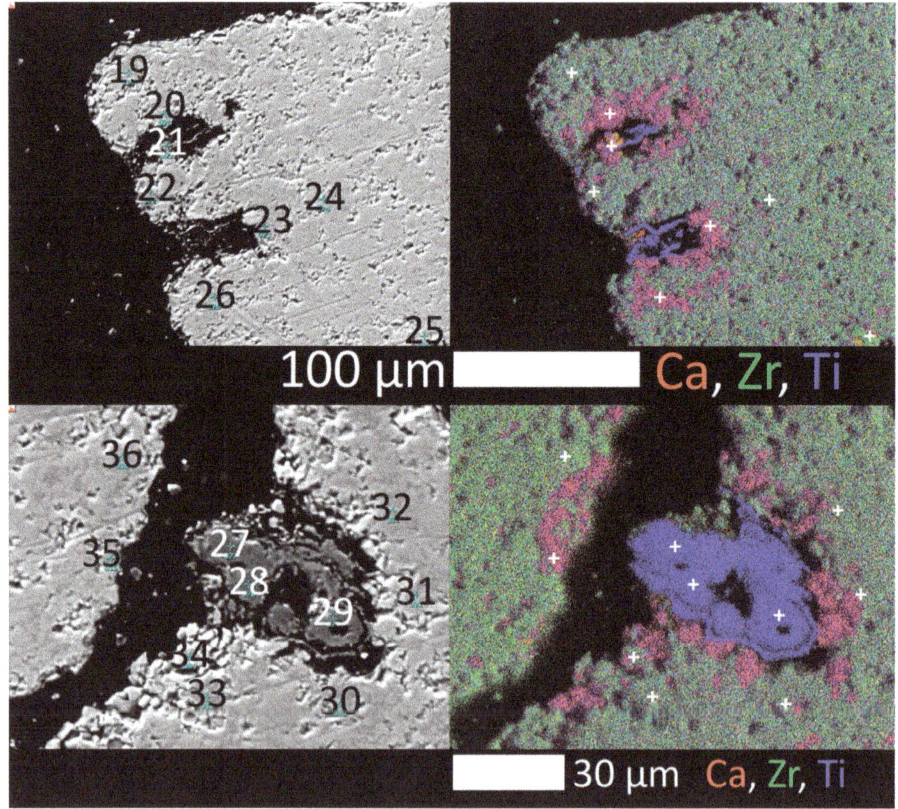

Figure 2. Rutile inclusions within zirconolite grains surrounded by perovskite. **Left**: backscattered electron (BSE) images, **right**: element map.

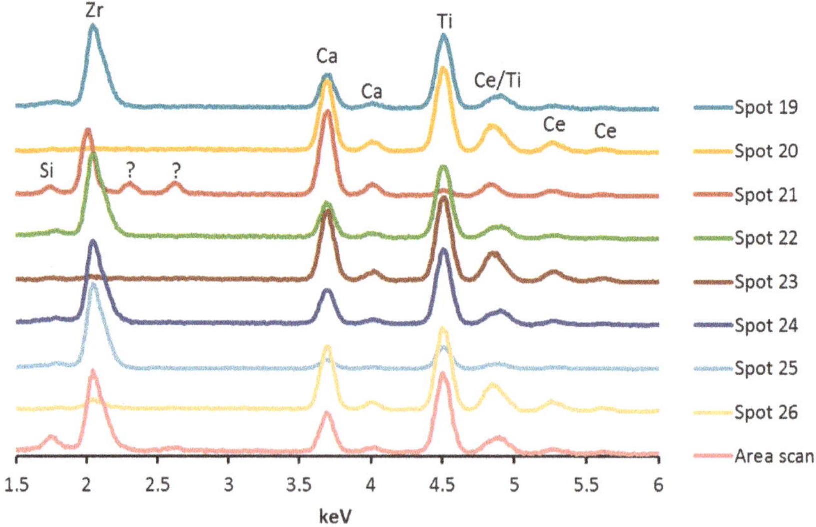

Figure 3. Spectra of spots analysed in the top half of Figure 2.

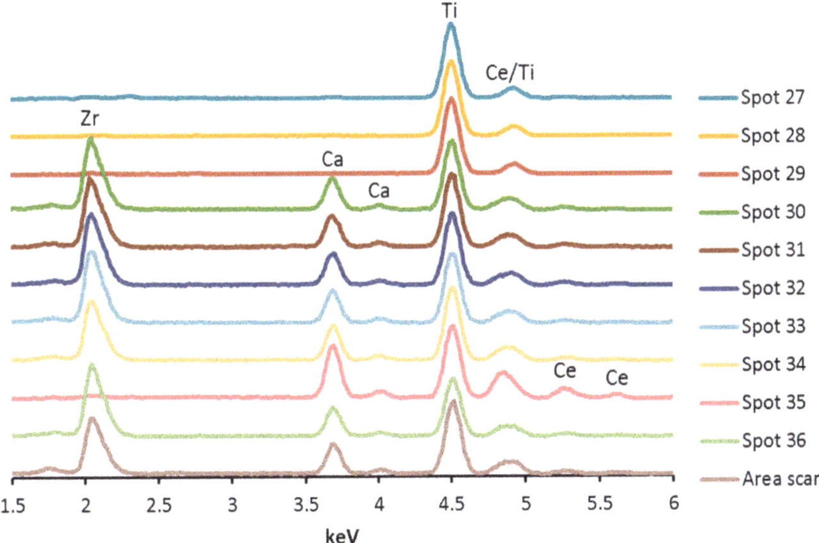

Figure 4. Energy dispersive X-ray spectroscopy (EDX) spectra of spots analysed in the bottom half of Figure 2.

It is difficult to resolve the cerium Lα peak from the titanium Kβ peak. For this reason, cerium was not included on any of the element maps. Closer examination shows the Ce Lβ peaks in some spectra allowing the presence of cerium to be confirmed. Of all the spots analysed on the feed sample, cerium was most prominent in the spectrum of spot 35 in Figure 4.

Eight oxygen atoms coordinate the calcium site in zirconolite, but the calcium site in perovskite is larger and coordinated by 12 oxygen atoms [19]. Hence, calcium within perovskite undergoes extensive isomorphous substitution with uranium, thorium and REEs [20]. Perovskite is commonly formed as a side-product in the synthesis of zirconolite and other titanate ceramics [14,21] and is an intentional phase in Synroc C, to host Sr-90. Studies on polyphase heterogeneous actinide titanate ceramics show that large lanthanide ions like Ce^{3+}/Nd^{3+} and trivalent actinides (Pu^{3+}, Am^{3+}, Cm^{3+}) favour the Ca site of perovskite over zirconolite. This explains the presence of cerium in the perovskite phase in this sample. The partition coefficient between zirconolite and perovskite was lower for larger cations [21].

Zirconolite is significantly more stable than perovskite [1,8]; thus, the formation of perovskite in synthetic samples intended for uranium sequestration should be minimised as much as possible. Pöml et al. [14] succeeded in synthesising a cerium doped zirconolite without detectable levels of perovskite by adding a stoichiometric excess of ZrO_2 during synthesis.

Along with the three major separate phases, rutile, Ce-perovskite and Ce-zirconolite, one of the spectra indicates the presence of a fourth minor Zr oxide phase (spot 25 in Figure 3). The boundaries between phases in the coarse zirconolite sample are clear and distinct unlike those observed in brannerite [22]. The boundaries between phases in this sample are clear and distinct unlike those observed in natural brannerite [22], as is apparent from EDX line analyses across a rutile inclusion. Rutile inclusions were typically surrounded by smaller perovskite inclusions though not all perovskite inclusions were associated with rutile.

3.2. Leaching Kinetics

Under similar leaching conditions, the synthetic zirconolite dissolved much more slowly than natural brannerite [9,12,23] and betafite [11,24,25]. Cerium extraction from zirconolite followed linear kinetics in sulphuric acid (Figure 5). After five hours of leaching, cerium extraction had yet to plateau.

Titanium dissolved at a slower rate than cerium but faster than zirconium. This suggests that zirconolite is not dissolving in significant amounts. Based on the observed leaching kinetics, perovskite is more susceptible to leaching than zirconolite. Calcium extraction kinetics were not included due to the significant levels of analytical error in measuring the calcium concentrations in solution. Apart from Ca, these results are consistent with earlier work that showed the typical order of elemental dissolution rates from zirconolite in acidic solutions is Ca > Ce > Ti > Zr [14].

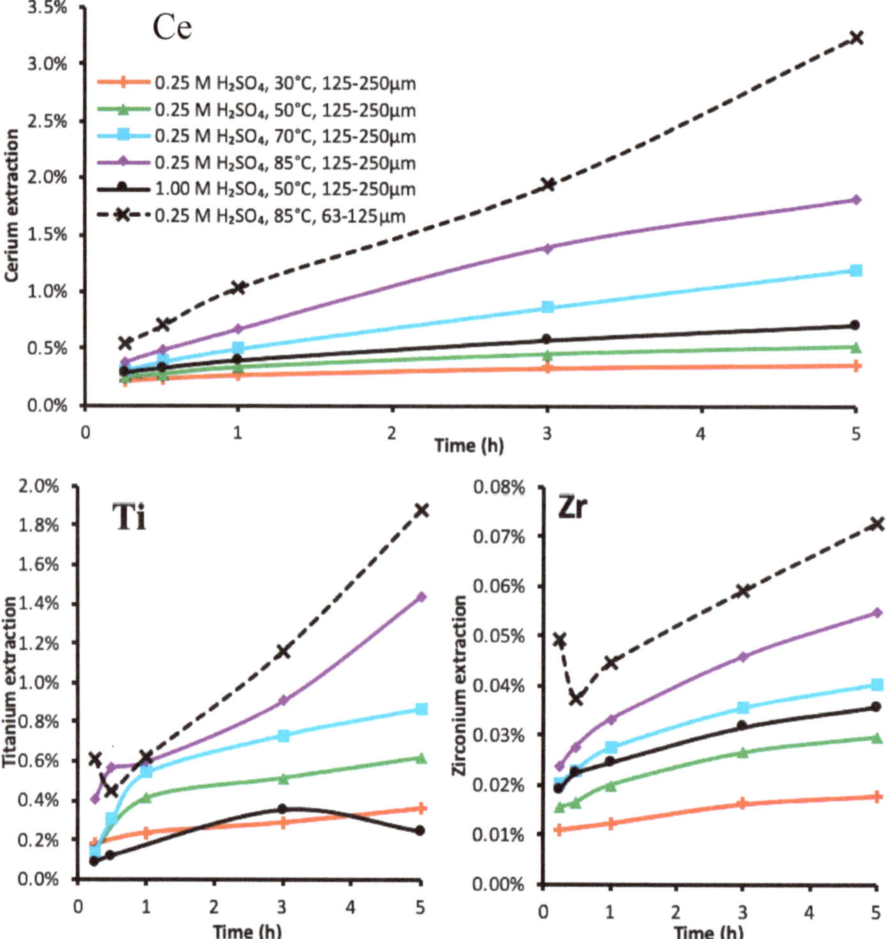

Figure 5. Leaching kinetics under various conditions in sulphuric acid media.

Similar trends were apparent during leaching in chloride media (Figure 6) to those observed for sulphate media. Cerium dissolved faster than titanium, which in turn dissolved faster than zirconium. While extraction rates were lower in chloride media compared to sulphate media at the same temperature and acid concentration, variations in acid concentration had a larger effect on the rate of dissolution in chloride media. Both of these behaviours have been observed when leaching brannerite in chloride and sulphate media over a wide range of temperatures and acid concentrations [10]. The order of uranium and titanium extraction from brannerite was approximately 0.5 with respect to H_2SO_4 while the order was approximately 1 with respect to HCl [9,10].

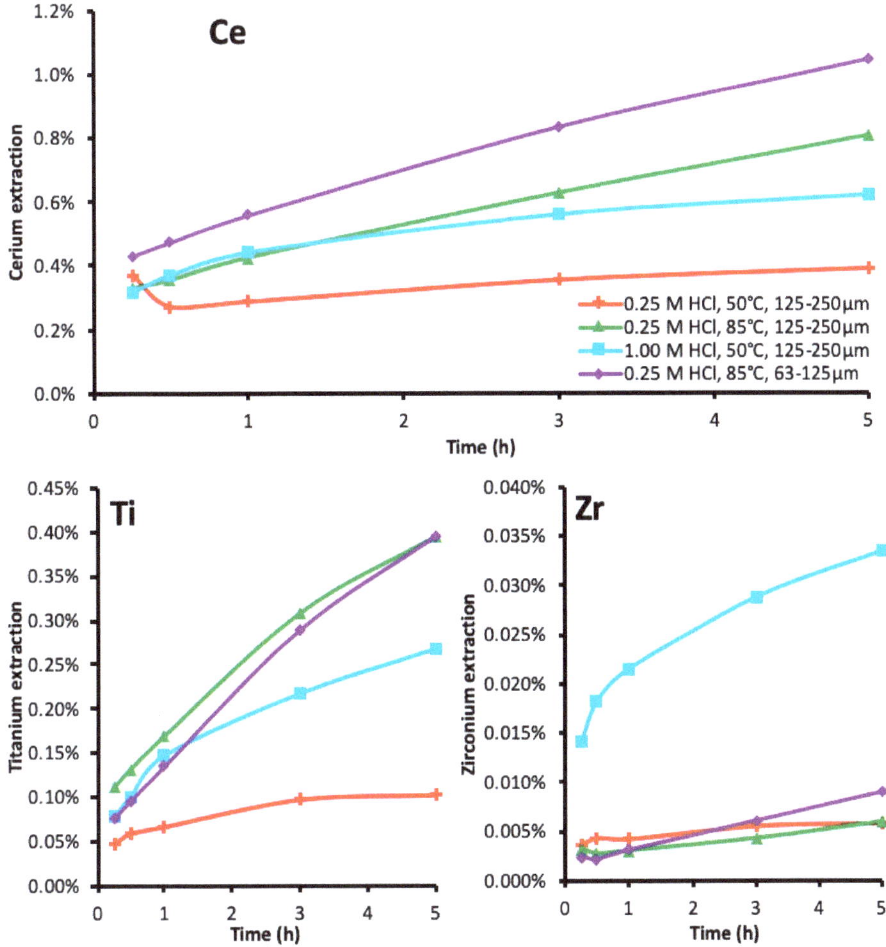

Figure 6. Leaching kinetics under various conditions in hydrochloric acid media.

After five hours of leaching, the extent of cerium dissolution was 3–6 times higher than that of titanium in hydrochloric acid and around 100 times higher than that of zirconium in 0.25 M HCl. In 1.00 M HCl, the Ce/Zr ratio decreased to approximately 20, with acid concentration having a significant effect on the dissolution rate of zirconium in chloride media. These trends match those observed in long term leaching studies on synthetic actinide waste forms such as zirconolite and pyrochlore at pH 2 in 0.01 M HNO$_3$ solution [26], and are in agreement with the relative solubility of the simple oxides of these elements (Figure 7; Reactions 1–5).

$$\text{Ca} \quad \mathbf{CaO + 2\,H^+ \rightarrow Ca^{2+} + H_2O} \qquad \text{(Reaction 1)}$$

$$\text{Ce} \quad (\text{III})\ 0.5\,Ce_2O_3 + 3\,H^+ \rightarrow Ce^{3+} + 1.5\,H_2O \qquad \text{(Reaction 2)}$$

$$\text{Ce} \quad (\text{IV})\ CeO_2 + 4\,H^+ \rightarrow Ce^{4+} + 2\,H_2O \qquad \text{(Reaction 3)}$$

$$\text{Ti} \quad TiO_{2(\text{rutile})} + 2\,H^+ \rightarrow TiO^{2+} + H_2O \qquad \text{(Reaction 4)}$$

$$\text{Zr} \quad ZrO_2 + 2\,H^+ \rightarrow ZrO^{2+} + H_2O \qquad \text{(Reaction 5)}$$

The relative rates of leaching were different in an alkaline environment (Figure 8). As with the acid leaching experiments, rates of dissolution in alkaline media were significantly slower than those observed for brannerite. Titanium dissolved faster than cerium, which dissolved much faster than zirconium. If the zirconolite was allowed to react for longer, it is expected that titanium would re-precipitate as titanium dioxide as observed with brannerite and Ti rich uranium ore [12,28]. Titanium is somewhat amphoteric and may dissolve as $Ti(OH)_5^-$ at high pH [29,30].

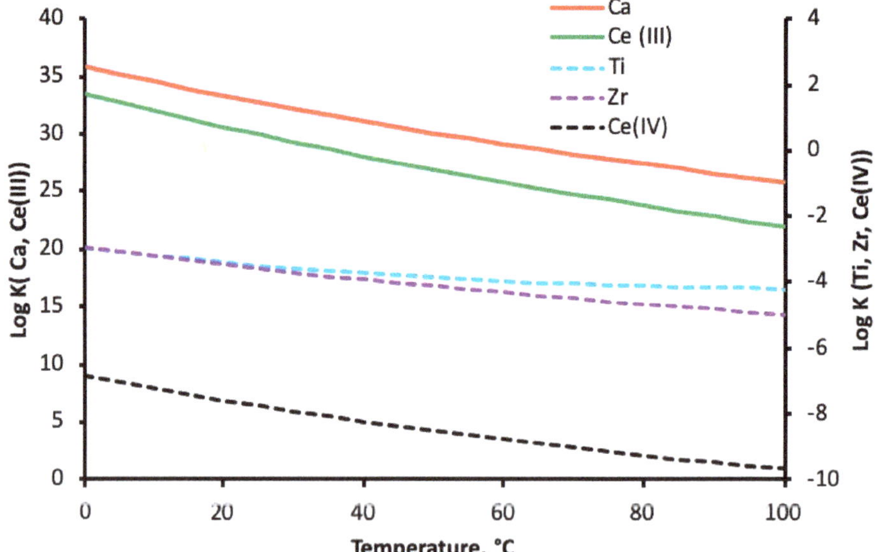

Figure 7. Solubilities of simple oxides of Ca, Ce, Ti and Zr from 0–100 °C as calculated using software [27]. Log K values for Reactions 1–5 were used.

While Ce_2O_3 will readily dissolve in acidic or neutral conditions, CeO_2 is far less soluble (Figure 7). Oxidising conditions may have caused cerium as Ce^{3+} or $CeOOH$ to be oxidised to insoluble CeO_2. Calculations [26] have indicated that this process is favourable (see Reactions 6 and 7). Increasing the pH will make Reaction 7 even more favourable.

$$0.5\,Ce_2O_3 + 0.5\,H_2O \rightarrow CeOOH_{(aq)}$$
$$\Delta_{rxn}G^{70°C} = -25.0\ kJ/mol \qquad \text{(Reaction 6)}$$

$$CeOOH_{(aq)} + Fe(CN)_6^{3-} + OH^- \rightarrow CeO_2 + Fe(CN)_6^{4-} + H_2O$$
$$\Delta_{rxn}G^{70°C} = -129.4\ kJ/mol \qquad \text{(Reaction 7)}$$

While the unreliability of the calcium assay data makes it impossible to determine its behaviour in solution, secondary calcite phases have been observed when leaching brannerite (~2% Ca) under similar conditions [12]. The leaching of calcium from perovskite in carbonate solutions forming anatase and calcite has been observed in natural titanium deposits [31]. The process may take place according to the following reactions:

$$CaTiO_3 + 2H_2CO_3 \rightarrow TiO_{2(anatase)} + Ca(HCO_3)_{2(aq)} + 2H_2O \qquad \text{(Reaction 8)}$$

$$Ca(HCO_3)_{2(aq)} \rightarrow CaCO_3 + CO_2 + H_2O \qquad \text{(Reaction 9)}$$

The overall process is described in Reaction 10:

$$CaTiO_3 + CO_{2(aq)} \rightarrow TiO_{2(anatase)} + CaCO_3, \ \Delta_{rxn}G^{70°C} = -48.4 \ kJ/mol \qquad \text{(Reaction 10)}$$

Clearly, this process is favourable under these conditions.

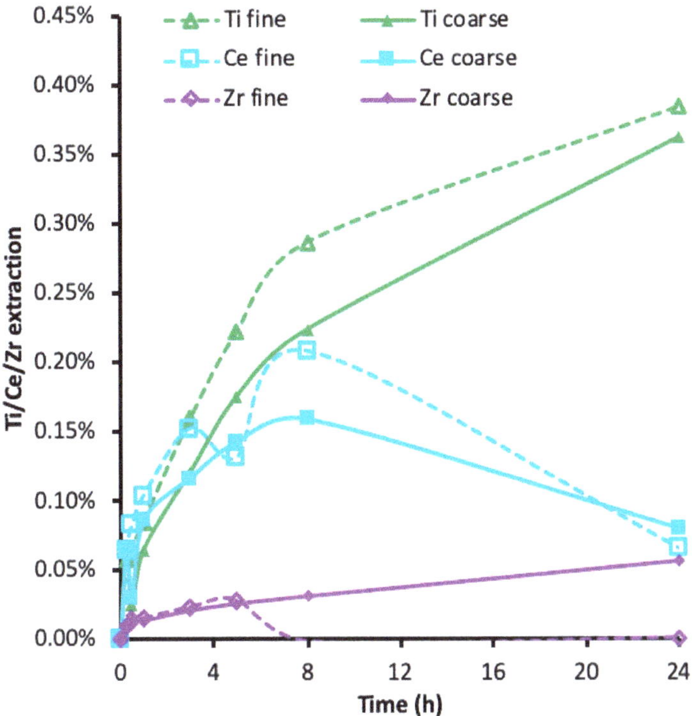

Figure 8. Extraction of titanium, zirconium and cerium from zirconolite in alkaline media at 70 °C.

3.3. Activation Energy

The rate of cerium dissolution showed a strong dependence on temperature. The average rate of dissolution between 1 and 5 h residence time was used to calculate the activation energy in the temperature range 30 to 85 °C, shown in Table 4. This can be considered to be the initial rate of extraction given the long periods over which zirconolite is known to dissolve [14,26]. Arrhenius plots for the sulphate leaching experiments showing data from tests conducted at 30, 50, 70 and 85 °C are shown in (Figure 9).

Table 4. Activation energy (kJ/mol) for the dissolution of Ce, Ti and Zr based on extraction rates from 1–5 h.

Element	H_2SO_4	HCl
Ce	41.5	35.5
Ti	29.8	50.2
Zr	21.4	15.4

Activation energies (Table 4) were also calculated for the hydrochloric acid leaching tests, though the results are less certain as chloride leaching was only done at two temperatures, 50 and 85 °C.

Arrhenius plots may have multiple regions corresponding to different rate determining steps [32], thus, such plots derived from only two temperature points may be unreliable.

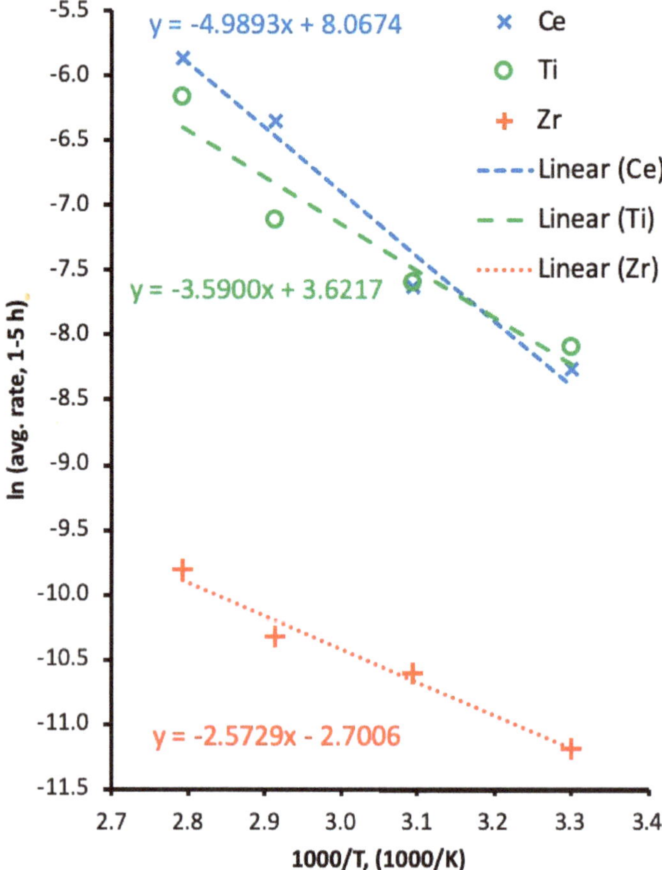

Figure 9. Arrhenius plot for Ce, Ti and Zr leaching in 0.25M H_2SO_4, based on the average rate of leaching from 1–5 h.

Omitting the outlying 85 °C point from the titanium calculation gives an activation energy of 21.4 kJ/mol in sulphate media, very close to the calculated activation energy for zirconium dissolution in the same media, possibly indicative of a similar dissolution mechanism.

Longer term leaching experiments over 14 days in 1 M HCl at 100–200 °C gave an activation energy of approximately 20 kJ/mol for the dissolution of cerium and titanium from synthetic zirconolite [14]. Leaching experiments with similar synthetic samples by Zhang et al. [33] between 25 and 75 °C and over a pH range of 2–12, showed that the activation energy for uranium release from zirconolite varied with pH when these calculations were repeated with data presented by Zhang et al. [33]. The activation energy for uranium release was typically 15–20 kJ/mol with the one outlier being the pH 4.1 tests which gave a calculated activation energy value of 37 kJ/mol.

Comparisons with other studies [9,10] showed that zirconolite underwent slower dissolution than brannerite or even betafite when leached under similar conditions (Figure 10).

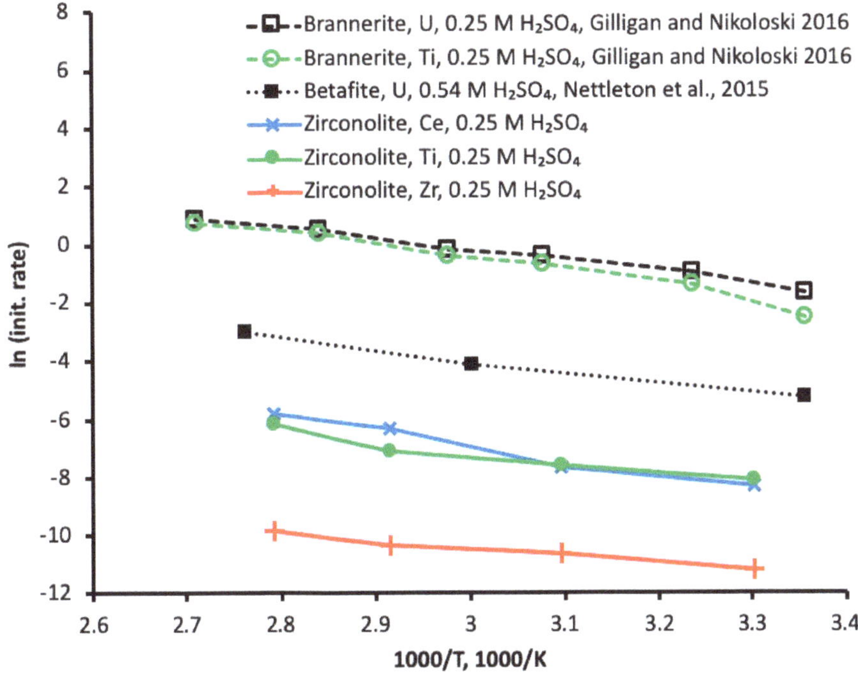

Figure 10. Arrhenius plots for the extraction of various elements from brannerite, betafite and zirconolite in sulphuric acid.

3.4. Leached Residue Characterisation

Unlike the brannerite studied previously, there were few apparent signs of corrosion after the leaching of zirconolite. Images, element maps and spectra were taken of zirconolite particles leached at the highest temperature in each lixiviant.

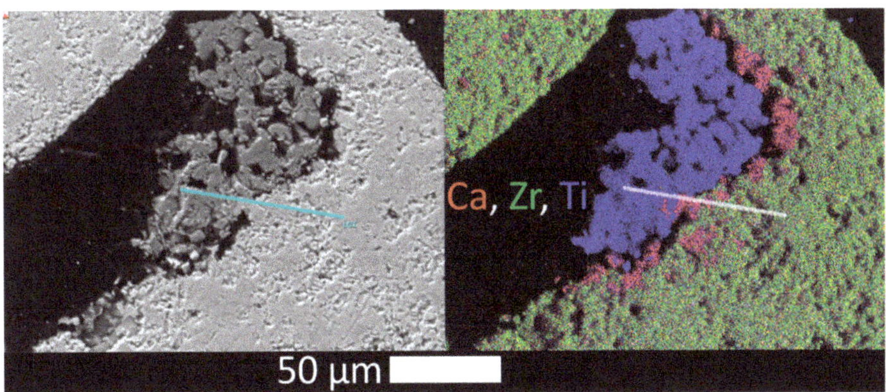

Figure 11. Grain boundary in zirconolite after leaching in 0.25 M HCl at 85 °C. **Left:** BSE image, **right:** element map.

Figure 11 shows a rutile inclusion surrounded by perovskite following chloride leaching. It is possible that the perovskite regions may have undergone some corrosion. There was no sign of pitting on the zirconolite (light green in all element maps) visible at the resolution of these images, though line analyses indicated that the outermost 2–5 μm were enriched in titanium and zirconium and depleted of calcium and cerium relative to the core of the particles.

These apparent changes in the distribution of elements in the solid phase are corroborated by the leaching kinetics data, which showed that the extraction of cerium was consistently higher than that of titanium and significantly higher than that of zirconium (Figures 5 and 6).

There were less visible signs of corrosion in the sulphate leaching system (Figure 12), although line EDX analyses indicate some selective leaching of calcium and cerium in the outermost 2–5 μm layer of zirconolite. Line A intersects a rutile inclusion and a perovskite grain, while line B runs across a protrusion of zirconolite corroded on both sides, both ends showing decreased Ca/Ce relative to Ti/Zr. There were no visible signs of corrosion on the rutile inclusions. Past experience with ilmenite [34] suggests minimal corrosion occurs in 0.25–1.00 M H_2SO_4 at 95 °C during 5 h of contact.

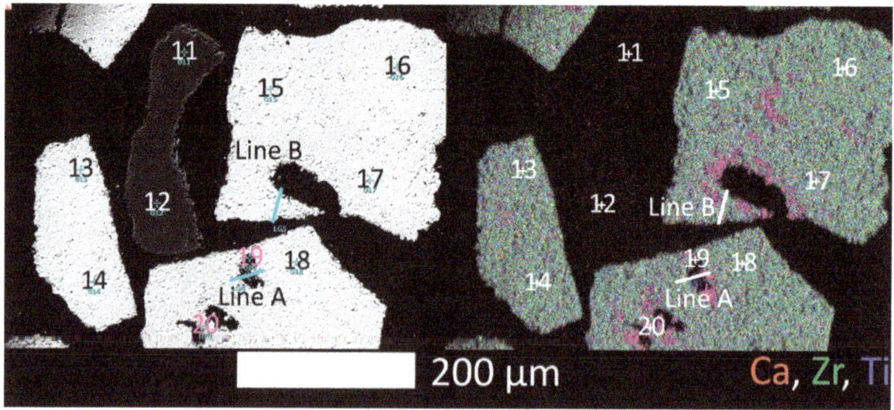

Figure 12. A large zirconolite particle with rutile inclusions after leaching in 0.25 M H_2SO_4 at 85 °C for 5 h Left: BSE image, right: element map.

There were minimal signs of corrosion in carbonate leaching media (Figures 13 and 14). Once again, this is consistent with the leaching kinetics (Figure 8) and past experience with brannerite [12,28].

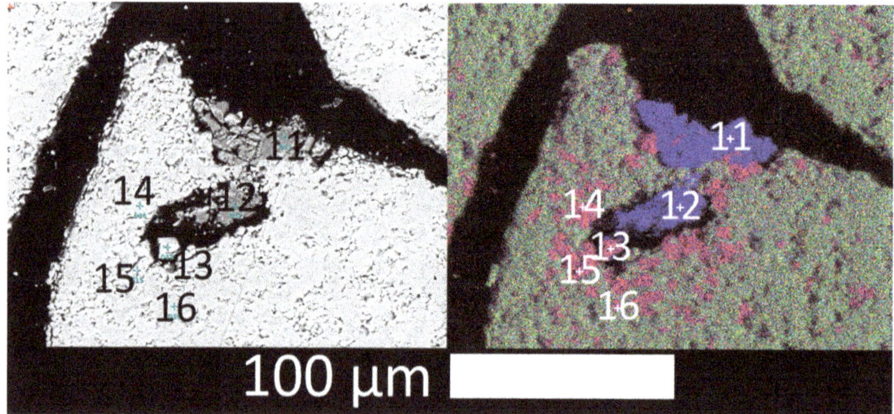

Figure 13. Backscattered electron SEM image (**left**) and Ca–Zr–Ti map (**right**) of zirconolite after 24 h of leaching in sodium carbonate at 70 °C with the locations of EDX analyses. Spectra are shown in Figure 14.

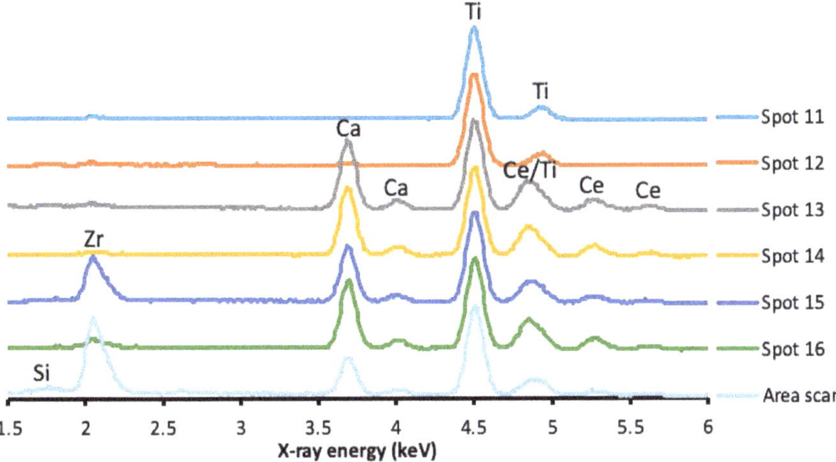

Figure 14. EDX spectra of rutile and perovskite inclusions in a zirconolite particle after leaching in sodium carbonate.

The alteration of zirconolite to secondary phases such as anatase and baddeleyite has been observed in earlier work [14,33]; however, these leaching tests typically ran over longer time periods of at least two weeks.

3.5. Reaction Mechanisms

Particle size had a clear effect on the rate of cerium dissolution in sulphate and chloride media. The rate at which various elements dissolve from zirconolite has been observed to be proportional to the surface area in contact with the lixiviant until saturation is reached and secondary phases begin to form [14]. The extraction rates of titanium and zirconium were less affected by particle size (Figure 5; Figure 6), as might be expected if they were forming secondary phases.

Calcium, aluminium and cerium dissolution were proportional to surface area, while zirconium and titanium were observed to plateau due to the formation of secondary solid phases on reaching saturation [14]. The reaction for zirconolite dissolution given by Pöml et al. [14] is:

$$(Ca_{1-x}Ce_x)Zr\left(Ti_{2-y}Al_y\right)O_7 + (6+2y)H^+$$
$$\rightarrow ZrO_2 + (2-y)TiO_2 + (1-x)Ca^{2+} + xCe^{4+} + yAl^{3+} + (3+y)H_2O \qquad \text{(Reaction 11)}$$

Titanium and zirconium precipitation does not occur until ZrO_2 and TiO_2 exceed saturation at the zirconolite-solution interface [14]. These experiments were not run for long enough for the zirconium and titanium concentrations in the bulk solution to plateau, though the outer 5 μm of zirconolite leached in sulphate media at 85 °C was enriched in titanium and zirconium indicating that saturation may have been reached at the solid-aqueous interface. The relative rates of extraction observed in this study matched those identified in longer term leaching studies [14,26].

It is under oxidising conditions that the similarities between cerium and plutonium break down. Cerium is oxidised to insoluble CeO_2 (Reaction 7), while insoluble PuO_2 can be oxidised further and remobilised as PuO_2^+ and PuO_2^{2+} complexes [35], similar to what is typically seen with uranium.

3.6. Crystallinity and Leachability

There are two reasons for the much lower extent of dissolution observed in the zirconolite leaching compared with earlier work with brannerite. Zirconolite is known to be highly chemically stable. This sample is also highly crystalline as is apparent from the XRD results (Figure 1). Crystalline phases are more refractory than metamict materials [2,23]. Even within the same mineral sample, heavily altered metamict zones are more susceptible to corrosion than less altered zones [22].

The process of metamictisation, by which a radioactive crystalline material gradually becomes amorphous from internal irradiation [36] decreases its chemical and physical stability [14]. However, if a metamict material is recrystallised by heating, the material becomes less soluble. This has been documented for brannerite [23], betafite [11] and synthetic zirconolite [26].

Synthetic titanates have been synthesised with plutonium-238 (half-life = 87.7 years) to study the rate and effects of radiation damage over the course of five years. Strachan et al. [26] showed that the degree of radiation damage has little to no effect on the chemical stability of zirconolite and pyrochlore. Similarly, studies of natural zirconolites have shown them to be highly chemically durable having survived 600 Ma or more in nature [2]. This means that measurements of the chemical stability of zirconolite based on recently prepared non-metamict samples are likely to be applicable to aged and metamict samples as well. Furthermore, this indicates that zirconolite is a good material to use for the long-term sequestration of radioisotopes.

The minor perovskite phase seemed to have undergone more corrosion than the surrounding zirconolite phase. This could be related to the formation of Ce^{3+}, which is easy to form, as observed in the fabrication, and is what stabilises perovskite. Perovskite should be avoided when synthesising zirconolite, as this may increase the proportion of soluble and mobile plutonium. Pöml et al. [14] achieved this by adding a stoichiometric excess of ZrO_2. However, this is likely to be less of a concern for the immobilisation of tetravalent actinide ions.

In addition to long term durability in the disposal environment, a wasteform for fissile material immobilisation must demonstrate proliferation resistance such that the fissile material cannot be retrieved by dissolution of the wasteform. When the United States Department of Energy was developing wasteform options for the disposition of surplus weapons grade plutonium, zirconolite, the initial choice of wasteform, was replaced by betafite (therein referred to as pyrochlore) on the grounds that the U–238 in the wasteform would guard against certain long term criticality events in the disposal environment [37]. This study, in conjunction with our previous work on betafite leaching, strongly indicates that the addition of depleted uranium to the wasteform is detrimental to proliferation resistance. Given the demonstrated durability of zirconolite, from both natural and synthetic samples, long term criticality risks in the disposal environment seem a remote possibility, and this supports the selection of zirconolite, above betafite, as the wasteform for disposition of surplus plutonium.

4. Conclusions

Synthetic zirconolite is significantly more stable than natural brannerite or betafite. The most intense conditions used in this study did not cause synthetic zirconolite to undergo significant leaching or visible corrosion despite the same conditions being sufficient for near complete dissolution of natural brannerite in under five hours [9,12,22].

There was some evidence for incongruent dissolution, as the outer 5 µm of some leached zirconolite particles were enriched in titanium and zirconium, indicating that these elements had exceeded saturation at the aqueous-solid interface.

Fine zirconolite dissolved faster than the coarse material, indicating that the rate of dissolution is related to surface area. In practice, the rate of dissolution could therefore be further minimised by forming the zirconolite waste ceramics into larger solid masses.

Author Contributions: Conceptualisation, A.N.N., E.R.M.; methodology, A.N.N., R.G., E.R.M. and J.S.; validation, A.N.N.; formal analysis, A.N.N., R.G.; investigation, A.N.N., R.G., E.R.M., J.S.; resources, A.N.N.; data curation, A.N.N., R.G.; writing—original draft preparation, R.G., A.N.N.; writing—review and editing, A.N.N., R.G; supervision, A.N.N.; project administration, A.N.N.; funding acquisition, A.N.N.

Funding: This research received no external funding.

Acknowledgments: The authors acknowledge Murdoch University for access to equipment.

Conflicts of Interest: The authors declare no conflict of interest.

References

1. Lumpkin, G.R. Ceramic waste forms for actinides. *Elements* **2006**, *2*, 365–372. [CrossRef]
2. Lumpkin, G.R. Alpha-decay damage and aqueous durability of actinide host phases in natural systems. *J. Nucl. Mater.* **2001**, *289*, 136–166. [CrossRef]
3. Gieré, R.; Williams, C.T. REE-bearing minerals in a Ti-rich vein from the Adamello contact aureole (Italy). *Contrib. Min. Pet.* **1992**, *112*, 83–100.
4. De Hoog, J.C.M.; van Bergen, M.J. Notes on the chemical composition of zirconolite with thorutite inclusions from Walaweduwa, Sri Lanka. *Mineral. Mag.* **1997**, *61*, 721–725. [CrossRef]
5. Rasmussen, B.; Fletcher, I.R. Zirconolite: A new U-Pb chronometer for mafic igneous rocks. *Geology* **2004**, *32*, 785–788. [CrossRef]
6. Seddio, S.M.; Jolliff, B.L.; Korotev, R.L.; Zeigler, R.A. Petrology and geochemistry of lunar granite 12032, 366–19 and implications for lunar granite petrogenesis. *Am. Mineral.* **2013**, *98*, 1697–1713. [CrossRef]
7. Gieré, R.; Williams, C.T.; Lumpkin, G.R. Chemical characteristics of natural zirconolite. *Schweiz. Miner. Petrogr. Mitt.* **1998**, *78*, 433–459.
8. Helean, K.B.; Navrotsky, A.; Vance, E.R.; Carter, B.; Ebbinghaus, B.; Krikorian, O.; Lian, J.; Wang, L.M.; Catalano, J.G. Enthalpies of formation of Ce-pyrochlore, $Ca_{0.93}Ce_{1.00}Ti_{2.035}O_{7.00}$, U-pyrochlore, $Ca_{1.46}U^{4+}_{0.23}U^{6+}_{0.46}Ti_{1.85}O_{7.00}$ and Gd-pyrochlore, $Gd_2Ti_2O_7$: Three materials relevant to the proposed waste form for excess weapons plutonium. *J. Nucl. Mater.* **2002**, *303*, 226–239. [CrossRef]
9. Gilligan, R.; Nikoloski, A.N. Leaching of brannerite in the ferric sulphate system. Part 1: Kinetics and reaction mechanism. *Hydrometallurgy* **2015**, *156*, 71–80. [CrossRef]
10. Gilligan, R.; Nikoloski, A.N. Leaching of brannerite in the ferric chloride system. *Hydrometallurgy* **2018**, *180*, 104–112. [CrossRef]
11. Nettleton, K.C.A.; Nikoloski, A.N.; Da Costa, M. The leaching of uranium from betafite. *Hydrometallurgy* **2015**, *157*, 270–279. [CrossRef]
12. Gilligan, R.; Nikoloski, A.N. Alkaline leaching of brannerite. Part 1: Kinetics, reaction mechanisms and mineralogical transformations. *Hydrometallurgy* **2017**, *169*, 399–410. [CrossRef]
13. Williams, C.T. The occurrence of niobian zirconolite, pyrochlore and baddeleyite in the Kovdor carbonatite complex, Kola Peninsula, Russia. *Mineral. Mag.* **1996**, *60*, 639–646. [CrossRef]
14. Pöml, P.; Geisler, T.; Cobos-Sabaté, J.; Wiss, T.; Raison, P.E.; Schmid-Beurmann, P.; Deschanels, X.; Jégou, C.; Heimink, J.; Putnis, A. The mechanism of the hydrothermal alteration of cerium- and plutonium-doped zirconolite. *J. Nucl. Mater.* **2011**, *410*, 10–23. [CrossRef]
15. Bellatreccia, F.; Della Ventura, G.; Williams, C.T.; Lumpkin, G.R.; Smith, K.L.; Colella, M. Non-metamict zirconolite polytypes from the feldspathoid-bearing alkalisyenitic ejecta of the Vico volcanic complex (Latium, Italy). *Eur. J. Mineral.* **2002**, *14*, 809–820. [CrossRef]
16. Bellatreccia, F.; Della Ventura, G.; Caprilli, E.; Williams, C.T.; Parodi, G.C. Crystal-chemistry of zirconolite and calzirtite from Jacupiranga, São Paulo (Brazil). *Mineral. Mag.* **1999**, *63*, 649–660. [CrossRef]
17. White, T.J.; Segall, R.L.; Hutchison, J.L.; Barry, J.C. Polytypic behaviour of zirconolite. *Proc. R. Soc. Lond. A* **1984**, *392*, 343–358. [CrossRef]
18. Reed, S.J.N. *Electron Microprobe Analysis and Scanning Electron Microscopy in Geology*, 2nd ed.; Cambridge University: Cambridge, UK, 2008.
19. Chang, L.L.; Howie, R.A.; Zussman, J. *Rock-Forming Minerals: Volume 5. Non-Silicates*; Deer, W.A., Howie, R.A., Zussman, J., Eds.; Longman: London, UK, 1975.
20. Wenk, H.R.; Bulakh, A.G. *Minerals, Their Constitution and Origin*; Cambridge University Press: Cambridge, UK, 2004.
21. Lumpkin, G.R.; Smith, K.L.; Blackford, M.G. Partitioning of uranium and rare earth elements in Synroc: Effect of impurities, metal additive, and waste loading. *J. Nucl. Mater.* **1995**, *224*, 31–42. [CrossRef]
22. Gilligan, R.; Deditius, A.P.; Nikoloski, A.N. Leaching of brannerite in the ferric sulphate system. Part 2: Mineralogical transformations during leaching. *Hydrometallurgy* **2016**, *159*, 95–106. [CrossRef]
23. Charalambous, F.A.; Ram, R.; Pownceby, M.I.; Tardio, J.; Bhargava, S.K. Chemical and microstructural characterisation studies on natural and heat treated brannerite samples. *Miner. Eng.* **2012**, *39*, 276–288. [CrossRef]

24. McMaster, S.A.; Ram, R.; Pownceby, M.I.; Tardio, J.; Bhargava, S.K. Characterisation and leaching studies on the uranium mineral betafite [(U,Ca)$_2$(Nb,Ti,Ta)$_2$O$_7$]. *Miner. Eng.* **2015**, *81*, 58–70. [CrossRef]
25. McMaster, S.A.; Ram, R.; Faris, N.; Pownceby, M.I.; Tardio, J.; Bhargava, S.K. Uranium leaching from synthetic betafite: [(Ca,U)$_2$(Ti,Nb,Ta)$_2$O$_7$]. *Int. J. Miner. Process.* **2017**, *160*, 58–67. [CrossRef]
26. Strachan, D.M.; Scheele, R.D.; Icenhower, J.P.; Buck, E.C.; Kozelisky, A.E.; Sell, R.L.; Elovich, R.J.; Buchmiller, W.C. *Radiation Damage Effects in Candidate Ceramics for Plutonium Immobilization*; Final Report, PNNL-14588; Pacific Northwest National Laboratory: Richland, WA, USA, 2004.
27. Roinne, A. *SC Chemistry Software*, version 7.1.1; Chemical Reaction and Equilibrium Modules. Outotec Research: Pori, Finland, 2012.
28. Gilligan, R.; Nikoloski, A.N. Alkaline leaching of brannerite. Part 2: Leaching of a high-carbonate refractory uranium ore. *Hydrometallurgy* **2017**, *173*, 224–231. [CrossRef]
29. Knauss, K.G.; Dibley, M.J.; Bourcier, W.L.; Shaw, H.F. Ti(IV) hydrolysis constants derived from rutile solubility measurements made from 100 to 300 °C. *Appl. Geochem.* **2001**, *16*, 1115–1128. [CrossRef]
30. Schmidt, J.; Vogelsberger, W. Aqueous Long-Term Solubility of Titania Nanoparticles and Titanium (IV) Hydrolysis in a Sodium Chloride System Studied by Adsorptive Stripping Voltammetry. *J. Solut. Chem.* **2009**, *38*, 1267–1282. [CrossRef]
31. Thompson, J.V. Titanium pigments from Colorado perovskite. In Proceedings of the SME Annual Meeting, Salt Lake City, UT, USA, 26 February–1 March 1990.
32. Gupta, C.K. *Chemical Metallurgy: Principles and Practice*; John Wiley & Sons: Weinheim, Germany, 2003.
33. Zhang, Y.; Hart, K.P.; Bourcier, W.L.; Day, R.A.; Colella, M.; Thomas, B.; Aly, Z.; Jostsons, A. Kinetics of uranium release from Synroc phases. *J. Nucl. Mater.* **2001**, *289*, 254–262. [CrossRef]
34. Gilligan, R. The Extractive Metallurgy of Brannerite: Leaching Kinetics, Reaction Mechanisms and Mineralogical Transformations. Ph.D. thesis, Murdoch University, Perth, Australia, 2017.
35. Langmuir, D. *Aqueous Environmental Geochemistry*; Prentice Hall: Portland, ME, USA, 1997.
36. Pabst, A. The metamict state. *Am. Mineral.* **1952**, *37*, 137–157.
37. Ebbinghaus, B.B.; Armantrout, G.A.; Gray, L.; Herman, C.C.; Shaw, H.F.; Van Konynenburg, R.A. *Plutonium Immobilization Project Baseline Formulation*; Lawrence Livermore National Laboratory: Livermore, CA, USA, 2000; UCRL-ID-133089, rev. 1. PIP-00-141.

 *metals*

Article

Synthesis of Nanosilica via Olivine Mineral Carbonation under High Pressure in an Autoclave

Srecko Stopic [1,*], Christian Dertmann [1], Ichiro Koiwa [2], Dario Kremer [3], Hermann Wotruba [3], Simon Etzold [4], Rainer Telle [4], Pol Knops [5] and Bernd Friedrich [1]

[1] IME Process Metallurgy and Metal Recycling, RWTH Aachen University, Intzestrasse 3,
    52056 Aachen, Germany; cdertmann@ime-aachen.de (C.D.); bfriedrich@ime-aachen.de (B.F.)
[2] Department of Applied Chemistry, College of Science and Engineering, Kanto Gakuin University,
    1-50-1, Mutsuurahigashi, Kanazawa-ku, Yokohama 236-8501, Japan; koiwa@kanto-gakuin.ac.jp
[3] AMR Unit of Mineral Processing, RWTH Aachen University, Lochnerstrasse 4-20, 52064 Aachen, Germany;
    kremer@amr.rwth-aachen.de (D.K.); wotruba@amr.rwth-aachen.de (H.W.)
[4] Department of Ceramics and Refractory Materials, GHI Institute of Mineral Engineering,
    RWTH Aachen University, Mauerstrasse 5, 52064 Aachen, Germany; etzold@ghi.rwth-aachen.de (S.E.);
    Telle@ghi.rwth-aachen.de (R.T.)
[5] Green Minerals, Rijksstraatweg 128, NL 7391 MG Twello, The Netherlands; pol@green-minerals.nl
*   Correspondence: sstopic@ime-aachen.de

Received: 24 May 2019; Accepted: 19 June 2019; Published: 24 June 2019

Abstract: Silicon dioxide nanoparticles, also known as silica nanoparticles or nanosilica, are the basis for a great deal of biomedical and catalytic research due to their stability, low toxicity and ability to be functionalized with a range of molecules and polymers. A novel synthesis route is based on CO_2 absorption/sequestration in an autoclave by forsterite (Mg_2SiO_4), which is part of the mineral group of olivines. Therefore, it is a feasible and safe method to bind carbon dioxide in carbonate compounds such as magnesite forming at the same time as the spherical particles of silica. Indifference to traditional methods of synthesis of nanosilica such as sol gel, ultrasonic spray pyrolysis method and hydrothermal synthesis using some acids and alkaline solutions, this synthesis method takes place in water solution at 175 °C and above 100 bar. Our first experiments have studied the influence of some additives such as sodium bicarbonate, oxalic acid and ascorbic acid, solid/liquid ratio and particle size on the carbonation efficiency, without any consideration of formed silica. This paper focuses on a carbonation mechanism for synthesis of nanosilica under high pressure and high temperature in an autoclave, its morphological characteristics and important parameters for silica precipitation such as pH-value and rotating speed.

Keywords: silica; synthesis; olivine carbonation; autoclave; precipitation

---

1. Introduction

As a result of the high nickel production costs associated with traditional pyrometallurgical techniques and the depletion of high-grade sulfide ores, renewed interest has developed concern on the production of nickel and cobalt by high pressure acid leaching (PAL) of nickel laterites. More than one third of the world's nickel is nowadays produced from laterite ores [1,2]. Laterites account for two thirds of the world's nickel resources. It is therefore likely that increasing amounts of nickel will be produced from laterites. Since laterite type ores naturally occur close to the surface, economical open pit mining techniques are employed to recover the ore after removal of the overburden [3]. The laterite ore consists of fresh saprolite such as $K_{0.4}(Si_{3.0}Al_{1.0})_{4.0}(Al_{2.0}Mg_{0.3})_{2.33}O_{10}(OH)_2$ and nontronite such as $Na_{0.3}(Fe^{3+})_2(Si,Al)_4O_{10}(OH)_2 \cdot nH_2O$. These silicate ores represent the various layers in the laterite bedrock. The limonite consists mainly of goethite. This continues to a nontronite rich zone. Saprolite is the next layer, which is distinguished from its rich magnesium silicate content. The lateritic ore mostly has a low level of Ni (1–3%), Co (max. 0.1%), Fe (20–30%) and high level of SiO_2 (more than 50%). The treatment of silicate based ores with different acids under an atmospheric pressure leads to formation of silica gel and breaking of leaching process. Silica gel represents an amorphous and porous form of silicon dioxide consisting of an irregular tridimensional framework of alternating silicon and oxygen atoms with nanometer-scale pores [4].

Similarly, high Si content in red mud and its slags produced by pyrometallurgical treatment for the Fe removal makes these secondary resources untreatable with conventional acid leaching routes due to the formation of silica gel. Alkan et al. [5] studied red mud and slags synthesized by electric arc furnace smelting, which contain rather moderate and extensive SiO_2. In the next step, the formed slag was exposed separately by red mud to dry digestion with sulfuric acid at room temperature aiming at selective Sc recovery without Ti and Si dissolution. An empirical dry digestion-leaching model was proposed for each starting material in a comparative manner in order to prevent the formation of silica gel using sulfuric acid.

The Eudialyte concentrate is a potential rare earth elements (REE) primary resource due to its good solubility in acid, low radioactivity and relatively high REE content (about 2%), but also contains more than 50% of silica. The treatment of the Eudialyte concentrate can produce silica gel during a treatment with some acids [6]. The main challenge is avoiding the formation of silica gel, which is non-filterable when using acid to extract REE. Ma et al. [7] have studied neural network modeling for the optimization of the extraction of rare earth elements from the Eudialyte concentrate by dry digestion and a subsequent leaching avoiding the formation of silica gel in the presence of the hydrochloric acid.

Development of ceramic nanoparticles such as silica, alumina and titania with improved properties has been studied with much success in several areas such as synthesis and surface science [8,9]. Advancement in nanotechnology has led to the production of nanosized silica, which has been widely used as filler in catalysis and glass industry. The silica particles extracted from natural resources contains metal impurities and are not favorable for advanced scientific and industrial applications.

The sol-gel process is widely applied to produce silica, glass, and ceramic materials due to its ability to form pure and homogenous products at mild conditions. The process involves hydrolysis and condensation of metal alkoxides ($Si(OR)_4$) such as tetraethylorthosilicate (TEOS, $Si(OC_2H_5)_4$) or inorganic salts such as sodium silicate (Na_2SiO_3) in the presence of mineral acid (e.g., HCl) or base (e.g., NH_3) as catalyst [10]. The synthesis of spherical hollow silica particles from sodium silicate solution with boric acid or urea as an additive was carried out by the ultrasonic spray pyrolysis method. This work dealt with the effect of four parameters (the concentration of the boric acid and urea, feed rate of reactant, reaction temperature and time) on particle size and standard deviation. As a result, the mean particle size and standard deviation decreased with increasing of all parameters except urea [11]. Ratanathavorn et al. [12] have studied silica nanoparticles synthesis by ultrasonic spray pyrolysis (USP) technique using tetraethylorthosilicate (TEOS) as a precursor in order to produce a fixative material for cream perfume fomulation. The results showed that the synthesis temperature of 500 °C provided the smallest size of silica nanoparticle, about 106 nm. The particle size decreased from 347 nm to 106 nm when the synthesis temperature increased from 300 °C to 500 °C.

The ultra-small hollow silica nanoparticles were synthesized using the prepared amorphous calcium carbonate (ACC) particles as a template. The ACC particles were firstly prepared by the carbonation method, which the procedure was conducted in the methanol solvent to form the $Ca(OCH_3)_2$ layers on the ACC particles. An effect of methanol concentration on the morphology of ACC particles was also investigated [13]. ACC particles were prepared by a carbonation method via bubbling CO_2 gas into calcium ions dispersing in methanol solution. An effect of methanol concentration on the $CaCO_3$ formation was investigated. The pH of the ACC preparation was studied in a range of 9.4 and 10. After that, ultra-small HSNPs were synthesized using the prepared ACC particles in the one-pot process. The results suggested that the synthesis of HSNPs using the ultra-small

ACC particles via the one-pot process is one of the most effective methods to produce ultra-small HSNP regarding to save energy and cost.

In a mineral carbonation process, silicate minerals can also be used as feedstock to form carbonates and H_4SiO_4, that are chemically stable in a geological timeframe. Silicate minerals usually are richer in alkaline earth metal content such as magnesium, sodium, and calcium. Common silicate minerals suitable for carbonation are forsterite (Mg_2SiO_4), antigorite ($Mg_3Si_2O_5(OH)_4$) and wollastonite ($CaSiO_3$) and their overall reaction rates are given in Equations (1)–(3):

$$Mg_2SiO_4(s) + 2CO_2(g) + 2H_2O(l) = 2MgCO_3(s) + H_4SiO_4(aq) + 89 \text{ kJ/mol} \tag{1}$$

$$Mg_3Si_2O_5(OH)_4(s) + 3CO_2(g) + 2H_2O(l) = 3MgCO_3(s) + 2H_4SiO_4(aq) + 64 \text{ kJ/mol} \tag{2}$$

$$CaSiO_3(s) + CO_2(g) + 2H_2O(l) = CaCO_3(s) + H_4SiO_4(aq) + 90 \text{ kJ/mol} \tag{3}$$

Stopic et al. [14] have shown the reaction path of direct forsterite carbonation in the aqueous solution without any deeper consideration of the formed silica particles as shown Equations (1)–(3) and Figure 1.

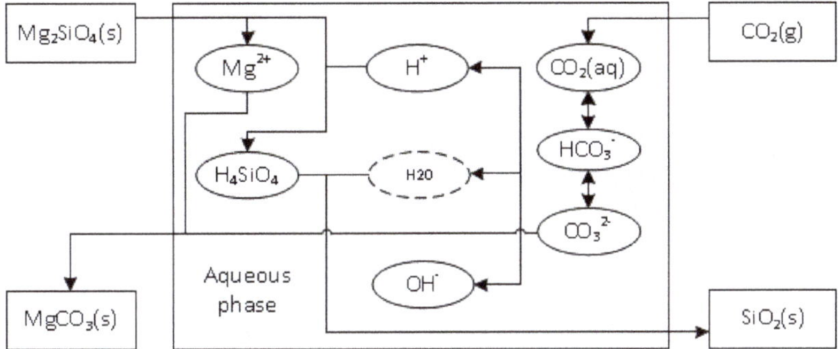

Figure 1. Reaction path of direct forsterite carbonation in the aqueous solution.

Although olivine is one mixed crystalline material (Mg, Fe)$_2$SiO$_4$, for simplicity, olivine consists only of Mg_2SiO_4, namely forsterite. First, gaseous carbon dioxide dissolves in the aqueous solution. Simultaneously, forsterite is dissolved in the aqueous solution (Equation (4)) forming aqueous silicic acid, then precipitates as amorphous silica (Equation (5)), which is a by-product, and lastly magnesium ions and carbonate form magnesite as shown with Equation (6):

$$Mg_2SiO_4(s) + 4\,H^+(aq) \xrightarrow{r_{Mg_2SiO_4}} 2\,Mg^{2+}(aq) + H_4SiO_4(aq) \tag{4}$$

$$H_4SiO_4(aq) \xrightarrow{r_{SiO_2}} SiO_2(s) + 2\,H_2O(l) \tag{5}$$

$$Mg^{2+}(aq) + CO_3^{2-}(aq) \xrightarrow{r_{MgCO_3}} MgCO_3(s) \tag{6}$$

The determination of process parameters such as temperature, pressure and pH for maximum overall conversion rates is elementary. Direct CO_2 sequestration at high pressure with olivine as a feedstock has already been performed in numerous studies at different temperatures and pressures with or without the use of additives such as carboxylic acid, and sodium hydroxide. It is reported that optimal reaction conditions are in the temperature range of 150–185 °C and in the pressure range of 135–150 bar [15]. Additives are reported to have a positive influence on carbonation rate, but without a study in detail. Optimal addition of additives are reported by Bearat et al. [16] in studies about the mechanism that limits aqueous olivine carbonation reactivity under the optimum sequestration reaction conditions observed as follows: 1 M NaCl + 0.64 M NaHCO$_3$, at 185 °C and P

(CO_2) about 135 bar. A reaction limiting silica-rich passivating layer forms on the feedstock grains, slowing down carbonate formation and raising process costs. Eikeland [17] reported that NaCl does not have significant influence on carbonation rate. The presented results show a conversion rate of more than 90% using a $NaHCO_3$ concentration of 0.5 M, without adding NaCl. Ideally, the solid phases exist as pure phases without growing together. In reality, different observations are made on the behavior of solid phases. Daval et al. [18] reported about the high influence of amorphous silica layer formation on the dissolution rate of olivine at 90 °C and elevated pressure of carbon dioxide. This passivating layer may be either built up from non-stoichiometric dissolution, precipitation of amorphous silica on forsterite particles or a combination of both. These previously mentioned results suggest that the formation of amorphous silica layers plays an important role in controlling the rate of olivine dissolution by passivating the surface of olivine, an effect that has yet to be quantified and incorporated into standard reactive-transport codes. In contrast to that, Oelkers et al. [19] and Hänchen [20] observed stoichiometric dissolution and no build-up of a passivating layer except during start-up of experiments. Furthermore, magnesite may precipitate on undissolved forsterite particles leading to a surface area reduction and therefore a reduction on forsterite dissolution rate, which was reported by Turri et al. [21]. In addition to this undesired intermixing of solids, they observed pure particles of magnesite to be predominant in the intermediate particle class, amorphous silica particles to be mainly present in the smallest particle class and unreacted olivine particles to be predominant in the largest particle class. This knowledge may be of value for subsequent separation of products such as magnesium carbonate and silica.

CO_2 sequestration with olivine as a feedstock was performed in a rocking batch autoclave at 175 °C and 100 bars in an aqueous solution and a CO_2-rich gas phase from 0.5 to 12 h. Turri et al. [21] showed maintainable recovery of separate fractions of silica, carbonates and unreacted olivine. Characterization of the recovered solids revealed that carbonates predominate in particle size range below 40 µm. The larger, residue fraction of the final product after carbonation consisted mainly of unreacted olivine, while silica is more present in the form of very fine spherical particles. An addition of sodium hydrogen carbonate at 0.64 M, oxalic acid at 0.5 M and ascorbic acid at 0.01 M was successfully applied in order to obtain maximal carbonation, what leads also to a complete formation of silica.

Our paper deals with the formation of magnesium carbonate and especially nanosilica using an olivine from Norway (40.1 MgO, and 48.7 wt % SiO_2) and with special attention on morphological characteristics of the obtained product and water solution after filtration, which was determined by structure and composition analysis (XRD, SEM; EDS; TEM and STEM).

2. Experimental Section

2.1. Materials

The samples used represent Steinsvik olivine from Norway as analyzed by a PW2404 XRF device (Malvern Panalytical B.V., Eindhoven, The Netherlands) and as shown in Table 1.

Table 1. Chemical composition of the investigated olivine from Norway (fraction between 20 and 63 µm) as analyzed by X-ray fluorescence (XRF) in wt %.

Component	SiO_2	Al_2O_3	Fe_2O_3	CaO	MgO	K_2O	MnO	Cr_2O_3	ZnO	NiO
in wt %	48.7	0.5	7.8	0.2	41.0	0.1	0.1	0.4	0.1	1.2

2.2. Procedures

The treatment of olivine was performed using the operations such as milling, sieving, carbonation in an autoclave, filtration and chemical analysis of solid and liquid sample shown at Figures 2 and 3. According to Reference 21 (Turri et al.) and Reference 14 (Stopic et al.) the carbonation tests have been carried out in the 1500 mL autoclave from Büchi Kiloclave Type 3E, Switzerland (as shown at

Figure 4) at 175 °C with 117 bar pure grade CO_2 in the presence and the absence of the additives such as sodium bicarbonate, oxalic acid and ascorbic acid in duration of 2–4 h. An amount ranging from 100 to 300 g sample has been added to 1000 mL solution with mixing rate 600 revolution per min in different experiments. After reaction, the liquid had very low contents of metal cations and was analyzed via the induced coupled plasma optical emission spectrometry ICP OES analysis (SPECTRO ARCOS, SPECTRO Analytical Instruments GmbH, Kleve, Germany). Characterization of the solid products was restricted to the X-ray powder diffraction XRD (Bruker AXS, Karlsruhe, Germany) and X-ray fluorescence XRF analyses using Device PW2404 (Malvern Panalytical B.V., Eindhoven, The Netherlands).

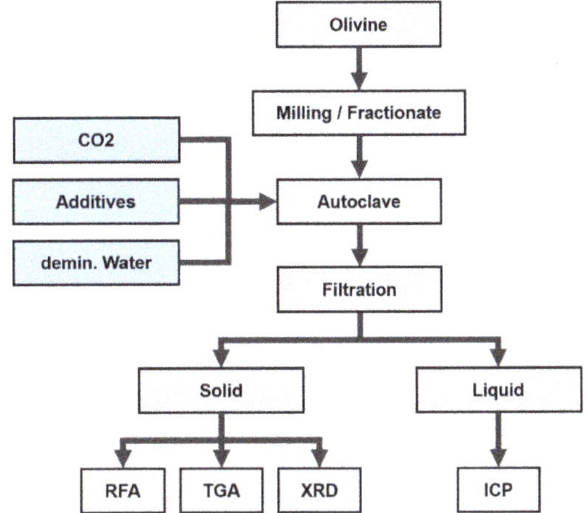

Figure 2. Procedure for experimental work and characterization of samples.

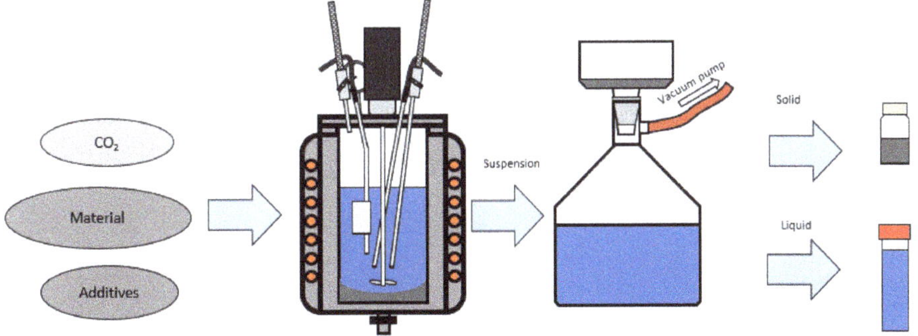

Figure 3. Carbonation process of olivine and sampling.

After milling the particle size fraction 20–63 µm was tested in an autoclave. The change of temperature and pressure during a heating from room temperature to 175 °C and a subsequent carbonation of 100 g olivine in 1000 mL water was followed in time. Carbonation was performed by injection of carbon dioxide from a bottle at the fixed temperature within 2 h. After this reaction time, the pressure was decreased to the atmospheric values and the solution was cooled to room temperature. After opening the cover of the autoclave the solution was filtrated as shown by Figures 2 and 3. Subsequent to drying of the solid residue, XRD analysis was performed for the product and an initial sample of olivine. X-ray powder diffraction patterns were collected by a Bruker-AXS D8 Advance diffractometer in Bragg–Brentano geometry, equipped with a copper tube coupled with a primary nickel filter providing Cu Kα1,2 radiation and LynxEye detector. The microstructure of the solid samples was examined using a scanning electron microscope (SEM)–JEOL6380 LV (JEOL Ltd., Tokyo, Japan). Energy dispersive X-Ray spectroscopy (EDS) was utilized by JSM-6000 (JEOL Ltd., Tokyo, Japan) to reveal elemental composition of the samples analyzed by SEM. The characterization of solid and liquid products was performed using the ICP-OES analysis (SPECTRO ARCOS, SPECTRO Analytical Instruments GmbH, Kleve, Germany).

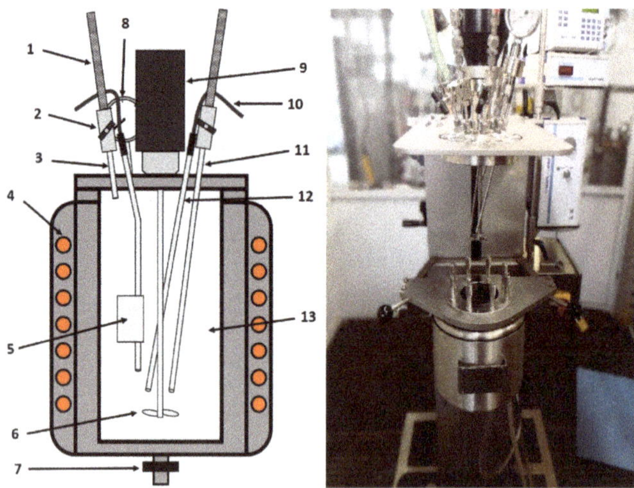

Figure 4. Sketch and picture of the autoclave: 1. Pressure pipe (Stahlflex); 2. needle valve; 3. tube for gas exhaust; 4. reactor shell with heating and cooling; 5. temperature sample head; 6. propeller mixer; 7. outlet valve; 8. analog manometer; 9. motor for the magnetic coupled stirrer; 10. cable for the measurement cutting site; 11. gas inlet; 12. testing rode for pressure; 13. working volume.

The values of temperature and pressure during carbonation are presented at Figure 5.

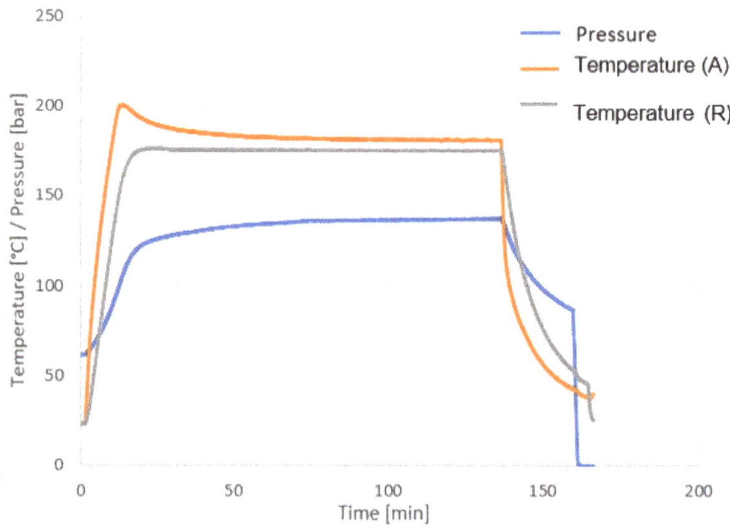

Figure 5. Pressure, temperature and time curves during work in autoclave (R-reaction; A-autoclave).

3. Results and Discussion

3.1. Product Characterization–Analysis via XRD and SEM of Solid Product after Carbonation

To evaluate the overall capability of the carbonation process, an experiment was performed using Norwegian olivine (20–63 μm) at 175 °C, 120 bar, 120 min, 600 rpm, in the presence of additives of sodium carbonate, oxalic and ascorbic acid considering the present mineralogical phases detected via XRD before and after the carbonation (Figure 6). A measurement range from 5–90° 2θ in 0.02° steps at 2 s per step are the chosen XRD parameters. XRD analysis was combined with a subsequent semi-quantitative evaluation of the mineral phase fractions. Table 2 provides information about the main mineral phase fractions as detected in the olivine samples within the given accuracy range. Forsterite, enstatite, lizardite and talc exhibited the most considerable mineralogical phase amounts. A decreased content of forsterite confirmed that the carbonation was successfully performed, what is indicated by 20–25% of magnesite in structure. The content of enstatite, lizardite and talc was not significantly changed, pointing out that these mineralogical phases show less reactivity within the performed process than forsterite.

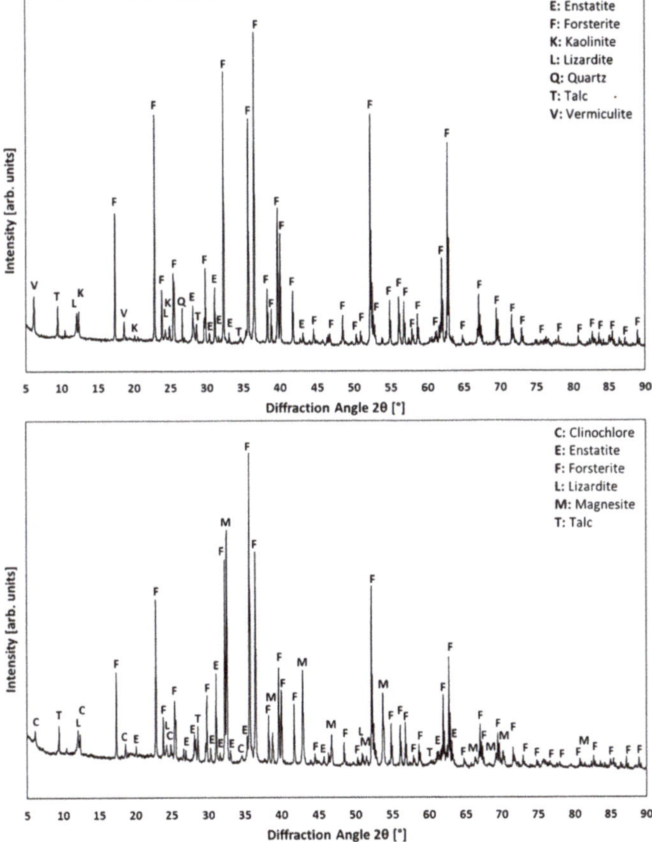

Figure 6. XRD analysis of Norwegian olivine samples; initial state (top) and carbonation product (bottom).

Table 2. The semi-quantitative XRD analysis before and after carbonation.

Mineral Phases	Olivine, Norway (20–63 μm)	
	Semi-Quantitative Composition in wt %	
	Before Carbonation	After Carbonation
Enstatite	5–10	5–10
Forsterite	75–80	50–55
Lizardite	≤5	≤5
Kaolinite	≤5	-
Talc	≤5	≤5
Magnesite	-	20–25

The presence of magnesite and silica was confirmed using the SEM analysis of the solid product, as shown in Figure 7.

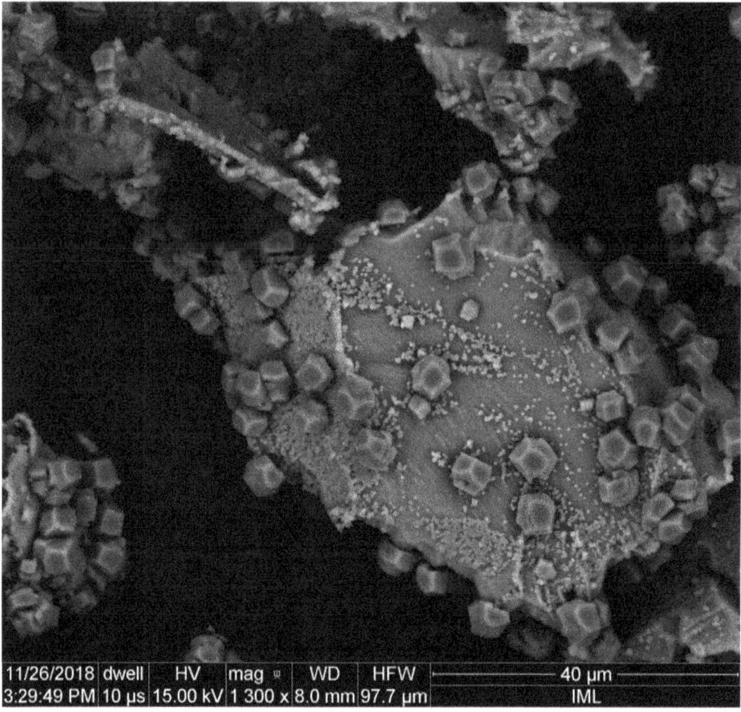

Figure 7. SEM analysis of initial olivine sample after carbonation.

As illustrated by Figure 7, SEM-analysis has confirmed that very small particles of SiO_2 and magnesite are formed as rhombohedrons or hexagonal prisms at the surface of partially carbonated magnesium silicate. The challenge of future work is a separation of the formed nanosilica particles from the product.

3.2. Product Characterization–Analysis of Precipitate from a Water

After filtration of the carbonated products, the white and yellow precipitate appeared during staying after 7–10 days, as shown in Figure 8. Then, this precipitate was separated from water solution and dried at 110 °C after the night. The obtained dried products were analyzed by XRD, SEM, EDS, and scanning transmission electron microscopy (STEM) by JEM-2100F (JEOL Ltd., Tokyo, Japan).

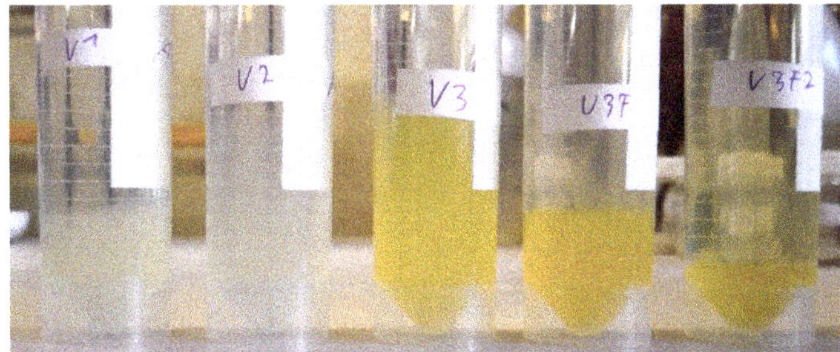

Figure 8. Precipitated solid residues from water solution after carbonation at 175 °C, 120 bar, s/L 1:10 (V1-600 rpm, 120 min; V2-600 rpm, 240 min; V3-1800 rpm, 120 min; V3F-1800 rpm, 40 min; V3F2-1800 rpm, 2 min).

As shown in Figure 9, an increase of stirring speed from 600 rpm to 1800 rpm in an autoclave leads to increased formation of products changing color from a white to yellow one. At the same way an increased reaction time leads to an increased production of solid residue. We suppose that an increased stirring speed has a positive influence for the separation of a formed silica rich layer at a non-reacted magnesium silicate. At the other side the pressure in an autoclave was increased from 120 to 170 bar with an increasing stirring speed from 600 to 1800 rpm, what is an additional support for the silica separation and precipitation from solution.

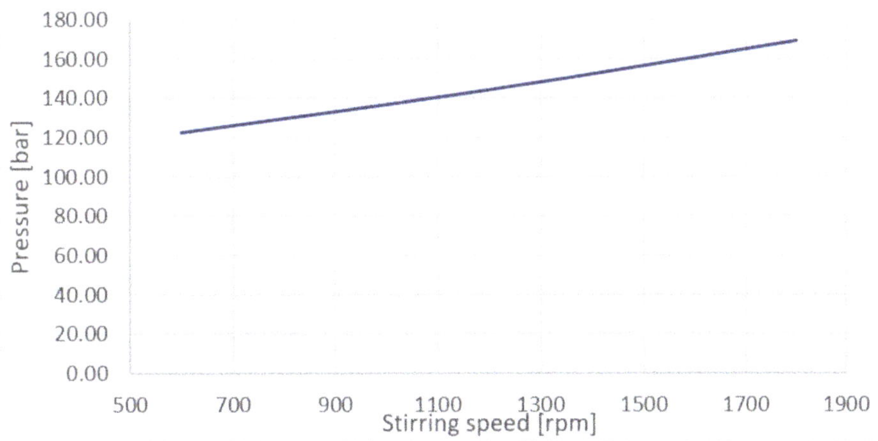

Figure 9. Relationship between operational pressure in an autoclave and stirring speed.

As shown in Table 3 and with Equation (7), the consumption of hydrogen ions, pH-Value was increased from the starting value 7.2 to maximal value of 8.57, forming the precipitate of SiO_2.

Table 3. The change of pH-Value of solution and pressure during a heating and carbonation.

Experiments	pH$_{Solution}$	pH$_{End}$	P$_{start\ for\ heating}$ to 175 °C [bar]	P$_{Start\ for\ reaction}$ [bar]	Stirring Speed [rpm]	P$_{End}$ [bar]
V1 (120 min)	7.20	8.32	61.70	121.60	600	137.20
V2 (120 min)	7.20	8.27	59.80	131.50	600	154.90
V3 (120 min))	7.20	8.57	59.60	162.80	1800	170.10
V3F (40 min)	7.20	8.38	62.60	152.70	1800	125.60
V3F2 (2 min)	7.20	8.35	62.40	159.70	1800	156.60

$$Mg_{1.8}Fe_{0.2}SiO_4 + H^+ \rightarrow 1.8\ Mg^{2+} + 0.2\ Fe^{2+} + SiO_2 + H_2O \tag{7}$$

The maximal pressure amounted 162.2–170.1 bar at 1800 rpm, what leads to an increased separation of nanosilica. Oelkers et al. [22] studied the products after carbonation of olivine via the reaction (8). They found that quartz is not stable at partial CO_2 pressures between 21.4 and 223.8 bar at temperatures from 120 to 200 °C. The spherical particles are containing high amounts of silicon and oxide atoms, which is confirmed by the EDS analysis (Figure 9).

$$0.5\ Mg_2SiO_4 + CO_2 \rightarrow MgCO_3 + 0.5\ SiO_2 \tag{8}$$

J. Götze and M. Göbbels [23] have mentioned that an increase of pH-value above six increases the solubility of silica. EDS Mapping was used for analysis of this dried final product drawn from water solution, as shown in Figure 10.

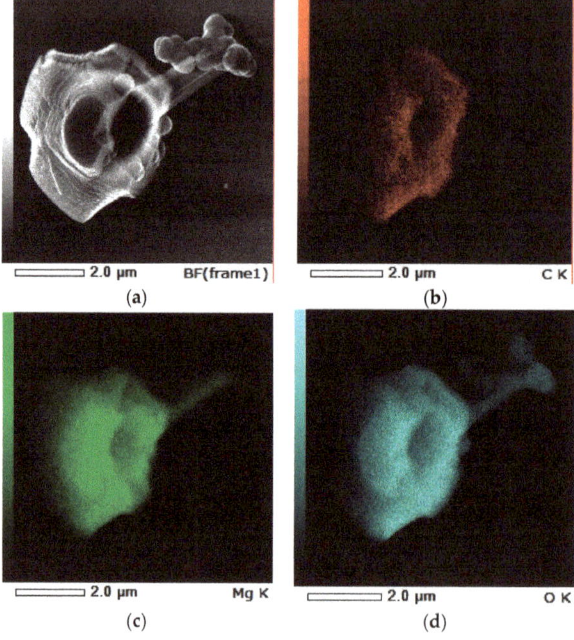

Figure 10. *Cont.*

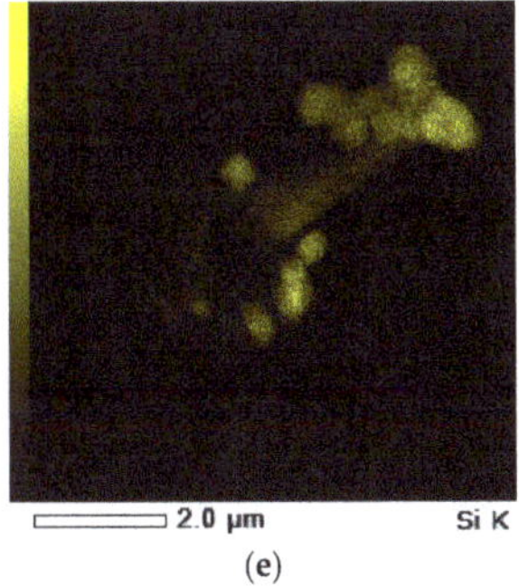

Figure 10. X-ray spectroscopy (EDS) mapping of final product obtained in the experiment V3F, (**a**) BF-SEM Image, (**b**) C K-carbon mapping, (**c**) Mg K-magnesium mapping, (**d**) O K-oxygen mapping, and (**e**) Si K-silicon mapping.

As shown in Figure 11, the TEM analysis of silicon oxide confirms that the formed silica particles have a spherical shape, with diameters of approximately 400–500 nm and the particles are amorphous.

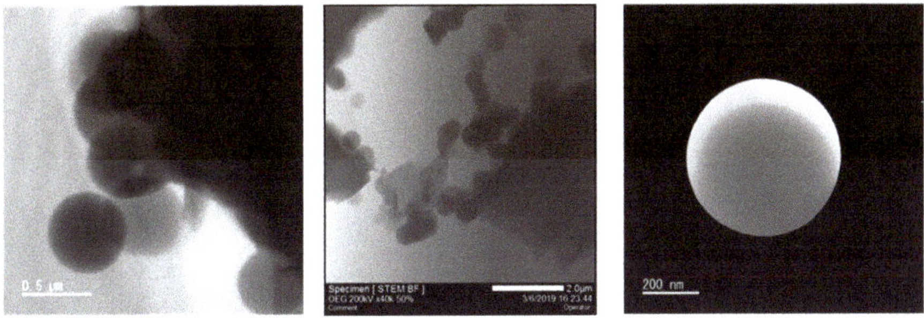

Figure 11. The TEM and STEM analysis of the obtained products.

Since an extensive share of the investigated material in Figure 10 shows low carbon content, there is a high possibility that this area is Mg_2SiO_4, what is previously shown in Figure 6. A new model for formation of nanosilica from magnesium silicate (forsterite) in used olivine was shown in Figure 12.

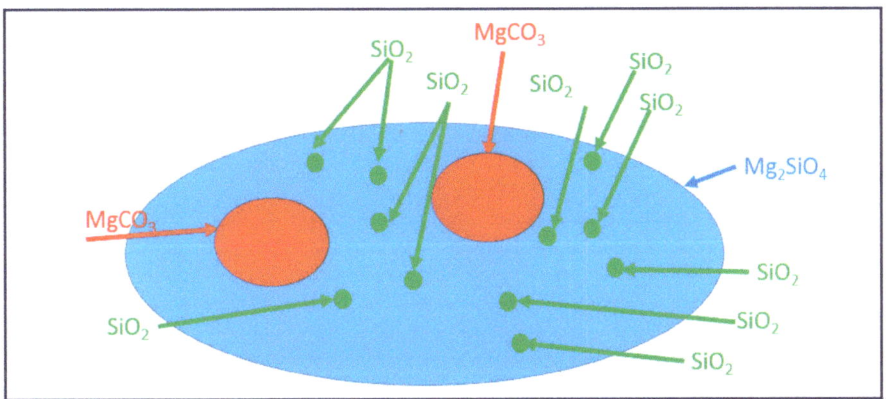

Figure 12. The proposed schematic model for the formation of SiO_2 from Mg_2SiO_4 in an autoclave.

We assumed that high stirring speed and high pressure in autoclave lead to the formation of ideally spherical silica particles together with larger fractions of magnesium carbonate. Weng et al. [24] found during a study of the kinetics and mechanism of mineral carbonation of olivine for CO_2 sequestration that the addition of sodium bicarbonate can dramatically increase the ionic strength and aid the dissolution of Si to temporarily aqueous H_4SiO_4 followed by decomposition to amorphous silica and consequently the removal of Si-rich layer. The aqueous silicon was not stable and can be decomposed into amorphous silica, which was extensively observed in the aqueous solution after carbonation and settled down for more than one month at room temperature.

3.3. Conclusions

Synthesis of nanosilica was studied via carbonation of olivine using size fraction between 20 and 63 μm with solid/liquid ratios of 1:10 at 175 °C and partial pressure of CO_2 more than 100 bar in an autoclave in the presence of additives such as sodium bicarbonate, oxalic and ascorbic acid. Under the above-mentioned conditions the ideally spherical particles of silica below 500 nm with amorphous grains were produced during carbonation. In comparison to ultrasonic spray pyrolysis, sol-gel and carbonation method via bubbling CO_2 gas into calcium ions dispersing in methanol solution under an atmospheric pressure, this synthesis was performed in a water solution in a closed reactor (autoclave) under higher pressure conditions above 100 bar avoiding the formation of silica gel, what blocks the metal extraction. An increase of stirring speed from 600 rpm to 1800 rpm raises the pressure from 120 to 170 bar and leads to an increase of silica production because of a removal of passivated silica formed layer at forsterite particles. The precipitated silicate particles were separated at pH-values between 8.32 and 8.57. In order to validate the first results in a 0.25 L-and 1.5 L autoclave, new scale up experiments will be performed in 10 L and 1000 L-autoclaves. Due to a good filterability of the carbonated product, separation of nanosilica from magnesite and magnesium carbonate shall be considered in future work in detail.

Author Contributions: S.S. conceptualized, managed the research, and co-wrote the paper. D.K. performed the preparation of the olivine materials (grinding, sieving) and co-wrote the paper. H.W. co-wrote the paper. C.D. participated in our experimental part, analyzed the data and co-wrote the paper. S.E. supervised the XRF- and XRD-analyses and co-wrote the paper with R.T. and B.F. supervised the personnel, provided funding and co-wrote the paper. I.K. performed the XRD, SEM, REM and STEM analysis of nanosilica. P.K. conceptualized the research and helped in the discussion of the morphological characteristics of obtained magnesium carbonate and nanosilica.

Funding: This research was funded by BMBF (Federal Ministry of Education and Research) in Berlin, grant number 033RCO14B (CO2MIN Project in period from 01.06.2017 to 31.05.2020).

Acknowledgments: For a continuous support and preparation of Figure 1 previously published in Metals in 2018, we would like to thank Andreas Bremen, AVT, RWTH Aachen University.

Conflicts of Interest: The authors declare no conflict of interest.

References

1. Cheng, C.J.; Urbani, M. Purification of laterite leach solutions by direct solvent extraction. In *Aqueous and Electrochemical Processing*; Kongoli, F., Itagaki, K., Yamauchi, C., Sohn, H.Y., Eds.; The Minerals, Metals, & Materials Society: San Diego, CA, USA, 2003; Volume 3, pp. 251–265.
2. Bergman, R.A. Nickel production from low-iron laterite ores: Process description. *CIM Bull.* **2003**, *96*, 127–138.
3. Zulfiqar, U.; Subhani, T.; Husain, S.W. Synthesis and characterization of silica nanoparticles from clay. *J. Asian Ceram. Soc.* **2016**, *4*, 91–96. [CrossRef]
4. Moskalyk, R.R.; Alfantazi, A.M. Nickel laterite processing and electrowinning practice. *Miner. Eng.* **2002**, *15*, 593–605. [CrossRef]
5. Alkan, G.; Yagmurlu, B.; Ditrich, C.; Gronen, L.; Stopic, S.; Ma, Y.; Friedrich, B. Selective silica gel scandium extraction from, iron–depleted red mud slags by dry digestion. *Hydrometallurgy* **2019**, *185*, 266–272. [CrossRef]
6. Ma, Y.; Stopic, S.; Friedrich, B. Hydrometallurgical treatment of a eudialyte concentrate for preparation of rare earth carbonate. *Johns. Matthey Technol. Rev.* **2019**, *63*, 2–13. [CrossRef]

7.   Ma, Y.; Stopic, S.; Gronen, L.; Obradovic, S.; Milivojevic, M.; Friedrich, B. Neural network modeling for the extraction of rare earth elements from eudialyte concentrate by dry digestion and leaching. *Metals* **2018**, *8*, 267. [CrossRef]

8.   Stopic, S.; Schroeder, M.; Weirich, T.; Friedrich, B. Synthesis of TiO_2 Core/RuO_2 shell particles using multistep ultrasonic spray pyrolysis. *Mater. Res. Bull.* **2013**, *48*, 3633–3635. [CrossRef]

9.   Košević, M.; Stopic, S.; Cvetković, V.; Schroeder, M.; Stevanović, J.; Panic, V.; Friedrich, B. Mixed RuO_2/TiO_2 uniform microspheres synthesized by low-temperature ultrasonic spray pyrolysis and their advanced electrochemical performances. *Appl. Surf. Sci.* **2019**, *464*, 1–9. [CrossRef]

10.   Ab Rahman, I.; Padavettan, V. Synthesis of silica nanoparticles by sol-gel: Size-dependent properties, surface modification, and applications in silica-polymer nanocomposites—A review. *J. Nanomater.* **2012**. [CrossRef]

11.   Kim, K.; Choi, K.; Yang, J. Formation of spherical hollow silica particles from sodium silicate solution by ultrasonic spray pyrolysis method. *Coll. Surf. Physicochem. Eng. Asp.* **2005**, *254*, 193–198. [CrossRef]

12.   Ratanathavorn, W.; Bouhod, N.; Modsuwan, M. Synthesis of silica nanoparticles by ultrasonic spray pyrolysis technique for cream perfume formulation development. *J. Food Health Bioenviron. Sci.* **2018**, *11*, 1–5.

13.   Nakashima, Y.; Takai, C.; Razavi-Khosroshahi, H.; Suthabanditpong, W.; Fuji, M. Synthesis of ultra-small hollow silica nanoparticles using the prepared amorphous calcium carbonate in one-pot process. *Adv. Powder Technol.* **2018**, *29*, 904–908. [CrossRef]

14.   Stopic, S.; Dertmann, C.; Modolo, G.; Kegler, P.; Neumeier, S.; Kremer, D.; Wotruba, H.; Etzold, S.; Telle, R.; Rosani, D.; et al. Synthesis of magnesium carbonate via carbonation under high pressure in an autoclave. *Metals* **2018**, *8*, 993. [CrossRef]

15.   Rahmani, O.; Junin, R.; Tyrer, M.; Mohsin, R. Mineral carbonation of red gypsum for CO_2 sequestration. *Energy Fuels* **2014**, *28*, 5953–5958. [CrossRef]

16.   Béarat, H.; McKelvy, M.; Chizmeshya, A.; Gormley, D.; Nunez, R.; Carpenter, R.; Squires, K.; Wolf, G. Carbon sequestration via aqueous olivine mineral carbonation: Role of passivating layer formation. *Environ. Sci. Technol.* **2006**, *40*, 4802–4808. [CrossRef]

17.   Eikeland, E.; Bank, A.; Tyrsted, C.; Jensen, A.; Iversen, B. Optimized carbonation of magnesium silicate mineral for CO_2 storage. *Appl. Mater. Interf.* **2015**, *7*, 5258–5264. [CrossRef] [PubMed]

18.   Daval, D.; Sissmann, O.; Menguy, N.; Salsi, G.; Gyot, F.; Martinez, I.; Corvisier, J.; Machouk, I.; Knauss, K.; Hellmann, R. Influence of anorphous silica lyer formation on the dissolution rate of olivine at 90°C and elevated pCO_2. *Chem. Geol.* **2011**, *284*, 193–209. [CrossRef]

19.   Oelkers, E. An experimental study of forsterite dissolution rates as afunction of temperature and aqueous Mg and Si concentrations. *Chem. Geol.* **2001**, *175*, 485–494. [CrossRef]

20.   Hänchen, M.; Prigiobbe, V.; Storti, T.; Seward, M.; Mazzotti, M. Dissolution kinetics of forsteritic olivine at 90–150 °C including effect of the presence of CO_2. *Geochim. Cosmochim. Acta* **2006**, *70*, 4403–4416. [CrossRef]

21.   Turri, L.; Muhr, H.; Rijnsburger, K.; Knops, P.; Lapicque, F. CO_2 sequestration by high pressure reaction with olivine in a rocking batch autoclave. *Chem. Eng. Sci.* **2017**, *171*, 27–31. [CrossRef]

22.   Oelkers, E.H.; Declercq, J.; Saldi, G.D.; Gislason, S.R.; Schott, J. Olivine dissolution rates: A critical review. *Chem. Geol.* **2018**, *500*, 1–19. [CrossRef]

23.   Götze, J.; Göbbels, M. *Introduction in Applied Mineralogy*; Springer: Berlin/Heidelberg, Germany, 2017; p. 270.

24.   Wang, F.; Dreisinger, D.; Jarvis, M.; Hitchins, T. Kinetics and mechanism of mineral carbonation of olivine for CO_2. *Miner. Eng.* **2019**, *131*, 185–197. [CrossRef]

 *metals*

Article

Synthesis of Tribological WS$_2$ Powder from WO$_3$ Prepared by Ultrasonic Spray Pyrolysis (USP)

Nataša Gajić [1,*], Željko Kamberović [2], Zoran Anđić [3], Jarmila Trpčevská [4], Beatrice Plešingerova [4] and Marija Korać [2]

[1] Innovation Center of the Faculty of Technology and Metallurgy in Belgrade Ltd., University of Belgrade, Karnegijeva 4, 11120 Belgrade, Serbia

[2] Faculty of Technology and Metallurgy, University of Belgrade, Karnegijeva 4, 11120 Belgrade, Serbia; kamber@tmf.bg.ac.rs (Ž.K.); marijakorac@tmf.bg.ac.rs (M.K.)

[3] Innovation Center of the Faculty of Chemistry in Belgrade Ltd., University of Belgrade, Studentski trg 12-16, 11000 Belgrade, Serbia; zoranandjic@yahoo.com

[4] Faculty of Materials, Metallurgy and Recycling, Technical University of Košice, Letná 9, 042 00 Košice, Slovensko; jarmila.trpcevska@tuke.sk (J.T.); beatrice.plesingerova@tuke.sk (B.P.)

* Correspondence: ngajic@tmf.bg.ac.rs; Tel.: +381-605-236-021

Received: 30 December 2018; Accepted: 22 February 2019; Published: 28 February 2019

Abstract: This paper describes the synthesis of tungsten disulfide (WS$_2$) powder by the sulfurization of tungsten trioxide (WO$_3$) particles in the presence of additive potassium carbonate (K$_2$CO$_3$) in nitrogen (N$_2$) atmosphere, first at lower temperature (200 °C) and followed by reduction at higher temperature (900 °C). In addition, the ultrasonic spray pyrolysis of ammonium meta-tungstate hydrate (AMT) was used for the production of WO$_3$ particles at 650 °C in air. The HSC Chemistry® software package 9.0 was used for the analysis of chemistry and thermodynamic parameters of the processes for WS$_2$ powder synthesis. The crystalline structure and phase composition of all synthesized powders were analyzed by X-ray diffraction (XRD) measurements. The morphology and chemical composition of these samples were examined by scanning electron microscopy (SEM) combined with energy dispersive X-ray analysis (EDX).

Keywords: tribology materials; tungsten disulfide; ultrasonic spray pyrolysis; tungsten trioxide

---

1. Introduction

According to current studies, nearly one-quarter of the world's total energy consumption originates from tribological contacts [1]. By improving the performance of current tribological materials, energy losses due to friction and wear could potentially be reduced by 40% in the long term (15 years) and by 18% in the short term (8 years) [1]. Hence, in the face of increasing global requirements for saving energy, there is no doubt that the search for efficient tribological materials with improved performances will continue in the coming years because the application conditions of future tribomechanical systems will undoubtedly be much more demanding than the current ones [2]. Today's market of tribological materials has unique requirements for the release of harmful substances (Cd, Pb, Cr, etc.) and the prevention of their migration into the environment during their use [3]. It is generally considered that these elements and their compounds, which can be introduced into food or water, are difficult to eliminate from the body and according to current studies can have carcinogenic effects [4]. For all of the above reasons, despite their excellent performances, tribological materials must also be acceptable from economic and ecological points of view. Among the various tribological materials, WS$_2$ has attracted a large amount of attention from researchers for its specific properties and extensive promising applications [5]. Even in small quantities, WS$_2$ contributes to the high performance of tribological materials and their specific properties (chemical stability in a wide temperature range, the possibility of revitalizing the surface that they protect, they are corrosion-resistant, inert, non-toxic, non-magnetic, and have lamellar structure, low shear strength, high oxidation and thermal degradation resistance, and so forth), which are of particular importance for the functioning of modern tribomechanical systems. WS_2 is applicable in various types of industries, such as automotive, aerospace, military, medical, and so forth.

WS_2 crystals have a hexagonal structure composed of a layer of tungsten atoms packed in between two layers of sulfur atoms, as shown in Figure 1a. The bonding between W-S layers is very strong and covalent, but layers of S atoms are loosely bound through weak van der Waals forces. This structure is responsible for the interlayer mechanical weakness with low shear strength, which results in a macroscopic lubricating effect. Figure 1b shows the lubrication mechanism of WS_2 single sheets.

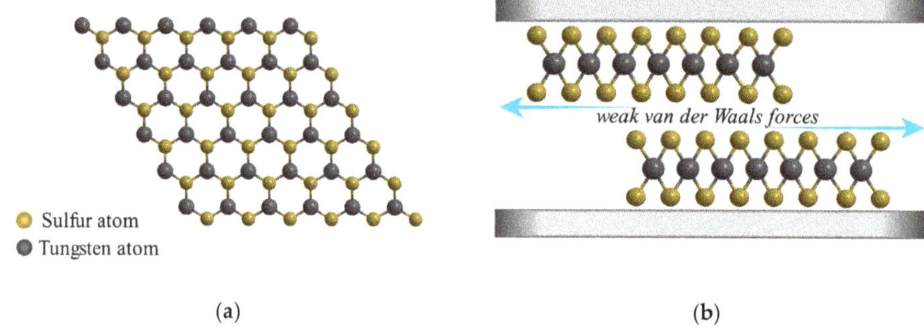

Figure 1. (**a**) Hexagonal structure of WS_2 (top view); (**b**) The relative sliding between two single layers of WS_2 (slide view).

So far, various methods of producing WS_2 of different size and morphology using WO_3 as precursor have been established, such as: gas–solid phase reaction [6–8], chemical vapor deposition (CVD) [9–11], hydrothermal method [12–14], mechanochemical activation method [15], and solid–solid phase reaction [16–18].

Gas–solid reactions present a very simple approach to generating WS_2. In most cases, WO_3 is reacted with a sulfur-containing compound for extended periods of time at high temperatures. For example, WS_2 has been synthesized in a tubular furnace through gas phase reactions between WO_3 particles and H_2/H_2S gases at a temperature of 840 °C for 30 min in Ar gas flow [6], by sulfidation of hexagonal WO_3 with H_2S/H_2 (15% H_2S) at different temperatures: 400, 500, and 800 °C for 4 h [7], and by sulfurization of WO_3 nanostructured thin film in a mixture of H_2S and Ar gas (10:90) at different partial pressure values [8]. However, this method involves exposure to extremely toxic and harmful H_2S at elevated temperatures.

WS_2 flakes have been synthesized successfully on SiO_2/Si substrate by the sulfurization of WO_3 powder at high temperatures by the CVD method [9–11]. Furthermore, there was no poisonous H_2S gas released during these experiments and WS_2 of a high purity was obtained. Although this method has many advantages, it is very demanding to coordinate the complicated relations among many process parameters.

Using the hydrothermal method [12–14], WS_2 has been synthesized by autoclaving a mixture of WO_3 and sulfur precursor, followed by washing and drying of the resulting product. Although various inexpensive precursors can be used for this method, the productivity of this process is low and additional thermal treatment is necessary because the obtained WS_2 has an amorphous structure.

Wu Z. et al. [15] have synthesized WS_2 nanosheets by novel mechanochemical activation methods in which a ball-milled mixture of WO_3 and S powder was annealed at 600 °C for 2 h in an atmosphere of Ar gas. This method seems to be environmentally advantageous and may be an alternative to the traditional route of synthesis, but it is a very complex process and a robust method for the production of particles with small dimension. Another difficulty arises from the fact that there is still a lack of clear interpretation of the exact reaction and activation mechanism.

Regarding solid–solid phase reaction, the synthesis of WS_2 powder was carried out by sulfurization of the WO_3 powder with thiourea in a N_2 atmosphere at 850 °C for 1 h in a horizontal tube furnace [16–18].

In this investigation the sulfurization of the WO_3 particles with S powder in the presence of additive K_2CO_3 in a nitrogen atmosphere was studied. In order to obtain an adequate precursor for synthesis, this research involved the production of WO_3 particles using an ultrasonic spray pyrolysis method. It is a simple and low-cost method, which in continuous operation can generate spherical, non-agglomerated submicron particles by using commercially available (inexpensive) precursors [19]. The as-prepared WO_3 particles were used for WS_2 synthesis without any post processing because they were free of impurities.

2. Experimental

2.1. Materials and Methods

The raw materials used for the experimental test performed in this work were: WO_3 obtained by ultrasonic spray pyrolysis (USP), commercial WO_3 (Chemapol, Prague, Czech Republic), sulfur (Solvay & CPC Barium Strontium GmbH & Co, Hannover, Germany, powder with characteristic size <45 μm, purity 99.95%), and K_2CO_3 (Zorka, Šabac, Serbia). Commercially available WS_2 powder (SpeedUP INTERNATIONAL, Belgrade, Serbia, with a minimum 99.4% WS_2 and WO_3) was used for comparative analysis. The production of precursor, WO_3 powder, as well as the synthesis of WS_2 powders were carried out in a rotary tilting tube furnace (ST-1200RGV).

The HSC Chemistry® software package 9.0 was used for the analysis of the chemistry and thermodynamic parameters of the processes for the synthesis of WS_2 powder [20]. The determination of appropriate conditions for WS_2 synthesis was crucial for this analysis. Therefore, the Gibbs free energy of theoretically suspected chemical reactions during the WS_2 synthesis process, the phase stability diagram for the W–O–S system, and the equilibrium composition of the species in the WO_3–S–K_2CO_3 system were calculated using HSC Chemistry software.

All obtained powders, as well as the commercial WS_2 powder, were subjected to analysis on a scanning electron microscope (SEM) equipped with energy dispersive X-ray analysis (EDX), a MIRA FE-SEM, from TESCAN Inc. In SEM images of the WO_3 powder, well-dispersed powder without any agglomerates was clearly seen and the determination of particle size distribution for the WO_3 powder was done using the Image Pro Plus Software. However, the particle size of the WS_2 powder was not identifiable using image analysis due to the poor dispersion of samples. Size measurements of the obtained WS_2 powders were thus performed using a laser particle size analyzer (Malvern Instruments, Malvern, UK). In addition, all samples were subjected to X-ray diffraction analyses using a Philips PW 1710, X-Pert Pro diffractometer (Co Kα radiation, generated at 40 kV and 30 mA). Measurements were carried out at an angle interval $10° < 2\theta < 119°$ with step $0.017°$. XRD results were analyzed by employing the Rietveld method with the help of PowderCell Software and the RIFRANE® programme.

2.2. Synthesis of WO$_3$ Powder

The apparatus for WO_3 synthesis consisted of an aerosol generator ("Profi Sonic", Prizma, Kragujevac), a horizontal reactor with a quartz glass tube (0.5 m diameter and 1.2 m length), a vacuum pump (VP125), and powder collectors (Figure 2).

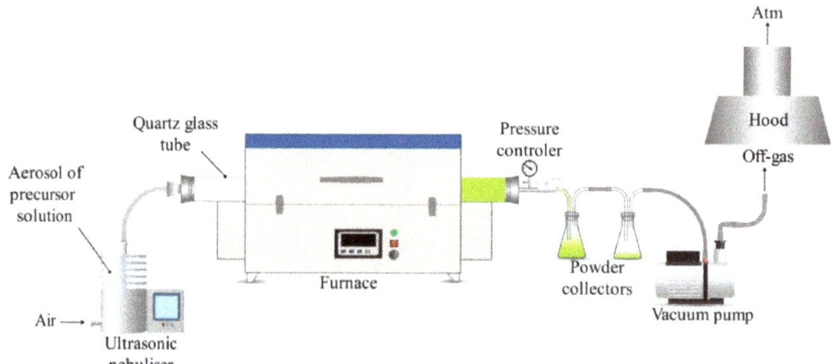

Figure 2. Schematic illustration of experimental setup for the ultrasonic spray pyrolysis (USP) synthesis of WO_3.

Firstly, the ammonium meta-tungstate hydrate (AMT—$H_{26}N_6O_{41}W_{12}$·aq) diluted in distilled water was put into the particle nebulizer to generate an aerosol. The concentration of AMT solution was 10 mmol/L. For this ultrasonic nebulizer system, the resonance frequency was set to 1.7 MHz. A vacuum pump and air with a flow rate of 5 L/min was used to introduce the generated aerosol droplets into the tubular reactor. Prior to the introduction, the temperature of the reactor was raised to 650 °C. The pressure of the system was adjusted using the reactor pressure controller. The calculated retention time of droplets in the reaction zone was estimated to be about one second. After thermal decomposition of the transported aerosol in the furnace, the formed WO_3 was partially collected in the bottles with water and alcohol. During the spray pyrolysis process, the evaporation of water from aerosol droplets increases the concentration of AMT in the droplets. Finally, AMT is thermally decomposed according to Equation (1):

$$2\,H_{26}N_6O_{41}W_{12} \rightarrow 24\,WO_3 + 12\,NH_3\uparrow + 8\,H_2O\uparrow + O_2\uparrow. \tag{1}$$

As shown in Figure 3, the mechanism of WO_3 particle formation proposed by Arutanti et al. [21] is comprised of different steps starting from an initial solution of AMT.

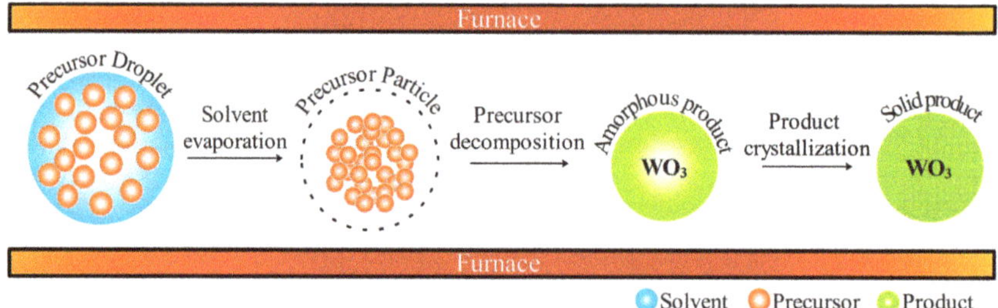

Figure 3. The mechanism of the formation of WO_3 particles in the spray pyrolysis method.

2.3. Synthesis of WS_2 Powder

A schematic illustration of the WS_2 synthesis process is shown in Figure 4.

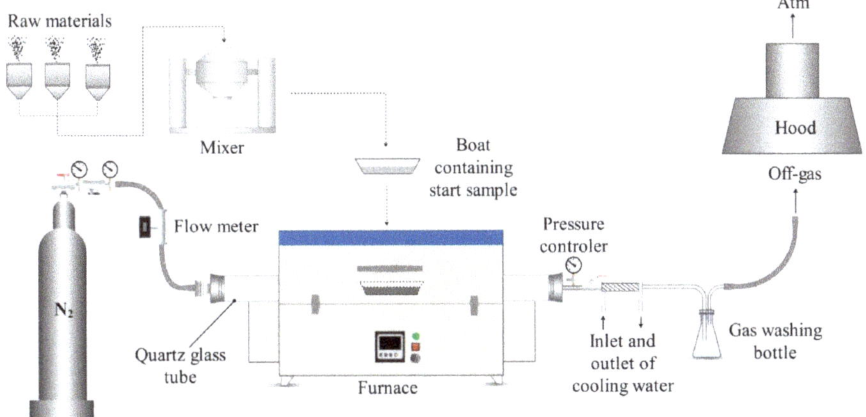

Figure 4. Schematic illustration of the experimental setup for the synthesis of WS_2 powder.

K_2CO_3 powder (5 wt.%) was added to the WO_3 and S powder mixture with a weight ratio of 60:40. The powders were mixed and ground for 15 min in the mixer. Then, the as-prepared mixture was transferred to a covered ceramic boat. The boat was placed in the center of the quartz tube and into the furnace. High-purity nitrogen gas was introduced through one side of the furnace, whilst the other side of the quartz tube was connected to a cooling system and outlet gas washing system. The total flow rate of the N_2 gas was fixed at 200 $cm^3 \cdot min^{-1}$ for all experimental conditions. Prior to heating the furnace, nitrogen gas was flushed constantly for about 30 minutes to remove residual air in the furnace. First, the furnace with the sample was heated at a lower temperature (200 °C) at a rate of 10 °C/min and maintained under these conditions for 2 h. Then, it was further heated at a rate of 5 °C/min followed by reduction at a higher temperature (900 °C). After 2 h, the furnace was turned off and allowed to cool down to room temperature. Nitrogen gas flow was stopped and the WS_2 powder was collected from the boat.

3. Result and Discussion

3.1. Thermodynamic Analysis

Results of the thermodynamic analysis of the process of WS_2 powder synthesis are discussed below. Using the assumption of raw materials for the composition of the synthesis process, the following chemical reactions (Equations (2)–(4)) were considered:

$$3WO_3 + K_2CO_3 + 7S = K_2O \cdot WO_3 + 2WS_2 + CO_2(g) + 3SO_2(g), \tag{2}$$

$$3WO_3 + K_2CO_3 + 5S = K_2O \cdot WO_3 + CO_2(g) + 2SO_2(g) + WO_2 + WS_3, \tag{3}$$

$$3WO_3 + K_2CO_3 + 4S = K_2O \cdot WO_3 + CO_2(g) + 2SO_2(g) + WO_2 + WS_2. \tag{4}$$

Figure 5 shows the calculated results of the Gibbs energy of reactions versus temperature from the reaction of Equations (2)–(4). The temperature range that was considered was up to 1000 °C.

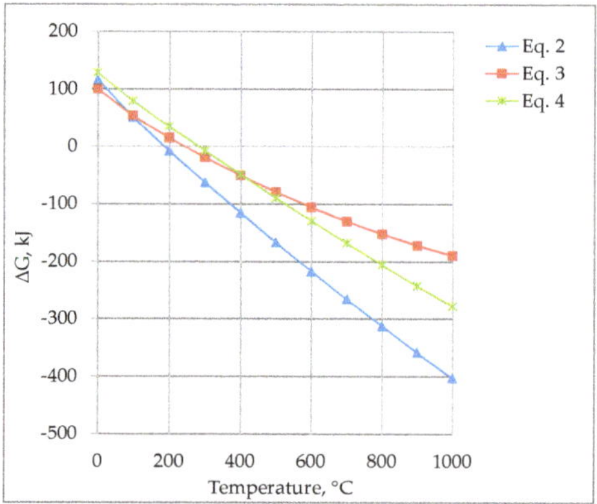

Figure 5. The change in Gibbs free energy (ΔG) versus temperature for the different possible reactions that take place during the WS_2 synthesis.

As shown in Figure 5, it can be clearly seen that the ΔG of all reactions has a negative value at temperatures higher than 300 °C, which means that all of the reactions are theoretically possible from the thermodynamic point of view. However, among them, the changes of the Gibbs energy of reaction presented with Equation (2) has a more negative value relative to Equations (3) and (4) (i.e., Equation (2) is dominant).

The equilibrium composition of the WO_3–K_2CO_3–S system at different temperatures was calculated using the HSC computer program based on the Gibbs energy minimization method, and the results are shown in Figure 6.

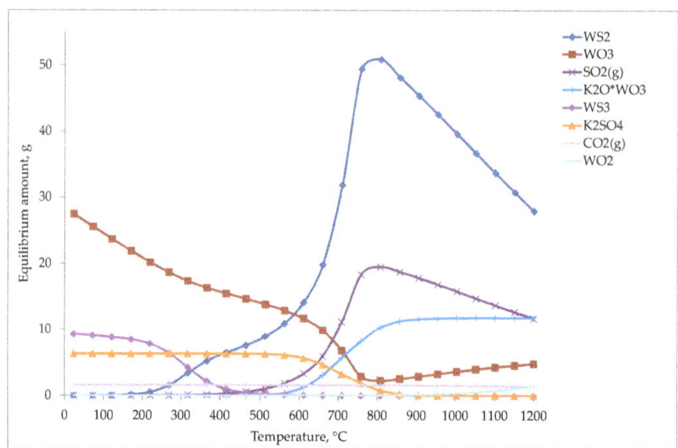

Figure 6. Equilibrium amount of WO3–S–K_2CO_3 reactant system in N_2 gas atmosphere.

The highest stability of WS_2 was achieved at approximately 800 °C when the stability of the reactants were decreasing. $K_2O \cdot WO_3$ and WO_2 phases also became stable at the same temperature. The main gaseous products of this reaction were gaseous SO_2 and CO_2.

In the analyzed system, the main products of the reactions at a temperature of 900 °C were WS_2, $K_2O \cdot WO_3$, WO_2, CO_2, and SO_2. These results were only qualitative.

By using thermodynamic software, a phase stability diagram for the W–O–S system for constant partial pressure of oxygen was constructed (Figure 7).

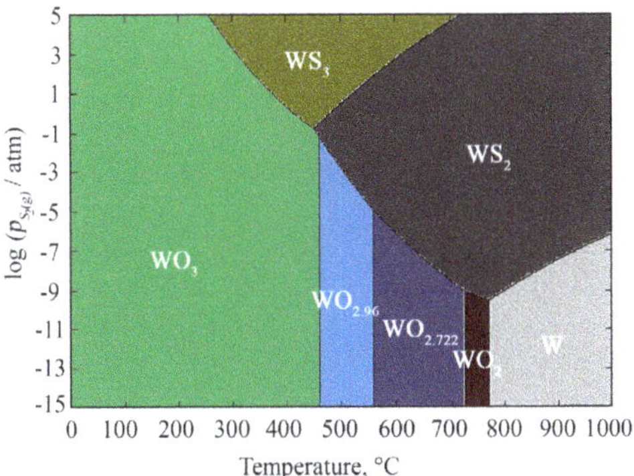

Figure 7. Temperature–partial pressures (Tpp) phase stability diagram for the W–O–S system at constant oxygen partial pressure.

Considering the logarithmic sulfur vapor pressure of 1 atm [22] and experimental temperature of 900 °C, it was found that the predominant stability area was of WS$_2$. At these temperatures, the vapor pressure of sulfur is such that it led to the saturation of the atmosphere with the sulfur vapor in the furnace. This meant that the amount of sulfur gas phase necessary to react with the WO$_3$ was sufficient to establish contact between the mentioned phases. Further, the diffusion of sulfur into WO$_3$ was enabled to form WS$_2$ powder.

The synthesis of WS$_2$ was performed in two stages: (i) initiation of synthesis at a low temperature (200 °C) followed by (ii) high-temperature (900 °C) reduction. The temperature of the first stage was selected to prolong the contact time of sulfur and WO$_3$ in the starting powder mixture, before sulfur self-ignition (232 °C) which promotes the transformation of sulfur to a gas phase and the evaporation of a large amount of S and/or SO$_2$. This loss of sulfur obstructs its diffusion into WO$_3$ and consequently obstructs the synthesis of WS$_2$. Thermodynamic analysis of the second stage indicated that the optimal temperature of the synthesis was above 800 °C. However, at lower temperatures oxides were formed and the chemical composition of the mixture was changed, and a lower level of crystallization occurred.

For that reason, the second stage of synthesis was carried out at 900 °C to prevent oxide formation and to increase the degree of crystallinity in the final WS$_2$ product.

3.2. Characterization of Synthesized WO$_3$ Powder

Figure 8a,b shows the SEM images of the synthesized WO$_3$ powder by USP method under different magnifications. The result of EDX analyses from the presented scanning surface is given in Figure 8c, which reveals that the sample consisted of the elements tungsten and oxygen, and no other elements were observed.

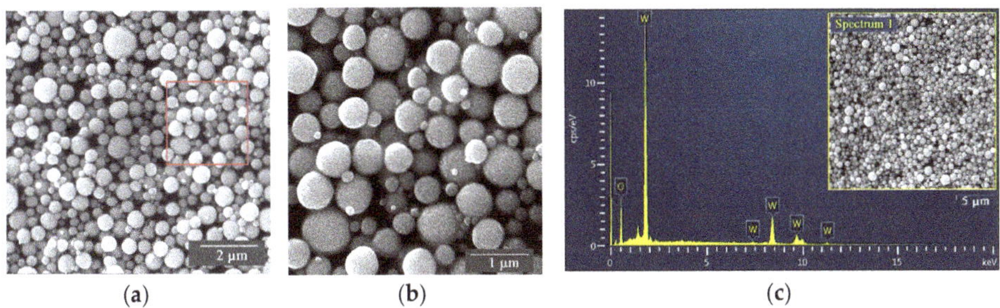

Figure 8. (**a**,**b**) SEM images and (**c**) EDX spectrum of WO$_3$ powder synthesized by USP.

The XRD pattern of the prepared WO_3 particles is shown in Figure 9. It is evident from the pattern that no diffraction peaks from other elements of compounds were found in the samples. Therefore, it is obvious that the as-prepared samples were composed of WO_3. The XRD pattern suggested that the prepared particles had two types of crystal structures: hexagonal and monoclinic.

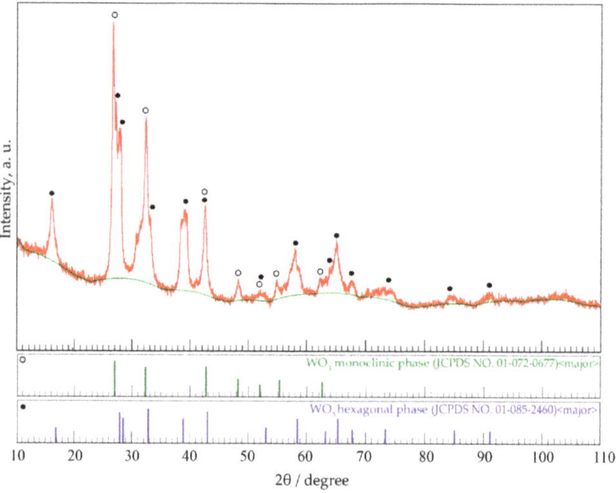

Figure 9. XRD pattern of WO_3 powder prepared by USP method.

The particle size distribution determination for the WO_3 powder prepared by USP is presented in Figure 10.

Size class [nm]			p3 [%]	Q3 [%]	1-Q3 [%]	q3 [%/nm]
	<	100	8.0	8.0	92.0	0.08
100	-	150	12.0	20.0	80.0	0.24
150	-	200	4.0	24.0	76.0	0.08
200	-	250	12.0	36.0	64.0	0.24
250	-	300	34.7	70.7	29.3	0.69
300	-	350	16.0	86.7	13.3	0.32
350	-	400	8.0	94.7	5.3	0.16
400	-	450	1.3	96.0	4.0	0.03
450	-	500	2.7	98.7	1.3	0.05
>	500		1.3	100.0	0.0	0.01

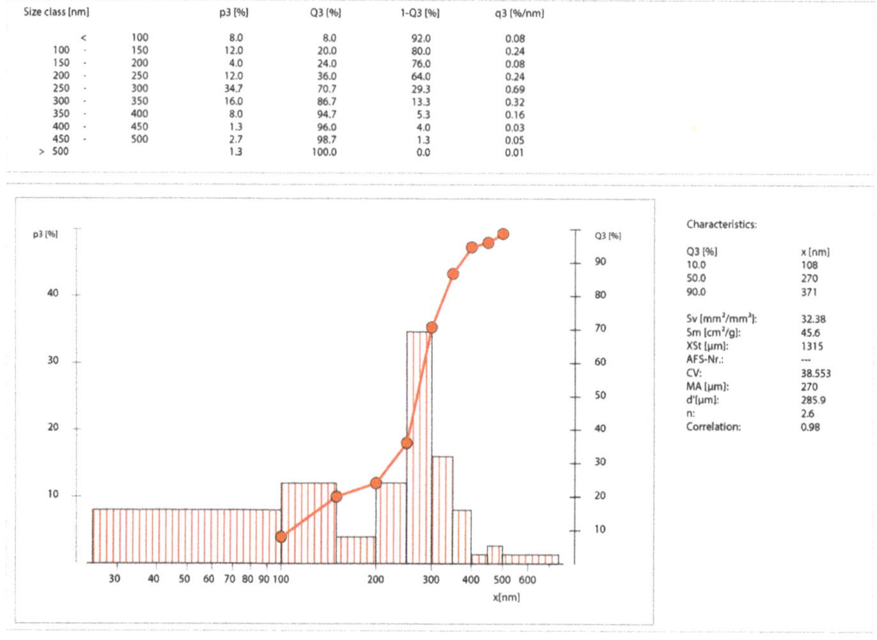

Figure 10. Particle size distribution for WO_3 powder prepared by USP method.

3.3. Characterization of WS$_2$ Powder

Figure 11a demonstrates an SEM image of powder obtained using the WO$_3$ precursor powder prepared by the USP method. The sample was composed of a large number of ultrathin WS$_2$ flakes. Furthermore, WO$_2$, WO$_3$, and K$_2$O·WO$_3$ phases with larger dimensions of particles were noticeable. The structure of the ultrathin flakes from the marked locations is presented more clearly in Figure 11b. The results demonstrate the presence of flower-shaped WS$_2$ particles composed of nanoflakes of 200–500 nm in length and 50 nm in mean thickness. A further EDX analysis of the sample presented in Figure 11c reveals that the product was composed of W, S, O, and K, which suggests that the powder was composed of mainly WS$_2$ (Spectrum 1) and WO$_2$, WO$_3$, and K$_2$O·WO$_3$ (Spectrum 2) phases, in accordance with the results presented in Figure 6.

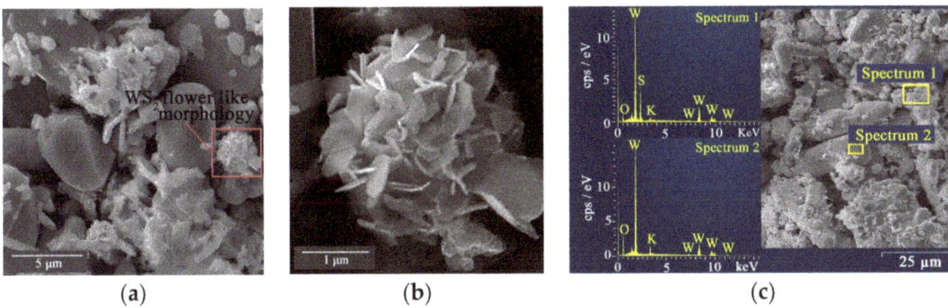

Figure 11. (**a**,**b**) SEM images and (**c**) EDX spectrum for WS$_2$ synthesized using WO$_3$ prepared by USP as precursor.

The results of the XRD analysis employing the Rietveld method are shown in Figure 12. These results revealed that WS$_2$ phase was predominant (present in two crystalized forms: hexagonal and rhombohedral) and was accompanied by the presence of WO$_2$ (9.80%), WO$_3$ (3.04%), and K$_2$O·WO$_3$ (1.56%) phases.

The obtained WS$_2$ particles were tested for their size and size distribution and the mean particle size was found to be around 950 nm (Figure 13).

WS$_2$ powder synthesized using WO$_3$ prepared by USP as precursor was selected as a lubrication additive in SF SAE 15W-40 motor oil. WS$_2$ powder was ultrasonically dispersed into the base oil for 10 min. In addition, a base oil without any additive was also prepared. Wear tests were conducted using a ball-on-disc configuration on a Bruker UMT-3 tribometer (Bruker, Billerica, MA, USA). This test method involves a ball-shaped upper specimen that slides against a rotating disk as a lower specimen under a prescribed set of conditions. Both the ball and the steel disc were cleaned with acetone and dried with a normal stream of air before the test. A normal load of 50 N and a linear sliding speed of 0.1 m/s were used for the experiments, for a sliding distance of 500 m. The results of base oil with additive showed that the average value of the friction coefficient was $\mu \sim 0.1$. It was concluded that the friction coefficient of the base oil ($\mu \sim 0.16$) was improved by adding WS$_2$. The mass ratio of WS$_2$/base oil in samples was 1.0 wt.%.

The generated WS$_2$ powder had plate-like particles, which were oriented differently when the starting material was commercial WO$_3$ powder (Figure 14a). It can also be clearly seen from Figure 14b that WS$_2$ particles were clustered together and exhibited evident agglomeration of up to 1 µm in size. The thickness of the WS$_2$ plate-like particles was approximately 100 nm and their lengths varied from 500 nm to 1 µm. Results of EDX analyses from marked locations showed the presence of W and S elements and a certain content of O and K elements which indicated the presence of WS$_2$ (Spectrum 2) and WO$_2$, WO$_3$, and K$_2$O·WO$_3$ (Spectrum 1) phases (Figure 14c).

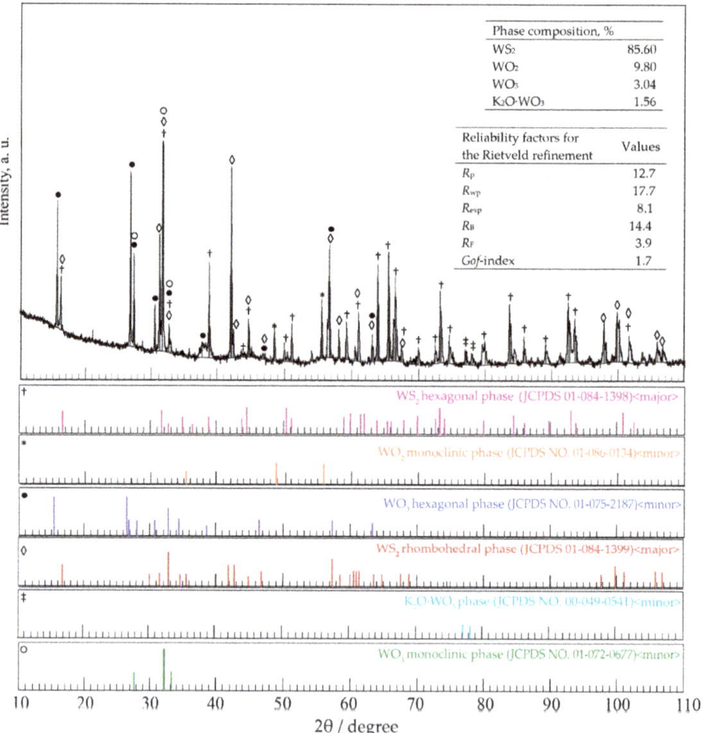

Figure 12. XRD pattern of WS$_2$ synthesized using WO$_3$ prepared by USP as precursor.

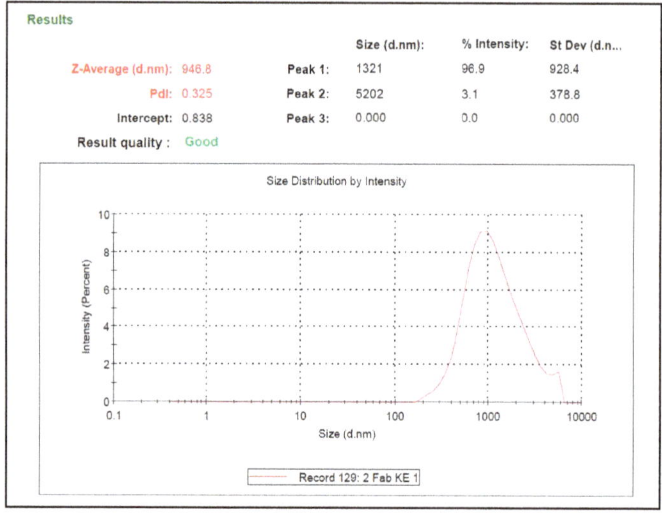

Figure 13. Particle size distribution by intensity of WS$_2$ particles synthesized using WO$_3$ prepared by USP as precursor.

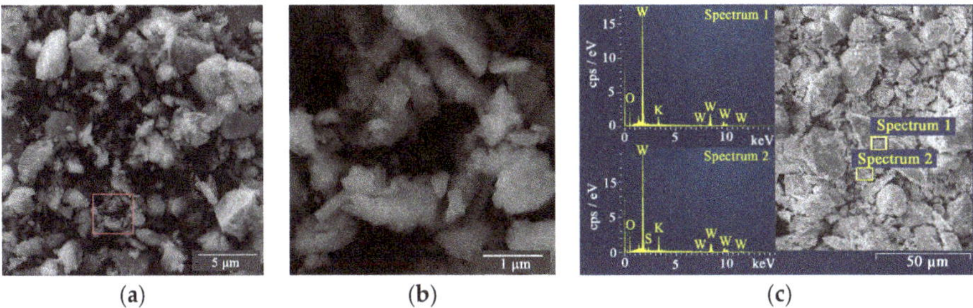

Figure 14. (**a**,**b**) SEM images and (**c**) EDX spectrum for WS_2 synthesized using commercial WO_3 as precursor.

The sample of WS_2 powder synthesized using commercial WO_3 as precursor was analyzed by XRD (Figure 15). Results showed the predominance of the WS_2 phase, present in two crystalized forms: hexagonal and rhombohedral. WO_3, WO_2, $K_2O\cdot WO_3$, and S phases were also identified, of which WO_3 was hexagonal and monoclinic. The strong and sharp diffraction peaks in the pattern indicated that the product was very highly crystallized.

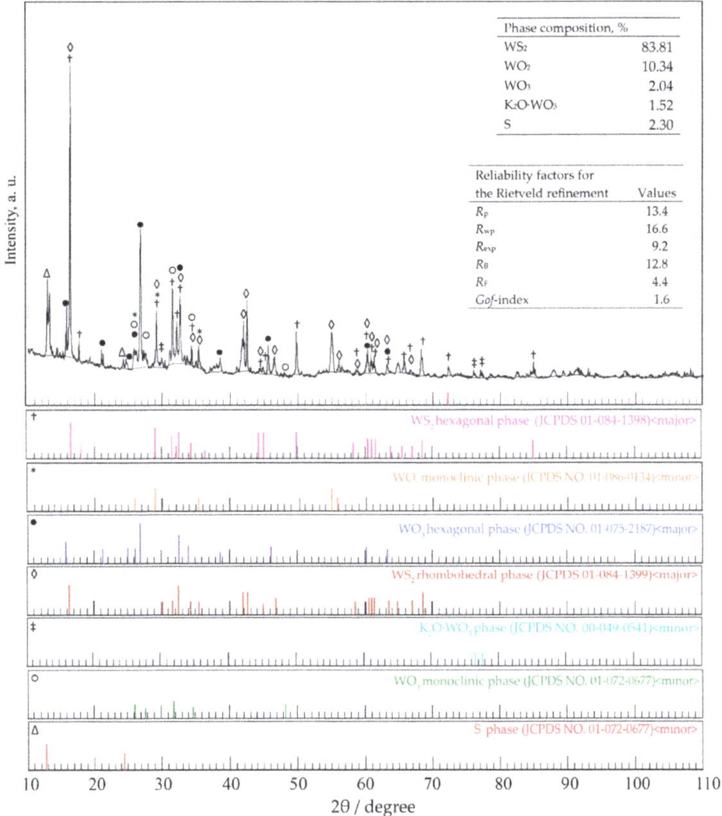

Figure 15. XRD pattern of WS_2 synthesized using commercial WO_3 as precursor.

The obtained WS_2 particles were tested for their size and size distribution and the mean particle size was found to be around 500 nm (Figure 16).

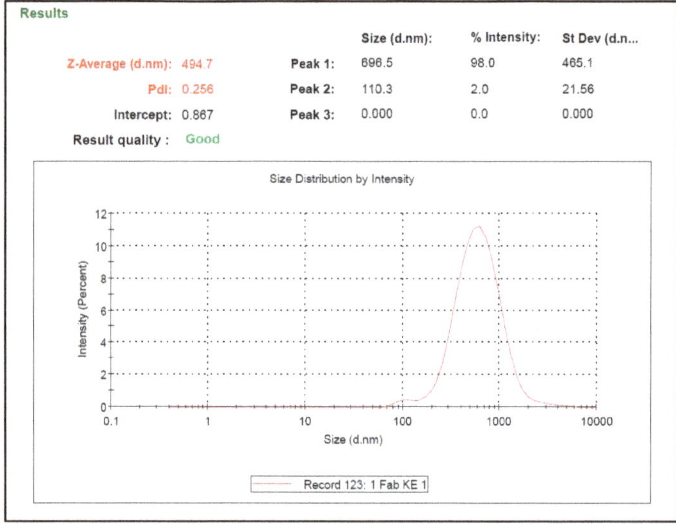

Figure 16. Particle size distribution by intensity of WS_2 particles synthesized using commercial WO_3 powder as precursor.

Commercial WS_2 powder as evidenced by SEM images consisted of nanoparticles with the presence of large agglomerates (Figure 17a). The uniform shape of particles can be seen in Figure 17b. On the basis of the EDX analysis as well as the high content of W and S, the presence of O was also confirmed (Figure 17c).

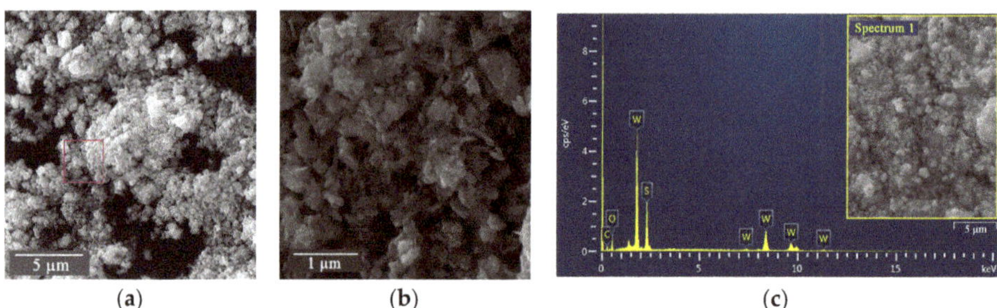

Figure 17. (a,b) SEM images and (c) EDX spectrum for commercial WS_2.

The XRD analysis of the commercial WS_2 powder indicated that there were major WS_2 and WO_3 phases and a minor $WO_3 \cdot 0.33H_2O$ phase, with low levels of crystallinity (Figure 18c). The presence of the amorphous phase was dominant, although the crystalline phase was also present.

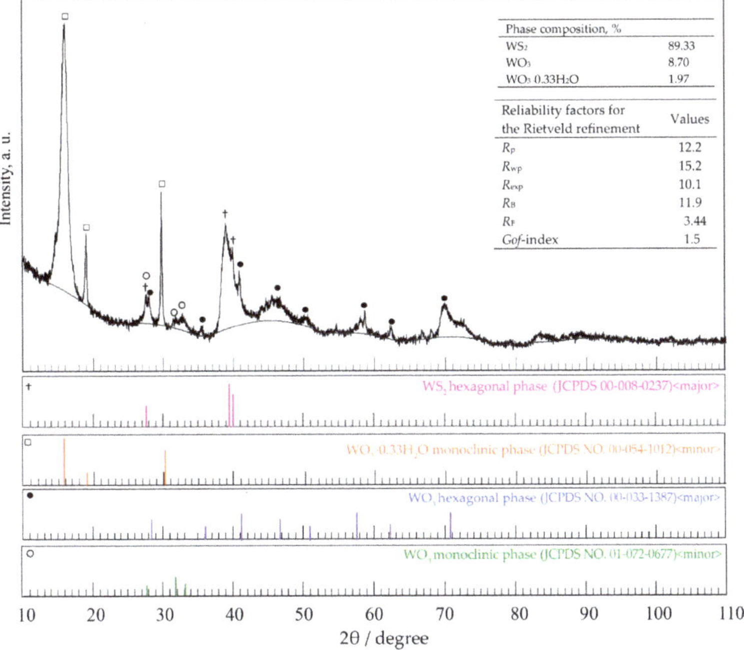

Figure 18. XRD pattern of commercial WS_2 powder.

Comparative analysis of the obtained powders with commercial WS_2 powder showed differences in particle size and agglomerates. The commercial WS_2 powder had the smallest particle size, but had more agglomerates compared to synthesized powders. Agglomerated powders are not good for tribological properties [23].

Particles of the synthesized WS_2 powders ranged from submicrometers to micrometers in size. The mixture of these two powders should provide better friction and wear performance between contacting surfaces, according to N. Wu et al [24].

Apart from the presence of sulfide as the main product, commercial WS_2 powders had a WO_3 (8.70%) phase, whereas synthesized WS_2 powders had $K_2O \cdot WO_3$ (1.52–1.56%), WO_2 (9.80–10.34%), and WO_3 (2.04–3.04%) phases. In order to obtain WS_2 powders that will provide low and stable friction coefficients, the presence of WO_3 should be avoided [25]. In synthesized powders, the WO_3 phase was less abundant relative to the levels found in commercial WS_2 powders. Both tungsten dioxides and tungsten disulfides exhibited similar lubrication performances [26]. The materials with low oxygen content were more resistant to wear [27]. In this respect, WO_2 in synthesized WS_2 powders led to a lower friction coefficient. In addition, the $K_2O \cdot WO_3$ phase in synthesized powders improved the thermal stability of obtained WS_2 powders [28].

4. Conclusions

In response to the requirements for improving performances of currently available tribological materials, this work demonstrated the synthesis of WS_2 powder using ultrafine WO_3 powder prepared by USP method. In addition, we synthesized WS_2 powder using commercial WO_3 powder as precursor.

The synthesis of WS_2 powder with the addition of K_2CO_3 as a fluxing agent reduced the WO_3 precursor and protected the sample against oxidation, which is one of the advantages of the applied method.

Further, the results of the XRD analysis of synthesized powders were in accordance with thermodynamic predictions. In the synthesized WS_2 powders, $K_2O \cdot WO_3$ and WO_2 phases were present in addition to the WS_2 phase. These phases are useful from the aspect of thermal stability and friction coefficient control of the powders.

As a result of this investigation, two sizes of WS_2 particles ranging from submicrometers to micrometers were obtained. Hence, future efforts are planned to investigate mixed micro/submicron lubrication systems composed of the synthesized WS_2 powders.

In summary, tribological WS_2 powder was successfully prepared by a facile and environment-friendly method.

Author Contributions: Conceptualization, Ž.K.; Formal analysis, J.T.; Investigation, N.G.; Methodology, Ž.K. and Z.A.; Resources, M.K.; Software, N.G. and B.P.; Supervision, Z.A.; Validation, J.T. and B.P.; Writing—original draft, Ž.K.; Writing—review & editing, M.K.

Funding: This research was funded by the Ministry of Education, Science and Technological Development of the Republic of Serbia, project No. 34033.

Acknowledgments: This paper was done with the financial support of the Ministry of Education, Science and Technological Development of the Republic of Serbia and it is a result of project No. 34033.

Conflicts of Interest: The authors declare no conflict of interest.

References

1. Holmberg, K.; Erdemir, A. Influence of tribology on global energy consumption, costs and emissions. *Friction* **2017**, *5*, 263–284. [CrossRef]
2. Stojanović, B.; Ivanović, L. Tribomechanical systems in design. *J. Balk. Tribol. Assoc.* **2014**, *20*, 25–34.
3. Wu, X.; Cobbina, S.; Mao, G.; Xu, H.; Zhang, Z.; Yang, L. A review of toxicity and mechanisms of individual and mixtures of heavy metals in the environment. *Environ. Sci. Pollut. Res.* **2016**, *23*, 8244–8259. [CrossRef] [PubMed]
4. Tchounwou, P.; Yedjou, C.; Patlolla, A.; Sutton, D. Heavy metal toxicity and the environment. In *Molecular, Clinical and Environmental Toxicology*; Springer: Basel, Switzerland; Heidelberg, Germany, 2012; pp. 133–164.
5. Wang, H.; Xu, B.; Liu, J. *Micro and Nano Sulfide Solid Lubrication*; Science Press: Beijing, China, 2012; ISBN 978-7-03-031785-8.
6. Therese, H.A.; Li, J.; Kolb, U.; Tremel, W. Facile Large Scale Synthesis of WS_2 Nanotubes from WO_3 Nanorods Prepared by a Hydrothermal Route. *Solid State Sci.* **2005**, *7*, 67–72. [CrossRef]
7. Huirache-Acuña, R.; Paraguay-Delgado, F.; Albiter, M.A.; Alvarez-Contreras, L.; Rivera-Muñoz, E.M.; Alonso-Núñez, G. Synthesis and characterization of WO_3 and WS_2 hexagonal phase nanostructures and catalytic test in sulfur remotion. *J. Mater. Sci.* **2009**, *44*, 4360–4369. [CrossRef]
8. Kumar, P.; Singh, M.; Gopal, P.; Reddy, G.B. Sulfurization of WO_3 nanorods into WS_2 as a function of H_2S/Ar partial pressure. *AIP Conf. Proc.* **2018**, *1953*, 030252.
9. Perkgoz, N. CVD growth and characterization of 2D transition metal dichalogenides, MoS2 and WS2. *Anadolu Univ. J. Sci. Technol. A Appl. Sci. Eng.* **2017**, *18*, 375–387. [CrossRef]
10. Liu, P.; Luo, T.; Xing, J.; Xu, H.; Hao, H.; Liu, H.; Dong, J. Large-Area WS_2 Film with Big Single Domains Grown by Chemical Vapor Deposition. *Nanoscale Res. Lett.* **2017**, *12*, 558. [CrossRef] [PubMed]
11. Hu, S.; Wang, X.; Meng, L.; Yan, X. Controlled synthesis and mechanism of large-area WS2 flakes by low-pressure chemical vapor deposition. *J. Mater. Sci.* **2017**, *52*, 7215–7223. [CrossRef]
12. Chen, X.; Wang, X.; Wang, Z.; Yu, W.; Qian, Y. Direct sulfidization synthesis of high-quality binary sulfides (WS2, MoS2, and V5S8) from the respective oxides. *Mater. Chem. Phys.* **2004**, *87*, 327–331. [CrossRef]
13. Shang, Y.; Xia, J.; Xu, Z.; Chen, W. Hydrothermal Synthesis and Characterization of Quasi-1-D Tungsten Disulfide Nanocrystal. *J. Dispers. Sci. Technol.* **2005**, *26*, 635–639. [CrossRef]
14. Shifa, T.A.; Wang, F.; Cheng, Z.; Zhan, X.; Wang, Z.; Liu, K.; Muhammad Safdar, M.; Sun, L.; He, J. Vertical-oriented WS_2 Nanosheet Sensitized by Graphene: An Advanced Electrocatalyst for Hydrogen Evolution Reaction. *Nanoscale* **2015**, *35*, 1–3.
15. Wu, Z.; Wang, D.; Zan, X.; Sun, A. Synthesis of WS_2 nanosheets by a novel mechanical activation method. *Mater. Lett.* **2010**, *64*, 856–858. [CrossRef]

16. Tang, H.; Zhang, X.; Li, L.; Li, C. Preparation Method of Graphite Alkene Like Tungsten Disulfide Nanometer Sheet. CN Patent 103641173 A, 19 March 2014.

17. Zhang, X.; Xu, H.; Wang, J.; Ye, X.; Lei, W.; Xue, M.; Tang, H.; Li, C. Synthesis of Ultrathin WS_2 Nanosheets and Their Tribological Properties as Lubricant Additives. *Nanoscale Res. Lett.* **2016**, *11*, 442. [CrossRef] [PubMed]

18. Zhang, X.H.; Tan, H.; Fan, Z.; Ge, M.Z.; Ye, X.; Xue, M. Synthesis and electrochemical performance of ultrathin WS_2 nanosheets. *Chalcogenide Lett.* **2017**, *14*, 419–423.

19. Stopić, S. *Synthesis of Metallic Nanosized Particles by Ultrasonic Spray Pyrolysis*; Shaker Verlag: Aachen, Germany, 2015; ISBN 978-3844035292.

20. Roine, A. *HSC Chemistry*® *v 9.0.*; Outotec Research Oy Center: Pori, Finland, 2016.

21. Arutanti, O.; Ogi, T.; Nandiyanto, A.B.; Iskandar, F.; Okuyama, K. Controllable crystallite and particle sizes of WO_3 particles prepared by a spray-pyrolysis method and their photocatalytic activity. *AIChE J.* **2013**, *60*, 41–49. [CrossRef]

22. Meyer, B. Elemental sulfur. *Chem. Rev.* **1976**, *76*, 367–388. [CrossRef]

23. Davim, J. (Ed.) *Tribology of Nanocomposites*; Springer: Berlin/Heidelberg, Germany, 2013; ISBN 978-3-642-33882-3.

24. Wu, N.; Hu, N.; Zhou, G.; Wu, J. Tribological properties of lubricating oil with micro/nano-scale WS_2 particles. *J. Exp. Nanosci.* **2018**, *13*, 27–38. [CrossRef]

25. Aldana, P.U. Tungsten Disulfide Nanoparticles as Lubricant Additives for the Automotive Industry. Ph.D. Thesis, Université de Lyon, Lyon, France, 2016.

26. Nian, J.; Chen, L.; Guo, Z.; Liu, W. Computational investigation of the lubrication behaviors of dioxides and disulfides of molybdenum and tungsten in vacuum. *Friction* **2017**, *5*, 23–31. [CrossRef]

27. Polcar, T.; Parreira, N.M.G.; Cavaleiro, A. Tungsten oxide with different oxygen contents: Sliding properties. *Vacuum* **2007**, *81*, 1426–1429. [CrossRef]

28. Çelikbilek Ersundu, M.; Ersundu, A.E.; Sayyed, M.I.; Lakshminarayana, G.; Aydin, S. Evaluation of physical, structural properties and shielding parameters for K_2O-WO_3-TeO_2 glasses for gamma ray shielding applications. *J. Alloys Compd.* **2017**, *714*, 278–286. [CrossRef]

Article

Influence of the Shape of Copper Powder Particles on the Crystal Structure and Some Decisive Characteristics of the Metal Powders

Ljiljana Avramović [1], Vesna M. Maksimović [2], Zvezdana Baščarević [3], Nenad Ignjatović [4], Mile Bugarin [1], Radmila Marković [1] and Nebojša D. Nikolić [5,*]

[1] Mining and Metallurgy Institute, Center for Technologies Development in Metallurgy, Zeleni bulevar 35, 19 210 Bor, Serbia; ljiljana.avramovic@irmbor.co.rs (L.A.); mile.bugarin@irmbor.co.rs (M.B.); radmila.markovic@irmbor.co.rs (R.M.)

[2] Vinča Institute of Nuclear Sciences, Department of Materials Science, University of Belgrade, 11 000 Belgrade, Serbia; vesnam@vin.bg.ac.rs

[3] Institute for Multidisciplinary Research, University of Belgrade, Kneza Višeslava 1a, 11 000 Belgrade, Serbia; zvezdana@imsi.bg.ac.rs

[4] Institute of Technical Sciences of the Serbian Academy of Science and Arts, Knez Mihailova 35/IV, 11 000 Belgrade, Serbia; nenad.ignjatovic@itn.sanu.ac.rs

[5] Institute of Chemistry, Technology and Metallurgy-Department of Electrochemistry, University of Belgrade, Njegoševa 12, 11 000 Belgrade, Serbia

* Correspondence: nnikolic@ihtm.bg.ac.rs; Tel.: +381-11-337-0390

Received: 13 November 2018; Accepted: 31 December 2018; Published: 9 January 2019

Abstract: Three different forms of Cu powder particles obtained by either galvanostatic electrolysis or a non-electrolytic method were analyzed by a scanning electron microscope (SEM), X-ray diffraction (XRD) and particle size distribution (PSD). Electrolytic procedures were performed under different hydrogen evolution conditions, leading to the formation of either 3D branched dendrites or disperse cauliflower-like particles. The third type of particles were compact agglomerates of the Cu grains, whose structural characteristics indicated that they were formed by a non-electrolytic method. Unlike the sharp tips that characterize the usual form of Cu dendrites, the ends of both the trunk and branches were globules in the formed dendrites, indicating that a novel type of Cu dendrites was formed in this investigation. Although the macro structures of the particles were extremely varied, they had very similar micro structures because they were constructed by spherical grains. The Cu crystallites were randomly oriented in the dendrites and compact agglomerates of the Cu grains, while the disperse cauliflower-like particles showed (220) and (311) preferred orientation. This indicates that the applied current density affects not only the morphology of the particles, but also their crystal structure. The best performance, defined by the largest specific surface area and the smallest particle size, was by the galvanostatically produced powder consisting of disperse cauliflower-like particles.

Keywords: copper; powder; electrolysis; hydrogen; SEM; XRD; PSD

---

1. Introduction

Copper powders have found a wide industrial application for a very long time [1]. Cu in a powder form is often used in the electrical and electronic industries due to its excellent electrical and thermal characteristics. The self-lubricating bearing is probably the most common application of Cu powder and about 70% of the total Cu powder production in a granular form is used for that purpose. This application takes advantage of the ability to produce a component with controlled interconnected and surface-connected porosity. Copper powders are also used in such nonstructural applications as brazing, cold soldering, and mechanical plating, as well as for medals and medallions, metal-plastic decorative products and a variety of chemical and medical purposes.

The Cu powder production methods can be divided into electrolytic and non-electrolytic (or chemical) methods. Processes such as ultrasonic spray pyrolysis [2], solvothermal synthesis [3], cementation [4], chemical reduction methods [5–8], the high-energy electrical explosion method [9], atomization [10], pyrolysis [11], polyol processes [12–14], hydrometallurgy [15], etc. belong to the group of non-electrochemical or chemical processes of synthesis. The chemical reduction method is often referred to as electroless deposition [15,16]. Various reducing agents, such as ascorbic acid [5,8,16], hydrazine hydrate [7], sodium borohydride [6], and formaldehyde [15], are used in the chemical reduction processes. Copper, in the form of nanoparticles, can be synthesized in ethylene glycol (EG) using copper sulphate as a precursor, and vanadium sulfate as an atypical reductant [17]. Metal nanoparticles can also be obtained by biosynthesis processes using the microorganisms, which has clear advantages compared to chemical synthesis methods [18]. The biosynthesis processes are environmentally friendly, no toxic chemicals or reagents are needed for these processes, and it is possible to synthesize particles that cannot be obtained using chemical synthesis methods.

Aside from the above-mentioned methods of synthesis, electrolysis is often used for Cu powder synthesis. The advantage of this method of synthesis can primarily be attributed to the fact that the shape and size of particles can be easily regulated by the choice of electrolysis regime and parameters [19]. The regimes, both constant (potentiostatic [20] and galvanostatic [21–23]) and periodically changing [19,24–26], are used for the Cu synthesis in powder form. The electrolysis parameters that affect the final shape of the particles, are: the type and composition of electrolytes, presence of additives, temperature, type of cathode, electrolysis time, etc. [19,21,25,27–30]. Special attention has been dedicated to the effect of the hydrogen evolution reaction as the parallel reaction to Cu electrolysis in the production range of Cu powder [20,24].

The shape of Cu powder particles is related to their synthesis method. The four various forms of Cu particles can be identified as: almost ideal micro spheres, irregular rough particles, disperse cauliflower-like particles and dendrites. Dendrites are the most commonly observed shape of particles and they can be obtained by both electrolysis and some chemical methods such as the galvanic replacement reaction method [31]. The very disperse cauliflower-like particles are formed by the electrolysis in conditions of strong hydrogen evolution as the parallel reaction [19,20]. Irregular particles are obtained by water atomization [32] while spherical particles are obtained by the gas-atomizing process [32]. Aggregates of non-uniform irregular particles are obtained by the polyol process [12]. Almost ideal Cu spheres can be also obtained by the polyol process [12]. Polyhedral, non-agglomerated monodispersed particles are obtained using ascorbic acid as the reducing agent [4]. The non-agglomerated almost spherical particles are obtained by ultrasonic spray pyrolysis [2].

Recently, a strong correlation between the morphologies of silver powder particles and their crystal structure was found [33,34]. Here, this type of investigation has been continued with the aim of establishing the existence of the same correlation for Cu powders. For this reason, the three completely various morphologies of Cu powder particles were analyzed and correlated with their crystal structure. Two of them are obtained by electrolysis under completely different conditions, without and with vigorous hydrogen evolution. The third type was commercially supplied, and on the basis of morphological characteristics, it is clear that it was obtained by some non-electrochemical method. In order to examine the effect of the shape of particles on the decisive characteristics that define the behavior of the powders as a collection of particles, the specific surface area (SSA) and particle size distribution (PSD) were also analyzed.

2. Materials and Methods

Electrolyte, formed by dissolution of 0.10 mol·dm^{-3} CuSO$_4$ in 0.50 mol·dm^{-3} H$_2$SO$_4$, was used for the electrolytic production of Cu powders. Cu powder was galvanostatically produced at the current densities of 14.4 and 384 mA·cm^{-2}. Hereinafter, these powders are denoted as Cu(14.4) for powder produced at 14.4 mA·cm^{-2} and Cu(384) for powder produced at 384 mA·cm^{-2}. Electrolysis was carried out at a temperature of 21.0 ± 0.5 °C in an open cell of cylindrical shape. The cylindrical Cu wire of the overall surface area 0.50 cm^2 was used as the working electrode. The counter electrode was Cu foil. Figure 1 shows a configuration of electrodes in the cell. The ultra-pure water and p.a. reagents (CuSO$_4$ × 5H$_2$O, H$_2$SO$_4$) were used for electrolyte preparation for the Cu powder synthesis. Powder was produced with the electricity amount of 10 mA·h·cm^{-2}, and removed from the electrode surface after this amount of electricity was reached.

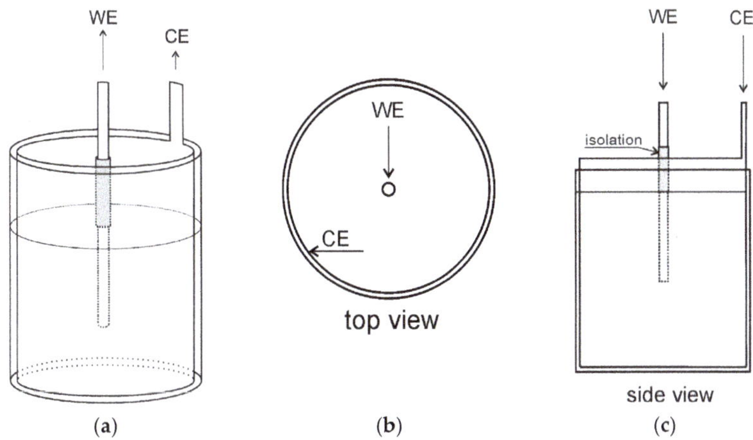

Figure 1. (a) Scheme of the cell used for production of the Cu powders by electrolysis. WE—working electrode; CE—counter electrode, (b) top view, (c) side view.

The electrolytically produced powder was compared to the commercially supplied powder, Sigma-Aldrich company (Saint Louis, MO, USA), product No. 326453 (powder (spheroidal), 14–25 μm, 99%). Hereinafter, this powder is denoted as Cu(CHEM).

A scanning electron microscope (SEM)—TESCAN Digital Microscopy company (model VEGA3, Brno, Czech Republic) was used for the morphological characterization of the produced particles.

The Rigaku Ultima IV diffractometer (Rigaku Co. Ltd., Tokyo, Japan) with CuK$_\alpha$ radiation was used to study the crystal structure of the produced powders. Four reflections in the 2θ range between 30° and 95° were recorded in order to determine the preferred orientation of powder particles. For this purpose, calculation of the Texture Coefficient, *TC*(*hkl*) and Relative Texture Coefficient, *RTC*(*hkl*), based on an analysis of data obtained by the X-ray diffraction (XRD) method, was made. It was made in the following way: using an intensity of each reflection (*hkl*) plane, the following ratios (in percentage) were calculated by applying Equation (1) [35]:

$$R(hkl) = \frac{I(hkl)}{\sum\limits_{i}^{4} I(h_i k_i l_i)} \times 100 \tag{1}$$

where *I*(*hkl*) is the intensity of each (*hkl*) reflection plane, while $\sum\limits_{i}^{4} I(h_i k_i l_i)$ represents the sum of intensities of all recorded reflection planes. Note: the values of intensities are given in cps.

Then, using the determined *R*(*hkl*) coefficients, the values of the Texture Coefficient, *TC*(*hkl*), were calculated by applying Equation (2):

$$TC(hkl) = \frac{R(hkl)}{R_s(hkl)} \tag{2}$$

where $R_s(hkl)$ is determined in the same way as presented in Equation (1), but taking into consideration the Cu standard (04-0836).

In this way, accurate quantitative information about the absolute intensity of each of these reflections were obtained determining the $TC(hkl)$ coefficients. Also, the intensity of each reflection plane in relation to the other reflection planes represents the relevant information, and is defined as the Relative Texture Coefficient, $RTC(hkl)$ according to Equation (3):

$$RTC(hkl) = \frac{TC(hkl)}{\sum\limits_{i}^{4} TC(h_i k_i l_i)} \times 100 \tag{3}$$

The $RTC(hkl)$ coefficient defines the intensity of the considered (hkl) orientation in relation to the standard which is included in the TC values.

A MALVERN Instruments MASTERSIZER 2000 (MALVERN Instruments Ltd., Malvern, Worcestershire, UK) device was used to determine the particle size distribution (PSD) and specific surface area (SSA) of the powders. Malvern Software (Version 5.60, MALVERN Instruments Ltd., Malvern, Worcestershire, UK) was used in order to obtain the SSA values.

Determination of the Average Current Efficiency of Hydrogen Evolution ($\eta_{I,av}(H_2)$)

In order to quantify the amount of hydrogen, generated in the galvanostatic electrolysis, the average current efficiency of hydrogen evolution, $\eta_{I,av}(H_2)$ was determined. For that purpose, the working and counter electrodes of Cu were placed in a burette which was positioned so that the overall volume of evolved hydrogen in the galvanostatic electrolysis remained in it. The surface area of the Cu working electrode, placed in a burette, was 0.63 cm^2. The current efficiency of hydrogen evolution in time t_i, $\eta_{I,i}(H_2)$, in %, is given by Equation (4):

$$\eta_{I,i}(H_2) = \frac{V(H_2)}{\mu(H_2)S_0 j t_i} 100 \tag{4}$$

where

$$\mu(H_2) = \frac{V}{nF} = \frac{24{,}120 \text{ cm}^3}{2 \times 26.8 \text{ Ah}} = 450 \frac{\text{cm}^3}{\text{Ah}} \tag{5}$$

and $V(H_2)$ is the volume of evolved hydrogen in time t_i, nF is the number of Faradays per mole of spent ions, V is the molar volume of gas at temperature of 21.0 °C (i.e., 24,120 cm^3), S_0 is the surface of working electrode, and j is the current density of electrolysis. The average values of current efficiency for the hydrogen evolution reaction, $\eta_{I,av}(H_2)$ are determined as $\eta_{I,av}(H_2) = (1/t)\int\limits_{0}^{t} \eta_{I,i}(H_2)\mathrm{d}t$, where t is time of electrolysis.

3. Results

3.1. Morphological Analysis of Cu Powders Produced by the Electrolytic and Non-Electrolytic Methods

The polarization curve for the copper electrodeposition from 0.10 mol dm^{-3} CuSO$_4$ in 0.50 mol dm^{-3} H$_2$SO$_4$ is shown in Figure 2. The plateau of the limiting diffusion current density was in the range of overpotentials between 300 and 750 mV, with a limiting diffusion current density value (j_L) of 9.6 mA cm^{-2}. In the galvanostatic regime of electrolysis, Cu in the powder form is formed at current densities larger than the limiting diffusion current density [19].

Morphologies of the Cu particles, obtained at current densities corresponding to 1.5 ($j = 14.4$ mA·cm^{-2}) and 40 ($j = 384$ mA·cm^{-2}) times larger values than the limiting diffusion current density, are shown in Figures 3 and 4, respectively. Considering the fact that the hydrogen evolution as the second reaction in the Cu electrolysis, commences inside the limiting diffusion current density plateau, a strong difference in the amount of generated hydrogen at these current densities, with strong consequences for the morphology, structure and decisive characteristics of the obtained powders, was expected. Accordingly, inspection of the presented microphotographs shows the strong effect of the applied current densities on the morphology of Cu powder particles.

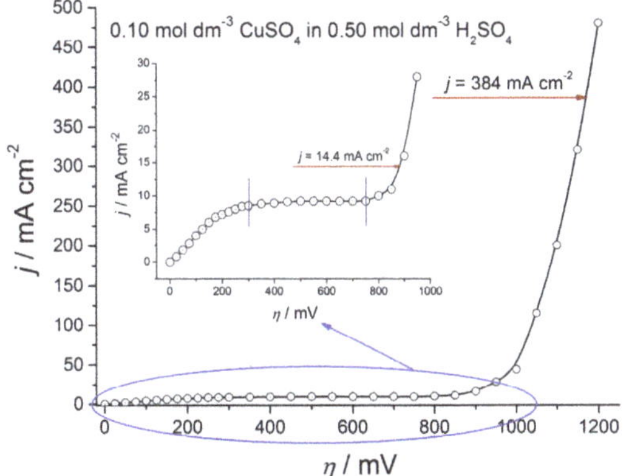

Figure 2. Polarization curve for the Cu electrodeposition from 0.10 mol·dm^{-3} CuSO$_4$ in 0.50 mol·dm^{-3} H$_2$SO$_4$.

Figure 3a shows the Cu deposit obtained immediately after finishing the process of electrolysis at 14.4 mA·cm^{-2}. The three morphological forms can be identified on this microphotograph: very branched 3D (three dimensional) dendritic forms, holes originating from detached hydrogen bubbles and cauliflower-like agglomerates of Cu grains. It can be seen that the shape of the dendritic forms (Figure 3a,b) was different from all of the so far observed forms of Cu dendrites. In previously observed forms of Cu dendrites, the tips of both the trunk and branches were sharp, while in this form they end with globules (Figure 3c,d). This form is observed for the first time in this investigation, and for this reason, it can be said that this shape represents a completely novel type of Cu dendrites. The size of the globules was from 3–5 μm for those in the branches, to about 10 μm at the tops of both branches and trunk. The size of the holes formed from the detached hydrogen bubbles, approached 100 μm. The typical cauliflower-like agglomerate of Cu grains, formed between the dendrites and holes is shown in Figure 3e. It consists of small agglomerates of Cu grains, surrounded by irregular channels. The size of the individual grains in these agglomerates was considerably smaller than the size of globules, and it was about 1 μm. The particles, obtained after removing the deposit from the electrode surface, are shown in Figure 3f,g. From a macromorphological point of view, no difference is observed between those on the electrode surface after the finished process of electrolysis and those after their removal. It is necessary to note the existence of channel structure inside the cauliflower-like particles (Figure 3f), which is a result of hydrogen evolution reaction.

(a) (b) (c)

Figure 3. *Cont.*

(d)   (e)   (f)

(g)

Figure 3. Morphologies of the powdered copper obtained at a current density of 14.4 mA·cm^{-2}: (**a**) appearance of electrode surface after the process of electrolysis, (**b**) dendrite, (**c**) tops of branches of dendrite, (**d**) top of trunk of dendrite, (**e**) cauliflower-like agglomerates of Cu grains formed among dendrites and holes, (**f**,**g**) particles obtained after removal from the electrode surface.

Figure 4a shows the SEM microphotograph of copper electrode, obtained at a current density of 384 mA·cm^{-2}. A typical honeycomb-like structure, constructed from holes formed from the detached hydrogen bubbles (Figure 4b) surrounded by the disperse cauliflower-like agglomerates of Cu grains (Figure 4c,d) was obtained. Increasing the current density from 14.4 to 384 mA·cm^{-2} led to an intensification of the hydrogen evolution reaction, which is manifest by the increase in the number of holes and by a decrease in their size. In this case, a hole size was about 70 μm. As a result of intensification of the hydrogen evolution, the cauliflower-like agglomerates of Cu grains were more disperse than those obtained at 14.4 mA·cm^{-2}. The size of the grains in these agglomerates was about 200 nm, that is, approximately five times smaller than those formed at 14.4 mA·cm^{-2}. The typical forms of Cu particles, obtained by removal of deposit from the electrode surface, are shown in Figure 4e,f. Inhibition of dendritic growth and the cauliflower-like character of the formed particles as a result of vigorous hydrogen evolution, are clearly visible in Figure 4e,f.

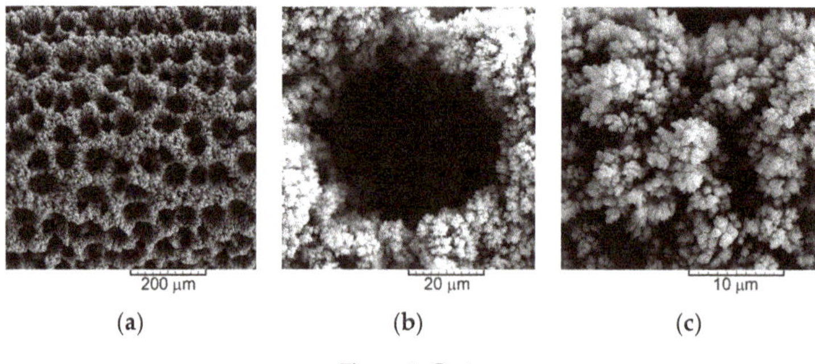

(a)   (b)   (c)

Figure 4. *Cont.*

(d)   (e)   (f)

Figure 4. Morphologies of the powdered copper obtained at a current density of 384 mA·cm^{-2}: (**a**) appearance of the electrode surface after the process of electrolysis (honeycomb-like structure), (**b**) hole formed from the detached hydrogen bubble, (**c**) disperse cauliflower-like agglomerates of Cu grains formed around holes, (**d**) typical agglomerate constructing disperse cauliflower-like agglomerates, (**e**,**f**) particles obtained after removal from the electrode surface.

Figure 5 shows the morphology of commercially supplied Cu powder. It can be seen in Figure 5 that this powder consists of agglomerates of approximately spherical (spheroidal) grains of various size. It can be seen that these agglomerates have a relatively compact structure. The size of these agglomerates was larger than 10 µm, and they were constructed from grain sized in the range of 1–10 µm. Also, some of them had very irregular shapes. Considering the fact that these forms have never been observed among electrolytically synthesized particles, it is clear that they are obtained in some non-electrochemical way. The irregular, relatively compact forms constructed from approximately spherical grains in a wide range of sizes confirmed this assumption.

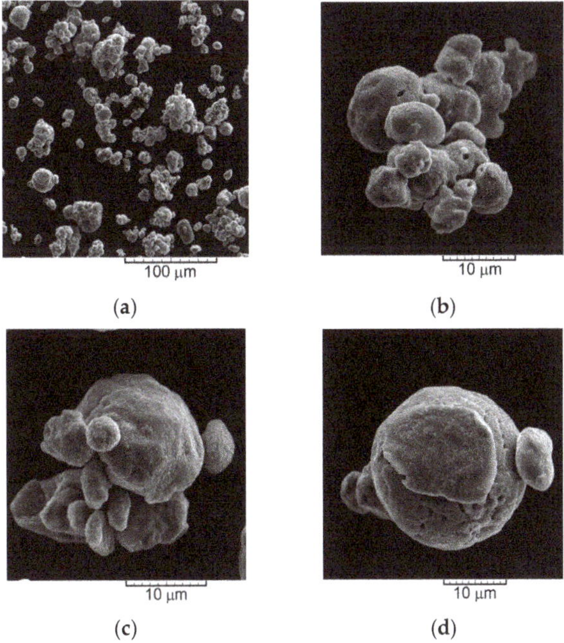

(a)   (b)

(c)   (d)

Figure 5. Typical copper particles obtained by chemical processes: (**a**) ×500, (**b**) ×4000, (**c**) ×4500, (**d**) ×3500.

3.2. Structural Analysis of Cu Powders Produced by the Electrolytic and Non-Electrolytic Methods

The XRD patterns of particles, obtained by removal the deposits obtained at current densities of 14.4 and 384 mA·cm^{-2}, and Cu standard (04-0836), are shown in Figure 6. The peaks at 2θ angles of 43.3°, 50.4°, 74.1° and 89.9° correspond to (111), (200), (220) and (311) crystal planes. The appearance of peaks at these angles confirms a face centered cubic (FCC) crystal lattice of Cu. A larger ratio of Cu crystallites oriented in the (111) plane observed in the both diffractograms, can be ascribed to the lower surface energy of this plane in relation to the other planes, because the surface energy (γ) values follow a trend: $\gamma_{111} < \gamma_{100} < \gamma_{110}$ [36,37]. The preferred orientation of the obtained particles was investigated by an analysis of the peak intensity ratios (111)/(200), (111)/(220) and (111)/(311), and by determination the *TC(hkl)* and *RTC(hkl)* coefficients.

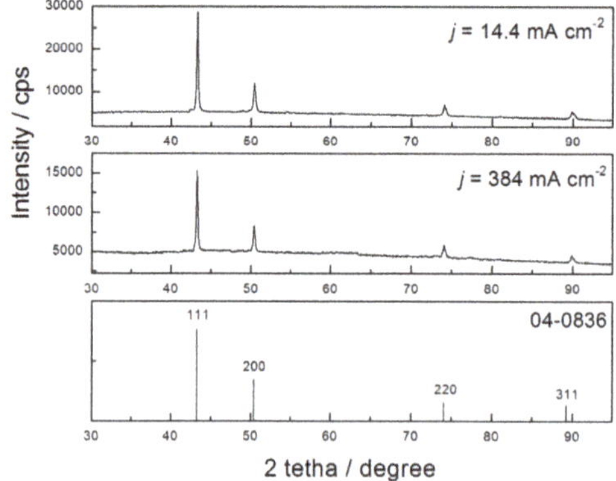

Figure 6. XRD patterns of Cu particles obtained at current densities of 14.4 and 384 mA·cm^{-2}, and Cu standard (04-0836).

The values of peak intensity ratios for the Cu particles produced at current densities of 14.4 mA·cm^{-2} (Cu(14.4)) and 384 mA·cm^{-2} (Cu(384)), and the same ratios for Cu standard are given in Table 1. It can be seen that (111)/(220) and (111)/(311) for the Cu(384) powder are considerably smaller than those for the Cu standard. On the other hand, for the Cu(14.4) powder, these ratios are still relatively close to the values for the Cu standard, indicating the random orientation of Cu crystallites in the particles shown in Figure 3. It is known [33] that the values of ratios considerably larger than those of the standard indicate the presence of (111) preferred orientation. However, there is no data about what happens with the crystal orientation when the peak intensity ratios are smaller than those of the standard. Therefore, the additional analysis of the crystal orientation of particles was carried out by determining the Texture Coefficient, *TC(hkl)*, and Relative Texture Coefficient, *RTC(hkl)*.

Table 1. Ratios of intensities of the diffraction peaks for the analyzed powders and Cu standard.

Type	Ratio of Intensities:		
of Powder Particles	(111)/(200)	(111)/(220)	(111)/(311)
Cu(14.4)	2.4	4.1	5.3
Cu(384)	1.9	2.6	3.4
Cu(CHEM)	2.5	5.2	6.0
Cu standard (04-0783)	2.2	5.0	5.9

The values of *TC(hkl)* and *RTC(hkl)* coefficients obtained for the electrolytically produced particles at current densities of 14.4 and 384 mA·cm^{-2} (Cu(14.4) and Cu(384), respectively) are given in Table 2. The values of *TC(hkl)* coefficients larger than 1 indicate the existence of preferred orientation [33–35].

On the other hand, since four main reflections were analyzed, the values of *RTC* coefficients larger than 25% indicate the existence of a preferred orientation.

Table 2. Texture calculations for Cu powders obtained by the galvanostatic regime at current densities (j) of 14.4 and 384 mA·cm^{-2} and for commercially available powder (j = 14.4 mA·cm^{-2}—14.4, j = 384 mA·cm^{-2}—384, commercially available powder—CHEM, s—Cu standard).

Plane	*R* (in %)			*R*$_s$ (in %)	*TC*			*RTC* (in %)		
(*hkl*)	$R_{14.4}$	R_{384}	R_{CHEM}		$TC_{14.4}$	TC_{384}	TC_{CHEM}	$RTC_{14.4}$	RTC_{384}	RTC_{CHEM}
(111)	54.1	45.3	57.3	54.6	0.99	0.83	1.05	23.6	17.3	26.6
(200)	22.5	24.1	22.3	25.1	0.90	0.96	0.89	21.5	20.0	22.5
(220)	13.1	17.2	10.9	10.9	1.20	1.58	1.00	28.6	32.9	25.3
(311)	10.3	13.4	9.5	9.4	1.10	1.43	1.01	26.3	29.8	25.6

The data analysis, presented in Table 2, shows that the values of the *RTC* coefficients for Cu(14.4) powder are about 25%, confirming the randomly orientated crystallites in the particles obtained at a current density of 14.4 mA·cm^{-2}. However, a different situation is observed when the particles produced at 384 mA·cm^{-2} were analyzed. The *RTC* coefficients for (220) and (311) planes are larger than those for the Cu standard, indicating the existence of a preferred orientation in these planes. Thus, it is clear that peak intensity ratios smaller than the values for the Cu standard (Table 1) indicate the preferred orientation in that plane.

Figure 7 shows the XRD pattern of chemically synthesized Cu powder (Cu(CHEM)) and Cu standard. The appearance of peaks at the same angles, as in the case of electrolytically produced Cu powders, confirms the face centered cubic (FCC) crystal lattice of Cu. Also, similar to the electrolytically produced powders, the Cu crystallites were predominately oriented in the (111) plane. The peak intensity ratios (111)/(200), (111)/(220) and (111)/(311) for this powder are included in Table 1. It can be seen in Table 1 that the values are very close to those for the Cu standard, confirming that the powder consists of randomly distributed spherical grains.

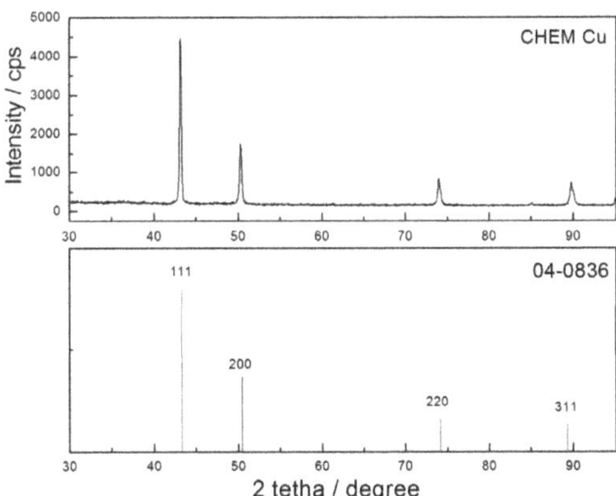

Figure 7. XRD pattern of chemically synthesized Cu powder, and Cu standard (04-0836).

The values of *TC*(*hkl*) and *RTC*(*hkl*) coefficients for this powder are shown in Table 2. The values of *TC* coefficients very close to 1, and *RTC* coefficients about 25%, clearly point to the random orientation of the Cu crystallites in these particles.

3.3. Analysis of the Specific Surface Area (SSA) and Particle Size Distribution (PSD) of the Considered Powders

To compare the powders obtained by the different methods of synthesis and under different conditions of electrolysis, the specific surface area (SSA) and particle size distribution (PSD), as two very important characteristics that describe the behavior of powders as a collection of the particles were analyzed. The values of the SSA for the analyzed powders are summarized in Table 3. It can be seen that the largest SSA value is obtained for the powder that consisted of disperse cauliflower-like particles obtained by electrolysis at 384 mA·cm^{-2}. The SSA of powder consisting of particles produced at this current density, was almost two time larger than the value for chemically synthesized powder. The smallest SSA value was obtained for the powder including the ordered dendritic structure galvanostatically produced at 14.4 mA·cm^{-2}.

Table 3. Values of specific surface area (SSA), particle size with the maximum volume ratio (PSMVR), and the average current efficiency for hydrogen evolution reaction ($\eta_{I,av}(H_2)$) for powders obtained by electrolysis at current densities of 14.4 mA·cm^{-2} (Cu(14.4)) and 384 mA·cm^{-2} (Cu(384)), and chemically synthesized powder (Cu(CHEM)).

The Type of Powder Particle	$\eta_{I,av}(H_2)$ %	SSA m^2 g^{-1}	PSMVR μm
Cu(14.4)	6.63	0.0389	23.2
Cu(384)	49.7	0.0997	10.9
Cu(CHEM)	/	0.0544	14.6

The PSD curves obtained for all three types of particles are shown in Figure 8. A uniform distribution of particles was obtained in all cases. The maximum volume ratios shift towards the smaller particle size in the same way as the increase of SSA. The maximum volume ratio for the powder produced at 384 mA·cm^{-2} corresponds to a particle size of 10.9 μm (this value as well as those obtained for the other powders are included in Table 3), and it is more than 50% smaller than the maximum volume ratio obtained at a current density of 14.4 mA·cm^{-2}, corresponding to a particle size of 23.2 μm. The presence of larger particles in powder produced at 384 mA·cm^{-2} (Figure 4e) is also identified in the PSD curve for this powder. For the chemically synthesized powder, the maximum volume ratio corresponds to a particle size of 14.6 μm, and it is situated between those for the electrolytically produced powders.

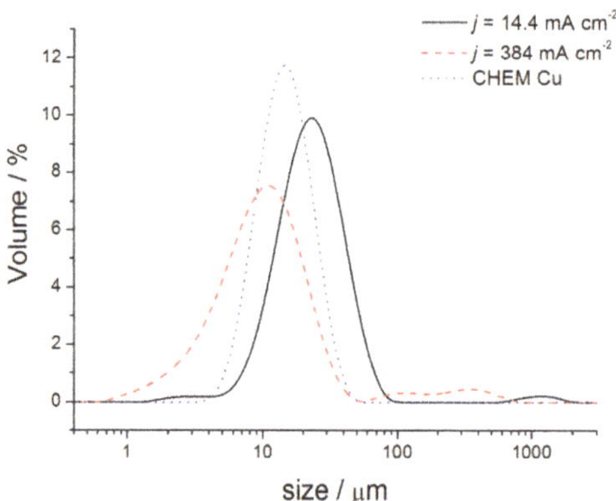

Figure 8. Particle size distribution (PSD) curves obtained for the particles produced at current densities of 14.4 and 384 mA·cm^{-2}, and chemically synthesized powder.

4. Discussion

Three different morphologies of Cu particles were analyzed in this study: the 3D dendrites, disperse cauliflower-like agglomerates of Cu grains and compact agglomerates of Cu grains. Two of these were obtained by the electrolysis processes: 3D dendrites and disperse cauliflower-like agglomerates of Cu grains. Hydrogen evolution, as a parallel reaction to the Cu electrolysis at high current densities and overpotentials, had a crucial role in creating the final morphology of these particles. The amount of produced hydrogen is quantified by determination of the average current efficiency for hydrogen evolution, as presented in Figure 9 for the Cu electrolysis at current densities of 14.4 and 384 mA·cm^{-2}.

Figure 9 shows the dependencies of the evolved hydrogen volume and current efficiency for the hydrogen evolution reaction in a time t_i on the electrolysis time obtained at current densities of 14.4 (Figure 9a) and 384 mA·cm^{-2} (Figure 9b). It is clear that increasing the current density leads to a strong intensification in hydrogen evolution. The values of the average current efficiency for the hydrogen evolution reaction, obtained at the current densities of 14.4 and 384 mA·cm^{-2}, are given in Table 3.

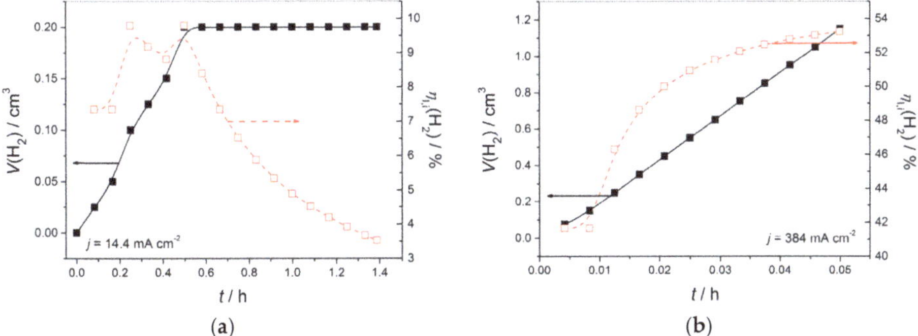

Figure 9. The dependencies of the evolved hydrogen volume and current efficiency for the hydrogen evolution reaction on the electrolysis time obtained at current densities of: (**a**) 14.4 mA·cm^{-2} and (**b**) 384 mA·cm^{-2}.

The amount of hydrogen that evolved at a current density of 14.4 mA·cm^{-2} was not sufficient to have any effect on the hydrodynamic conditions in the near-electrode layer. The proof for this is the rare holes formed from detached hydrogen bubbles and a somewhat non-uniform electrode surface determined by very branched 3D dendrites and cauliflower-like agglomerates of Cu grains (Figure 3a,b). As already mentioned, the 3D dendrites shown here represent a novel type of dendrite observed for the first time in this investigation. Formation of this type of dendrites can be explained from an electrochemical point of view in the following way: decrease of potential occurs with the electrolysis time in the galvanostatic regime of electrolysis. In this case, the potential at the end of electrolysis processes, corresponding to an amount passed electricity of 10 mA·h·cm^{-2}, was about 275 mV. This potential corresponds to the mixed activation-diffusion control at which formation of globules like those shown in Figure 3 occurs [19,38].

On the other hand, a very uniform honeycomb-like structure was formed at a current density of 384 mA·cm^{-2} (Figure 4). The amount of evolved hydrogen at this current density was vigorous enough to have a strong effect on the hydrodynamic conditions in the near-electrode layer. Namely, hydrogen generated during electrolysis causes a strong stirring of electrolyte in the near-electrode layer, leading to a decrease in diffusion layer thickness, an increase in the limiting diffusion current density, and a decrease in the degree of diffusion control in the electrodeposition process. The mechanism for the formation of the very disperse cauliflower-like particles, shown in Figure 4, was completely different to the one responsible for the formation of the dendritic particles (Figure 3). The concept of "effective overpotential" can be applied to the formation of this particle type [19,39]. According to this concept, when hydrogen evolution is vigorous enough, the electrodeposition process occurs at an overpotential that is effectively lower than the specified one, and this overpotential is denoted as "effective" in the deposition process. From the morphological point of view, this means that the morphologies of metal deposits become similar to those obtained at some lower overpotentials at which the hydrogen evolution does not occur or is very slow. Formation of the cauliflower-like particles instead of dendrites confirms that there is a lower degree of diffusion control at this current density than at 14.4 mA·cm^{-2}.

The third type of particles are irregular compact agglomerates of Cu grains (Figure 5). On the basis of the macromorphology, which is determined by the compact irregular shapes, it is clear that the particles were formed by some non-electrochemical method of synthesis.

The XRD analysis showed that the chemically obtained particles and those obtained by electrolysis at 14.4 mA·cm^{-2} were randomly oriented. Unlike these, the electrolytically obtained particles at 384 mA·cm^{-2} showed (220) and (311) preferred orientation. The different preferred orientation of the particles produced by the galvanostatic regime of electrolysis can be explained from the electrochemical point of view using the basic laws related to the metal electrocrystallization processes [19]. Namely, the (111) plane is a "slow-growing" one, while the other planes ((200), (220) and (311)) are the "fast-growing" planes [33,40]. In the growth process, the "fast-growing" planes disappear, while the "slow-growing" (111) plane survives. This explains the predominant orientation of Cu crystallites in the (111) plane in particles obtained at both current densities. On the other hand, a nucleation rate depends on the potential according to Equation (6) [19]:

$$J = K_1 \exp\left(-\frac{K_2}{E^2}\right) \tag{6}$$

where J is the nucleation rate, K_1 and K_2 are the constants independent of potential, and E is the potential. According to Equation (6), the nucleation rate increases with increasing the potential of electrolysis. At current density of 14.4 mA·cm^{-2}, a potential response was in the (275–900) mV range, while at 384 mA·cm^{-2}, this response was in the (1100–1250) mV range. Therefore, the nucleation rate was considerably larger at 384 mA·cm^{-2} than at 14.4 mA·cm^{-2}, causing the formation of a larger number of nuclei in the initial stage of electrolysis at 384 mA·cm^{-2} than at 14.4 mA·cm^{-2}. With the increased number of initially formed nuclei, there is an increased probability that the larger number of the "fast-growing" planes survive the growth process. In this way, the larger number of Cu crystallites oriented in (220) and (311) planes in the particles produced at 384 mA·cm^{-2} than at 14.4 mA·cm^{-2} can be explained.

The correlation between the morphology of the particles and crystal structure is also observed for some types of chemically synthesized particles. Namely, Cu nanowires synthesized under hydrothermal conditions using oleyl amine (OLA) and glucose as a reducing agent showed an XRD pattern with a higher intensity (200) peak than (111) peak [41]. The larger ratio of Cu crystallites oriented in the (200) plane than in (111) plane can be ascribed to the existence of a nanocubic structure [42], where a relatively large number of nanocubes coexist with the nanowires synthesized using OLA.

It is clear from the above consideration that electrolysis processes have certain advantages in the metal powder production over other methods. The advantages of this method include the easy control of the shape and size of particles by the choice of parameters of electrolysis. On the other hand, the particles obtained by chemical processes show a tendency towards aggregation [43], as observed here, which is avoided in the electrolysis processes by a vigorous hydrogen evolution reaction that can prevent this process. It is clear that the "current of hydrogen" formed through the interior of cauliflower-like particles (Figure 4e,f) prevents this aggregation.

5. Conclusions

Three types of particles with completely different shapes were analyzed by SEM, XRD and PSD. Two of them were obtained by electrolysis: the 3D dendrites and very disperse cauliflower-like agglomerates of the Cu grains. The third type of particle is a commercially available powder consisting of relatively compact agglomerates of the Cu grains. On the basis of the shape of particles, it was concluded that these particles were obtained by some non-electrochemical method.

Copper powders were synthesized by the galvanostatic regime of electrolysis at current densities of 14.4 and 384 mA·cm^{-2} at a temperature of 21.0 $\pm$ 0.5 °C using 0.10 mol·dm^{-3} CuSO$_4$ in 0.50 mol·dm^{-3} H$_2$SO$_4$. These current densities were selected to enable formation of either dendritic or cauliflower-like particles.

The dendrites obtained in this investigation ended in globules, which is different to the usual shape of Cu dendrites which have sharp tips at the top of the trunk and branches. The formation of globules at the tops of both the trunk and branches indicated the formation of a novel type of the Cu dendrites.

The Cu crystallites in the particles were randomly oriented (in both the electrolytic-produced powder at the current density of 14.4 mA·cm^{-2} and chemically synthesized powder) or showed (220) and (311) random orientation (in the galvanostatically produced powder at 384 mA·cm^{-2}). The change in preferred orientation for the galvanostatically produced powder was discussed on the basis of the theory of metal electrocrystallization.

The specific surface area of the Cu powders increased in the following order: SSA(14.4) < SSA(CHEM) < SSA(384).The particle size corresponding to the maximum volume ratio decreased similarly.

On the basis of comprehensive analysis of the three various forms of Cu powder particles, it is concluded that the powder obtained by electrolysis at the current density of 384 mA·cm^{-2} is more advantageous owing to it having the largest SSA and smallest particle size.

Author Contributions: L.A. performed the formation of copper powders by an electrolytic procedure; V.M.M. performed the XRD analysis and discussion of the corresponding data; Z.B. performed the SEM characterization of copper powders; N.I. performed the PSD analysis of copper powders; R.M. performed the preparation of powders for the SEM, XRD and PSD analysis; M.B. provided the reagents and equipment for the electrochemical experiments; and N.D.N. conceived and wrote the paper.

Funding: This research received no external funding.

Acknowledgments: This work was supported by the Ministry of Education, Science, and Technological Development of the Republic of Serbia under the research Project: "Electrochemical Synthesis and Characterization of Nanostructured Functional Materials for Application in the New Technologies" (Project No. 172046).

Conflicts of Interest: The authors declare no conflict of interest.

References

1. Copper and Copper Alloy Powder Metallurgy Properties and Applications. Available online: https://www.copper.org/resources/properties/129_6 (accessed on 13 November 2018).
2. Stopić, S.; Dvorak, P.; Friedrich, B. Synthesis of spherical nanosized copper powder by ultrasonic spray pyrolysis. *World Metall.* **2005**, *58*, 195–201.
3. Meshram, P.; Sinha, M.K.; Sahu, S.K.; Khan, P.; Pandey, B.D.; Mankhand, T.R. Solvothermal synthesis of high value copper powder from copper bleed solution of an indian copper smelter. *Powder Technol.* **2013**, *233*, 335–340. [CrossRef]
4. Jhajharia, R.; Jain, D.; Sengar, A.; Goyal, A.; Soni, P.R. Synthesis of copper powder by mechanically activated cementation. *Powder Technol.* **2016**, *301*, 10–15. [CrossRef]
5. Wu, S. Preparation of Fine Copper Powder using ascorbic acid as reducing agent and its application in MLCC. *Mater. Lett.* **2007**, *61*, 1125–1129. [CrossRef]
6. Sinha, A.; Das, S.K.; Vijaya Kumar, T.V.; Rao, V.; Ramachandrarao, P. Synthesis of nanosized copper powder by an aqueous route. *J. Mater. Synth. Process.* **1999**, *7*, 373–377. [CrossRef]

7.	Wu, S.P.; Meng, S.Y. Preparation of micron size copper powder with chemical reduction method. *Mater. Lett.* **2006**, *60*, 2438–2442. [CrossRef]

8.	Wu, S.P.; Ding, X.H. Preparation of fine copper powder with chemical reduction method and its application in MLCC. *IEEE Trans. Compon. Packag. Technol.* **2007**, *30*, 434–438. [CrossRef]

9.	Lee, S.; Lee, D.; Lee, S.-Y.; Cho, S.-S.; Uhm, S. Fabrication of nano-sized copper powders in liquid media via high-energy electrical explosion method: Use of high purity copper recovered from waste jelly-filled cable as raw material. *Mater. Test.* **2016**, *58*, 161–164. [CrossRef]

10.	Neikov, O.; Naboychenko, S.; Mourachova, I.; Gopienko, V.; Frishberg, I.; Lotsko, D. Production of copper and copper alloy powders. In *Handbook of Non-Ferrous Metal Powders: Technologies and Applications*; Elsevier: Oxford, UK, 2009; Chapter 16; pp. 331–332.

11.	Rosenband, V.; Gany, A. Preparation of nickel and copper submicrometer particles by pyrolysis of their formats. *J. Mater. Process. Technol.* **2004**, *153*, 1058–1061. [CrossRef]

12.	Chokratanasombat, P.; Nisaratanaporn, E. Preparation of ultrafine copper powders with controllable size via polyol process with sodium hydroxide addition. *Eng. J.* **2012**, *16*, 39–46. [CrossRef]

13.	Amit, S.; Sharma, B.P. Preparation of copper powder by glycerol process. *Mater. Res. Bull.* **2002**, *37*, 407–416. [CrossRef]

14.	Chow, G.M.; Kurihara, L.K.; Kemner, K.M.; Schoen, E.; Elam, W.T.; Ervin, A.; Keller, S.; Zhang, Y.D.; Budnick, J.; Ambrose, T. Structural, morphological, and magnetic study of nanocrystalline cobalt-copper powders synthesized by the polyol process. *J. Mater. Res.* **1995**, *10*, 1546–1554. [CrossRef]

15.	Djokić, S.S. Production of Metallic Powders from Aqueous Solutions without an External Current Source. In *Electrochemical Production of Metal Powders, Series: Modern Aspects of Electrochemistry*; Djokić, S.S., Ed.; Springer: New York, NY, USA, 2012; Volume 54, pp. 369–398.

16.	Djokić, S.S.; Djokić, N.S. Electroless deposition of metallic powders. *J. Electrochem. Soc.* **2011**, *158*, D204–D209. [CrossRef]

17.	Reverberi, A.P.; Salerno, M.; Lauciello, S.; Fabiano, B. Synthesis of copper nanoparticles in ethylene glycol by chemical reduction with vanadium (+2) salts. *Materials* **2016**, *9*, 809. [CrossRef] [PubMed]

18.	Park, T.J.; Lee, K.G.; Lee, S.Y. Advances in microbial biosynthesis of metal nanoparticles. *Appl. Microbiol. Biotechnol.* **2016**, *100*, 521–534. [CrossRef] [PubMed]

19.	Popov, K.I.; Djokić, S.S.; Nikolić, N.D.; Jović, V.D. *Morphology of Electrochemically and Chemically Deposited Metals*; Springer: New York, NY, USA, 2016; pp. 1–368.

20.	Nikolić, N.D.; Pavlović, Lj.J.; Pavlović, M.G.; Popov, K.I. Morphologies of electrochemically formed copper powder particles and their dependence on the quantity of evolved hydrogen. *Powder Technol.* **2008**, *185*, 195–201. [CrossRef]

21.	Orhan, G.; Hapci, G. Effect of electrolysis parameters on the morphologies of copper powder obtained in a rotating cylinder electrode cell. *Powder Technol.* **2010**, *201*, 57–63. [CrossRef]

22.	Orhan, G.; Gezgin, G.G. Effect of electrolysis parameters on the morphologies of copper powder obtained at high current densities. *J. Serb. Chem. Soc.* **2012**, *77*, 651–665. [CrossRef]

23.	Ostanina, T.N.; Rudoi, V.M.; Patrushev, A.V.; Darintseva, A.B.; Farlenkov, A.S. Modelling the dynamic growth of copper and zinc dendritic deposits under the galvanostatic electrolysis conditions. *J. Electroanal. Chem.* **2015**, *750*, 9–18. [CrossRef]

24.	Nikolić, N.D.; Branković, G.; Pavlović, M.G. Correlate between morphology of powder particles obtained by the different regimes of electrolysis and the quantity of evolved hydrogen. *Powder Technol.* **2012**, *221*, 271–277. [CrossRef]

25.	Nekouei, R.K.; Rashchi, F.; Joda, N.N. Effect of organic additives on synthesis of copper nano powders by pulsing electrolysis. *Powder Technol.* **2013**, *237*, 554–561. [CrossRef]

26.	Popov, K.I.; Pavlović, Lj.J.; Ivanović, E.R.; Radmilović, V.; Pavlović, M.G. The effect of reversing current deposition on the apparent density of electrolytic copper powder. *J. Serb. Chem. Soc.* **2002**, *67*, 61–67. [CrossRef]

27.	Pavlović, M.G.; Pavlović, Lj.J.; Maksimović, V.M.; Nikolić, N.D.; Popov, K.I. Characterization and morphology of copper powder particles as a function of different electrolytic regimes. *Int. J. Electrochem. Sci.* **2010**, *5*, 1862–1878.

28.	Nekouei, R.K.; Rashchi, F.; Ravanbakhsh, A. Copper nanopowder synthesis by electrolysis method in nitrate and sulfate solutions. *Powder Technol.* **2013**, *250*, 91–96. [CrossRef]

29. Nekouei, R.K.; Rashchi, F.; Amadeh, A.A. Using design of experiments in synthesis of ultra-fine copper particles by electrolysis. *Powder Technol.* **2013**, *237*, 165–171. [CrossRef]

30. Lou, W.; Cai, W.; Li, P.; Su, J.; Zheng, S.; Zhang, Y.; Jin, W. Additives-assisted electrodeposition of fine spherical copper powder from sulfuric acid solution. *Powder Technol.* **2018**, *326*, 84–88. [CrossRef]

31. Bakthavatsalam, R.; Ghosh, S.; Biswas, R.K.; Saxena, A.; Raja, A.; Thotiyl, M.O.; Wadhai, S.; Banpurkar, A.G.; Kundu, J. Solution chemistry-based nano-structuring of copper dendrites for efficient use in catalysis and superhydrophobic surfaces. *RSC Adv.* **2016**, *6*, 8416–8430. [CrossRef]

32. Copper Powder. Available online: http://www.cnpcpowder.com/products/copper/ (accessed on 13 November 2018).

33. Avramović, Lj.; Pavlović, M.M.; Maksimović, V.M.; Vuković, M.; Stevanović, J.S.; Bugarin, M.; Nikolić, N.D. Comparative morphological and crystallographic analysis of electrochemically- and chemically-produced silver powder particles. *Metals* **2017**, *7*, 160. [CrossRef]

34. Avramović, Lj.; Ivanović, E.R.; Maksimović, V.M.; Pavlović, M.M.; Vuković, M.; Stevanović, J.S.; Nikolić, N.D. Correlation between crystal structure and morphology of potentiostatically electrodeposited silver dendritic nanostructures. *Trans. Nonferr. Met. Soc. China* **2018**, *28*, 1903–1912. [CrossRef]

35. Berube, L.P.; Esperance, G.L. A quantitative method of determining of the degree of texture of zinc electrodeposits. *J. Electrochem. Soc.* **1989**, *136*, 2314–2315. [CrossRef]

36. Sivasubramanian, R.; Sangaranarayanan, M.V. Electrodeposition of silver nanostructures: From polygons to dendrites. *CrystEngComm* **2013**, *15*, 2052–2056. [CrossRef]

37. Xia, Y.; Xiong, Y.; Lim, B.; Skrabalak, S.E. Shape-controlled synthesis of metal nanocrystals: Simple chemistry meets complex physics? *Angew. Chem. Int. Ed.* **2009**, *48*, 60–103. [CrossRef] [PubMed]

38. Popov, K.I.; Živković, P.M.; Jokić, B.; Nikolić, N.D. The shape of the polarization curve and diagnostic criteria for the metal electrodeposition process control. *J. Serb. Chem. Soc.* **2016**, *81*, 291–306. [CrossRef]

39. Nikolić, N.D.; Popov, K.I.; Pavlović, Lj.J.; Pavlović, M.G. The effect of hydrogen codeposition on the morphology of copper electrodeposits. I. The concept of effective overpotential. *J. Electroanal. Chem.* **2006**, *588*, 88–98. [CrossRef]

40. Nikolić, N.D.; Maksimović, V.M.; Branković, G. Morphological and crystallographic characteristics of electrodeposited lead from the concentrated electrolyte. *RSC Adv.* **2013**, *3*, 7466–7471. [CrossRef]

41. Ravi Kumar, D.V.; Kim, I.; Zhong, Z.; Kim, K.; Lee, D.; Moon, J. Cu(II)–alkyl amine complex mediated hydrothermal synthesis of Cu nanowires: Exploring the dual role of alkyl amines. *Phys. Chem. Chem. Phys.* **2014**, *16*, 22107–22115. [CrossRef] [PubMed]

42. Yang, H.-J.; He, S.-Y.; Chen, H.-L.; Tuan, H.-Y. Monodisperse copper nanocubes: Synthesis, self-assembly, and large-area dense-packed films. *Chem. Mater.* **2014**, *26*, 1785–1793. [CrossRef]

43. Lu, L.; Sevonkaev, I.; Kumar, A.; Goia, D.V. Strategies for tailoring the properties of chemically precipitated metal powders. *Powder Technol.* **2014**, *261*, 87–97. [CrossRef]

Article

Preparation of Vanadium Oxides from a Vanadium (IV) Strip Liquor Extracted from Vanadium-Bearing Shale Using an Eco-Friendly Method

Yiqian Ma [1,*], Xuewen Wang [2], Srecko Stopic [1], Mingyu Wang [2], Dario Kremer [3], Hermann Wotruba [3] and Bernd Friedrich [1]

[1]   Institute of Process Metallurgy and Metal Recycling (IME), RWTH Aachen University, Intzestraße 3, 52056 Aachen, Germany; sstopic@ime-aachen.de (S.S.); bfriedrich@ime-aachen.de (B.F.)

[2]   School of Metallurgy and Environment, Central South University, Lushan South Road 932, Changsha 410083, China; wxwcsu@163.com (X.W.); wmydxx@163.com (M.W.)

[3]   Department of Mineral Processing, RWTH Aachen University, Lochnerstraße 4, 52064 Aachen, Germany; kremer@amr.rwth-aachen.de (D.K.); wotruba@amr.rwth-aachen.de (H.W.)

*   Correspondence: yma@ime-aachen.de; Tel.: +49-2419-5228

Received: 5 November 2018; Accepted: 23 November 2018; Published: 27 November 2018

Abstract: In the traditional vanadium precipitation process, the use of ammonium salts can produce serious pollution problems from the ammonia waste-water and the ammonia gas generated during the processing. In this reported study, an eco-friendly technology was investigated to prepare vanadium oxides from a typical vanadium (IV) strip liquor, obtained after the hydrometallurgical treatment of a vanadium-bearing shale. Thermodynamic analysis demonstrated that $VO(OH)_2$ could be prepared as a precursor over a suitable solution pH range. Experimental results showed that by adjusting the pH to around 5.6, at room temperature, 98.6% of the vanadium in the strip liquor was formed into hydroxide, in 5 min. After obtaining the $VO(OH)_2$, it was washed with dilute acid to minimize the level of impurities. VO_2 and V_2O_5 were then produced by reacting the $VO(OH)_2$ with air or argon, in a tube furnace. The XRD analyses of the products showed that VO_2 had been produced in air and V_2O_5 had been produced in argon. The purity of the VO_2 was 98.82% after calcining for 2 h at 550 °C, in argon flow, at a rate of 50 mL/min. It was found that the purity of the V_2O_5 was 98.70%, using the same reaction conditions in air. Compared to the traditional precipitation method that uses ammonium salt, followed by calcination, this proposed method is eco-friendly and employs less quantities of reagents and energy, and two types of products can be produced. Consequently, this process could promote the sustainable development of the vanadium chemical industry.

Keywords: vanadium precipitation; vanadium oxides; vanadium-bearing shale; vanadium strip liquor

1. Introduction

Vanadium is used in the preparation of alloys, catalysts, medicines, redox batteries, and ceramics, in the chemical industry, which makes it an essential metal in modern industry [1–6]. Although it enjoys a variety of applications, most vanadium is used in large quantities as an additive in the production of steel, primarily in the form of V_2O_5. As such, it can significantly improve the performance of steel [7]. The low-valence vanadium oxides (VO_2 and V_2O_3) are suitable for the manufacture of high vanadium-containing ferrovanadium and vanadium nitride [8,9]. In addition, VO_2 has a metal-semiconductor phase change property, which makes it suitable for use in photoelectric and magnetic thermally-sensitive materials, for a variety of applications, including smart window layers, memory layers, thermal sensors, and electrical and infrared light switching devices [10–12].

Currently, the common process used for producing V_2O_5 from various raw material sources, including vanadium-titanomagnetite, spent catalyst, fly ash from oil industry, and vanadium-bearing shale, can be summarized as pretreatment $\rightarrow$ leaching $\rightarrow$ purification $\rightarrow$ precipitation $\rightarrow$ calcination [13–17]. After obtaining a purified and enriched vanadium solution, ammonium precipitation with ammonium salts ensures a high precipitation efficiency and high-purity of the final product (V_2O_5). The precipitated ammonium polyvanadate is then decomposed into V_2O_5, at ~400 °C [18–20]. To produce VO_2, the common method is to further reduce the V_2O_5, at high temperature [21,22]. However, the use of ammonium salts causes serious pollution problems when ammonia waste-water and ammonia gas are produced during the processing [6]. The ammonia nitrogen wastewater has posed a great threat to the purity of water resources and treatment of the waste-water, to meet the strict discharge requirements, is costly [23,24]. Some new technologies, such as microbial reduction and membrane distillation, have been developed to treat the ammonia nitrogen waste-water [25,26], but simply avoiding the use of ammonium salts, altogether, in the process would be more reasonable. For example, recently, Liu et al. [27] studied a vanadium oxide electrowinnng process, from an alkaline solution. Zhang et al. [28] reported an eco-friendly technology of hydrothermal hydrogen reduction, to prepare a pure vanadium oxide product from a vanadium enriched solution. Both of these methods produced vanadium oxide directly from a vanadium solution, based on the characterization of the solutions. Autoclave and electro-decomposition units were utilized in their studies.

Vanadium-bearing shale, which is also called stone coal, accounts for 87% of the resource reserves of vanadium in China, and is widely distributed over many provinces [6]. This shale is characterized by its low vanadium-content and complex, polymetallic ores. Extraction of vanadium from this resource has been studied, extensively, over the past decade, especially, for the leaching and purification process [15,29,30]. Acid leaching, followed by solvent extraction, is the most widely applied process for treating this material and this approach has been successfully industrialized [31,32]. There are two mature solvent extraction methods for the treatment of acid leach solution—extraction of V(IV) using an acidic organophosphorus extractant, such as D2EHPA [31,33], and extraction of V(V) using an amine extractant, such as N235 [32]. After stripping, a purified vanadium liquor is obtained. In the case of the V(IV) strip liquor, it is first oxidized, if the ammonium precipitation method is to be used, and the precipitation process requires a temperature of ~90 °C. V_2O_5 is produced after calcination of the precipitate (ammonium polyvanadate).

The description of the conventional processes lends credence to the significance of developing an eco-friendly method for the vanadium precipitation process for the vanadium extraction industry. In this reported work, a novel technology for preparing vanadium oxides from a V(IV) strip liquor was studied to surmount the shortcomings of the conventional methods for vanadium oxide production. The experimental approach proposed here would be simpler than others described in the literature [27,28], and the direct preparation of a low-valence vanadium oxide was also attempted. Based on the solution's chemistry, the precursor was initially prepared and two types of products were obtained, using different atmospheres.

2. Thermodynamic Analysis

The potential-pH diagram of the V-H_2O system at 298 K is shown in Figure 1. These thermodynamic data were obtained from the literature [34]. As can be seen from this diagram, various species, in different valence states, exist in a solution of vanadium, based on its pH. Furthermore, Figure 2 shows the V-H_2O system, at low pH and the electrode potentials of $E^\theta(Fe^{3+}/Fe^{2+}) = 0.77$ V, $E^\theta(Fe^{2+}/Fe) = -0.46$ V [35]. It is known that Fe(II) cannot co-exist with V(V) at low pH, because there is no overlap in their respective predominant-areas of the diagram. In the acid leach solution of the vanadium-bearing shale, Fe(III) is the major impurity that can adversely affect the V(IV) extraction by acidic organophosphorus extractants, like D2EHPA. Therefore, prior to solvent extraction of V(IV), addition of Fe powder to the solution would reduce the Fe(III) to Fe(II), thereby, avoiding the adverse effects of Fe(III) and ensuring that vanadium is extracted from the solution in the form of V(IV). Correspondingly, the vanadium in the strip liquor is at the +4 state.

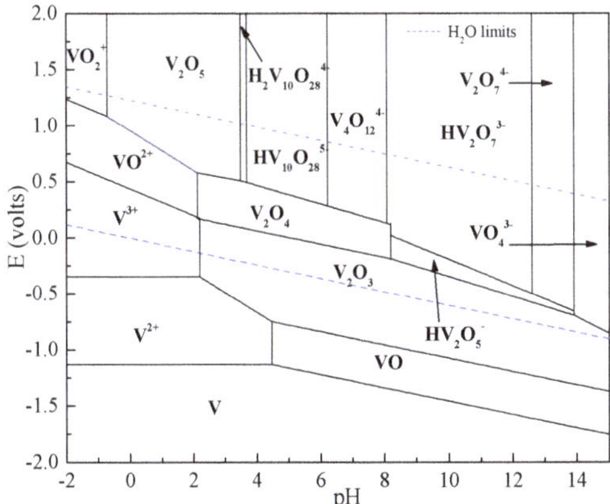

Figure 1. *E*-pH diagram of the system V-H$_2$O at 298.15 K (activity of vanadium is 1.0).

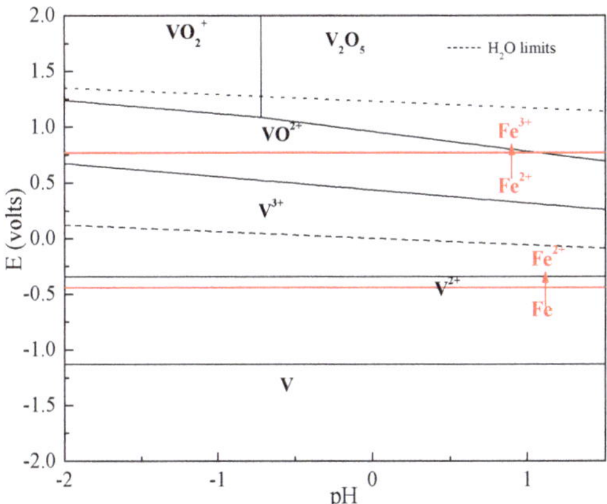

Figure 2. *E*-pH diagram of the V-H$_2$O and Fe-H$_2$O systems at 298.15 K and low pH (Activities: Vanadium 1.0, iron 1.0).

The pH value of the solution directly affects the nature of the chemical species of the vanadium in the solution. Table 1 lists the chemical reactions and $\Delta_r G^\theta$ in the V(IV)-S-H$_2$O system at 278.15 K. After calculating, the reaction equilibrium constants and the relationship between the pH and logV(IV) of the different species, were obtained. Using these data, the activity-pH diagrams for V(IV)-water-sulfur systems was plotted (Figure 3). As can be seen from these data, the activity of V(IV), in solution, would be the lowest when the pH of the solution was around 5.0. Therefore, VO(OH)$_2$ can be obtained through hydrolysis, by adjusting the pH of the strip liquor.

Table 1. Chemical reactions and $\Delta_r G^\theta$ in the V(IV)-S-H$_2$O system at 278.15 K ($\Delta_r G^\theta$ data were obtained from the literature [35]).

Chemical Reactions	$\Delta_r G^\theta$ (KJ·mol^{-1})
$VOSO_4 + H^+ = VO^{2+} + HSO_4^-$	-1.61
$HV_2O_5^- + 2SO_4^{2-} + 5\,H^+ = 2VOSO_4 + 3H_2O$	-114.92
$HV_2O_5^- = V_4O_9^{2-} + H_2O$	-1512.22
$V_2O_4 + 2SO_4^{2-} + 4H^+ = 2VOSO_4 + 2H_2O$	-68.1
$HV_2O_5^- + H^+ = V_2O_4 + H_2O$	-46.82
$V_4O_9^{2-} + 2H^+ = 2V_2O_4 + H_2O$	-90.38

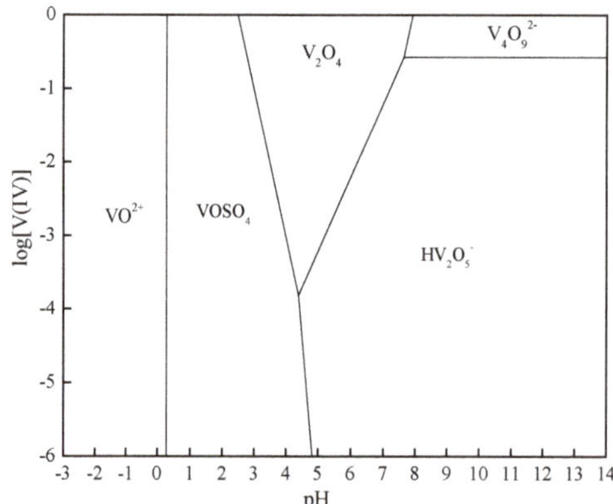

Figure 3. Activity-pH diagram for V(IV)-S-H$_2$O system at 298.15 K (activity of sulfur 1.0).

3. Materials and Methods

3.1. Materials and Analysis

The strip V(IV) liquor was obtained from a demo plant test, which was part of a project concerned with the extraction of vanadium from a decarburized vanadium-bearing shale. The vanadium-bearing shale containing 0.65 wt.% V$_2$O$_5$, 6.51 wt.% Al$_2$O$_3$, and 2.57 wt.% Fe$_2$O$_3$ was supplied by Datang Huayin Electric Power Co., Ltd. (Changsha, China). The H$_2$SO$_4$ and NaOH used in the study were of analytical grade, and all aqueous solutions were prepared using distilled water. The air and Argon gas used in this work were 99.9% pure and were supplied by Changsha Zhanyuan Gas Co., Ltd. (Changsha, China).

Vanadium was titrated with ammonium ferrous sulphate. The contents of other elements were measured by inductively-coupled plasma emission spectroscopy (ICP) with a PS-6 PLASMA SPECTROVAC, BAIRD (Waltham, MA, USA). Thermal gravimetric analysis (TGA), (Pyris 1 TGA, Perkin Elmer, Waltham, MA, USA) was used to identify the chemical reactions that occurred in the samples during the calcination processes. The products were characterized by X-ray diffraction (XRPD, PANalytical X'PERT-PRO diffractometer, Malvern Panalytical, Eindhoven, The Netherland).

3.2. Procedures

3.2.1. Source of the V(IV) Strip Liquor

The process flowchart for preparing the strip V(IV) liquor from the vanadium-bearing shale is shown in Figure 4. As shown, the sulfuric acid roasting, after water leaching, achieved a high leaching efficiency of vanadium. This technology and the detailed parameters of the process have been reported by our group [36]. The main composition of the acid leach solution is listed in Table 2, where it can be seen that it was a typical acid vanadium leach solution, with impurities like iron and aluminum. After removing the Al using alum, iron powder was used as the reductant to reduce the iron in solution from Fe(III) to Fe(II), avoiding the interference of Fe(III) in the V(IV) extraction. This principle was detailed in the previous thermodynamic analysis. After this initial treatment, Na$_2$CO$_3$ is added to the mixture to adjust the pH of the leach solution to about 2.5. Four stages of solvent extraction was then conducted, under the following conditions—the kerosene solution with 20% (*v/v*) D2EHPA and 5% (*v/v*) TBP, *A/O* phase ratio of 1:1, and an equilibrium time of 5 min. At last, the loaded organic phase was stripped by 1.0 mol/L H$_2$SO$_4$, with four stages, at an *O/A* phase ratio of 5:1 and equilibrium time of 5 min. The extraction and stripping process can be expressed by Equations (1) and (2):

$$\text{Extraction: VO}^{2+}_{(aq)} + 2(\text{HA})_{2(o)} \rightleftharpoons \text{VOA}_2 \cdot 2\text{HA}_{(o)} + 2\text{H}^+_{(aq)} \tag{1}$$

$$\text{Stripping: } VOA_2 \cdot 2HA_{(o)} + 2H^+{}_{(aq)} \rightleftharpoons VO^{2+}{}_{(aq)} + 2(HA)_{2(o)} \tag{2}$$

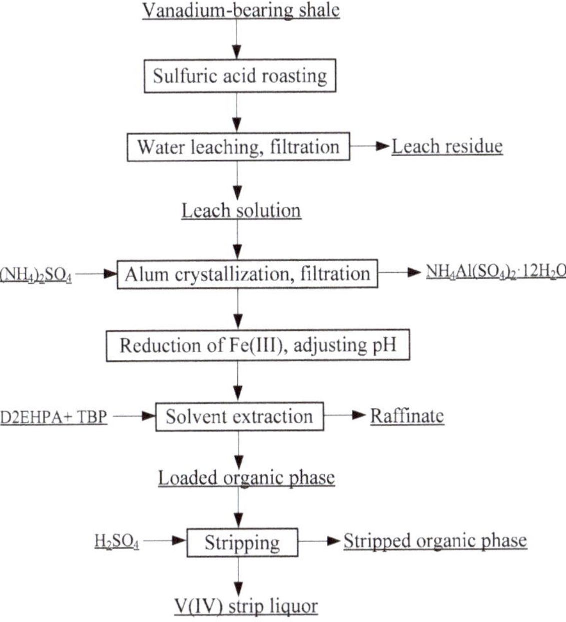

Figure 4. Proposed flowchart for preparation of the V(IV) strip liquor from vanadium-bearing shale.

Table 2. Chemical composition of the acid leach solution and the V(IV) strip liquor.

Concentration (g/L)	V	Fe(II)	Fe(III)	Al	Mg	K	Na	Ca	SO_4^{2-}
Acid leach solution	2.31	2.71	6.84	25.79	1.09	7.44	1.01	1.11	185
V(IV) strip liquor	22.1	0	0.009	0.01	0.008	0.006	<0.001	0.005	98

A blue band of V(IV) liquor, with a pH of about 0, containing 22.1 g/L vanadium was obtained. This band of V(IV) liquor was the raw material used to prepare the vanadium oxides and its composition is shown in Table 2. The results in Table 2 show that impurities had been removed in the process of enriching the vanadium, during the solvent extraction process.

3.2.2. Preparation of VO_2 and V_2O_5

Figure 5 shows the proposed flowchart for the preparation of vanadium oxides from the V(IV) strip liquor. Before calcination, the $VO(OH)_2$ precursor was prepared. Hydrolysis precipitation was used to precipitate the vanadium from the strip liquor. The precipitation percentage was affected by the final pH value and the temperature. Another issue that arose in this process was the control of the impurities in the hydrolysis products. There were few impurity ions in the strip liquor (Table 2), but the few Fe(III) could co-precipitate with $VO(OH)_2$. It was due to the oxidation of Fe(II) by air and then the co-extraction with V(IV). Fortunately, the vanadium oxide products can have a small amount of iron, if they are applied in the steel industry. On the other hand, the hydrolysis product contained some inclusion of Na^+, as NaOH was the neutralizer. Na^+ should be eliminated by washing the precipitate. After obtaining a relatively pure $VO(OH)_2$, a tube furnace (OTL 1200, Nanjing Nanda Instrument Co., Ltd. Nanjing, China) was used to calcine the vanadium hydroxide. The maximum working temperature of the tube furnace was 1100 °C and the maximum heating rate was 10 °C/min.

The Schematic drawing of the experimental apparatus is shown in Figure 6. The reactions for the overall process can be expressed as follows:

$$2VO^{2+} + 2OH^- \rightarrow VO(OH)_2 \downarrow \tag{3}$$

$$4VO(OH)_2 + O_2 \xrightarrow[550\ ^\circ C]{Air} 2V_2O_5 + 4H_2O \tag{4}$$

$$VO(OH)_2 \xrightarrow[550\ ^\circ C]{Air} VO_2 + H_2O \tag{5}$$

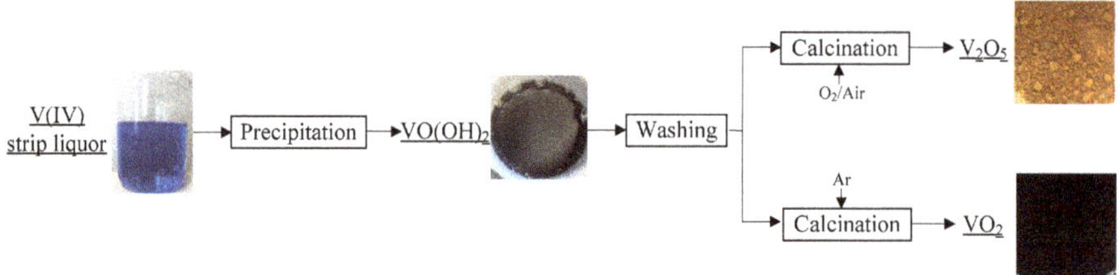

Figure 5. Proposed flowchart for preparation of vanadium oxides from the V(IV) strip liquor.

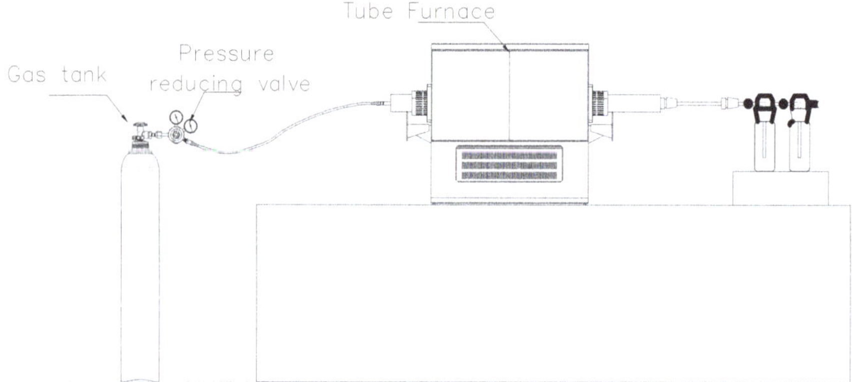

Figure 6. Schematic drawing of experimental apparatus.

4. Results and Discussion

4.1. Effect of pH on Vanadium Precipitation

As shown in Figure 3, The V(IV) in the strip liquor can be hydrolyzed and precipitated by neutralization. The effect of the final solution pH on the V(IV) precipitation was determined by varying the pH of the solution from 4.2 to 10.7 and the results are shown in Figure 7.

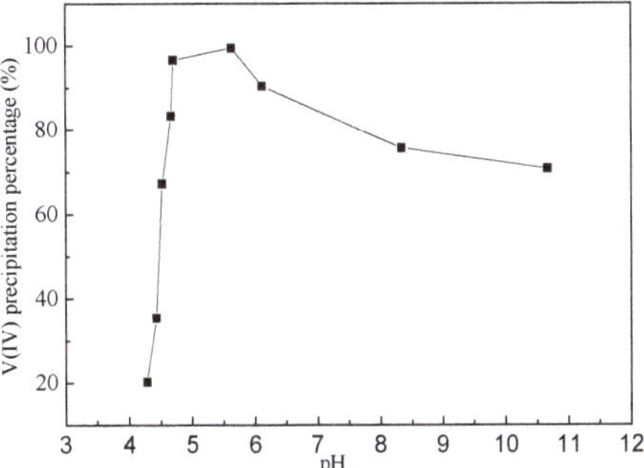

Figure 7. Effect of final pH on the vanadium precipitation percentage.

These experiments were conducted at room temperature for 30 min. As can be seen, the V(IV) precipitation yield increased with an increase in pH from 4.2 to 5.6, and it attained 99.5% at a pH of 5.6, then decreased. This effect was consistent with the data depicted in the activity-pH diagram for the V(IV)-water-sulfur system. The V(IV) was initially precipitated as $VO(OH)_2$, but this would dissolve slowly when the pH was increased further, converting the $VO(OH)_2$ to $HV_2O_5^-$. Therefore, the optimum solution pH for V(IV) precipitation was around 5.6.

4.2. Effect of Time on Vanadium Precipitation

The effect of precipitation time on the V(IV) precipitation was evaluated at room temperature, after adjusting the pH to the optimum value of 5.6. Some samples were taken at various times, during the reaction, and the vanadium content of the solution after precipitation was determined. As shown in Figure 8, the precipitation of the V(IV) product was quite rapid. The precipitation yield reached 98.6% in 5 min, and no further increase was noted, with a prolonged reaction time. Therefore, 5 min of reaction was considered to be sufficient for the full precipitation of the desired product, after adjusting the solution pH to ~5.6.

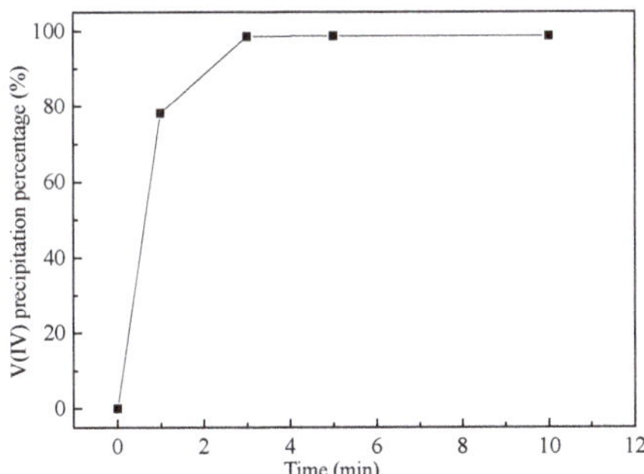

Figure 8. Effect of reaction time on the concentration of vanadium precipitate.

4.3. Effect of Temperature on Vanadium Precipitation

The effect of temperature on the vanadium precipitation is shown in Figure 9. As can been seen from these results, the V(IV) precipitation yield was comparable at temperatures below 40 °C, but higher temperatures resulted in a decrease of the precipitation yield. This effect could be due to the re-dissolution of the precipitation at these higher temperatures. Therefore, it was concluded that room temperature was the optimum temperature for this process.

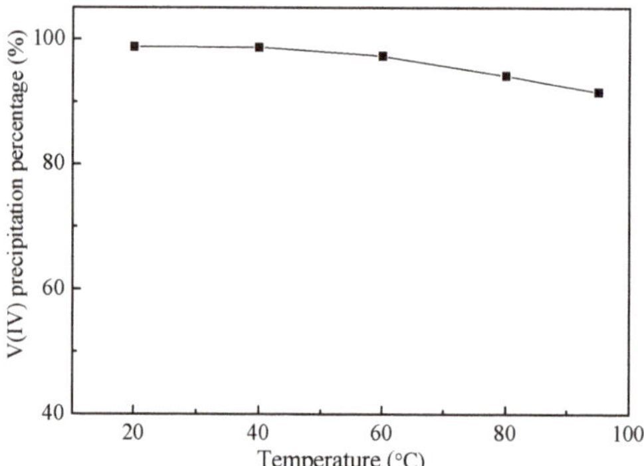

Figure 9. Effect of temperature on the vanadium precipitation percentage.

4.4. Removal of Na⁺ by Washing

The resulting precipitate always contained some Na^+, which needed to be removed before the calcining process could be conducted. It was found that washing the precipitate, with acidic water (pH ~2.0 with H_2SO_4), was effective in removing the unwanted Na^+, and this washing needed to be conducted with stirring. Figure 10 shows the effect of washing on the liquid/solid ratio of the precipitate.

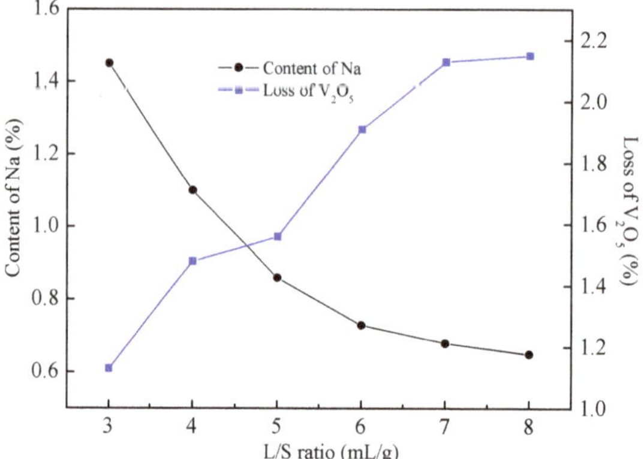

Figure 10. Effect of liquid to solid ratio on the washing of Na^+.

As can be seen from the results in Figure 10, the Na^+ content in the $VO(OH)_2$ decreased with the increase in the liquid/solid ratio, but the loss of vanadium increased, correspondingly. To meet the composition requirements of the products, an acidic water/precipitate weight ratio of 6/1 (mL/g) was used to wash out the Na^+, and then a relative pure precursor, $VO(OH)_2$ was obtained.

4.5. Preparation of VO_2 and V_2O_5 by Calcination

Thermal (TG-DTA) analysis was utilized to study the behavior of $VO(OH)_2$, during calcination in inert (argon) atmosphere and oxygen (air) atmosphere. Figure 11 shows the TG-DTA curves of the precursor in argon atmosphere.

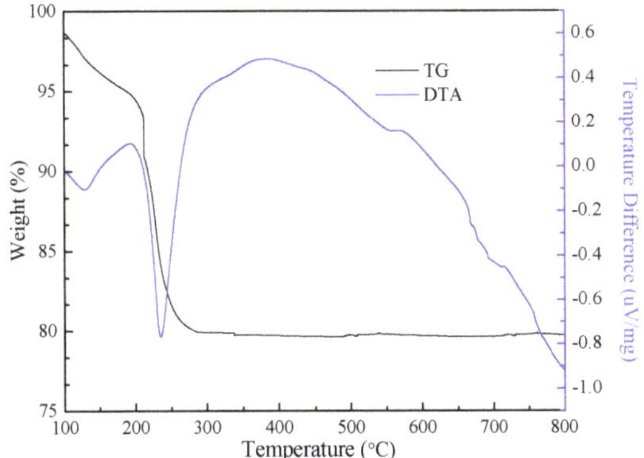

Figure 11. Thermal (TG-DTA) analysis of the VO(OH)$_2$ in argon atmosphere.

These results show that there was a small weight loss below 200 °C that was due to the loss of physically-adsorbed water, on the precipitate. A large weight loss occurred from 215 °C to 325 °C, which was accompanied by an evident endothermic peak (DTA), centered at about 238 °C. This result was attributed to the decomposition of the VO(OH)$_2$. No additional weight change and endothermic peaks could be observed in the thermogram, after the temperature was greater than 325 °C. We can infer from these results that VO$_2$ was produced and it gradually crystallized. As a comparative result, Figure 12 shows the thermal (TG-DTA) curves of the VO(OH)$_2$ calcined in an air atmosphere. Below 270 °C, the precipitate behaved similarly to the precipitate calcined in argon. The decomposition of the VO(OH)$_2$ also produced an endothermic peak (DTA) centered at about 238 °C, accompanied by a fast weight loss in the sample (TG). However, a small weight increase occurred from 275 °C to 410 °C, which was accompanied by a small exothermic peak (DTA). This was attributed to the oxidizing reaction, where the product transformed from VO$_2$ to V$_2$O$_5$. In addition, the DTA curve, in air, resulted in an obvious endothermic peak, centered at about 690 °C. This was considered to be attributable to the melting point of V$_2$O$_5$ being 690 °C.

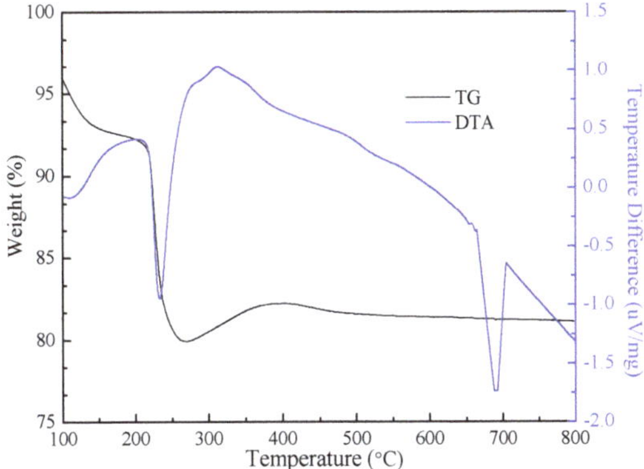

Figure 12. Thermal (TG-DTA) analysis of the VO(OH)$_2$ in air atmosphere.

Based on the results of the thermal (TG-DTA) analysis, the VO(OH)$_2$ calcined in the tube furnace in argon and air atmosphere resulted in VO$_2$, in air, and V$_2$O$_5$, in Ar. In the case of the VO$_2$, the calcining was conducted with a temperature heating rate of 10 °C/min and an argon flow rate 50 mL/min. The VO$_2$ powder product was obtained after 2 h of calcination at 550 °C. In the same manner, the V$_2$O$_5$

powder was produced, using the same conditions, with argon substituted for air. The two products were then characterized by XRD and the results are shown in Figures 13 and 14.

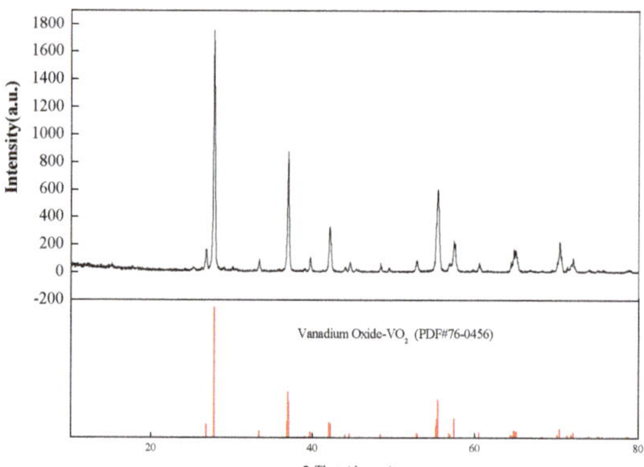

Figure 13. XRD patterns of the product obtained after calcination in argon.

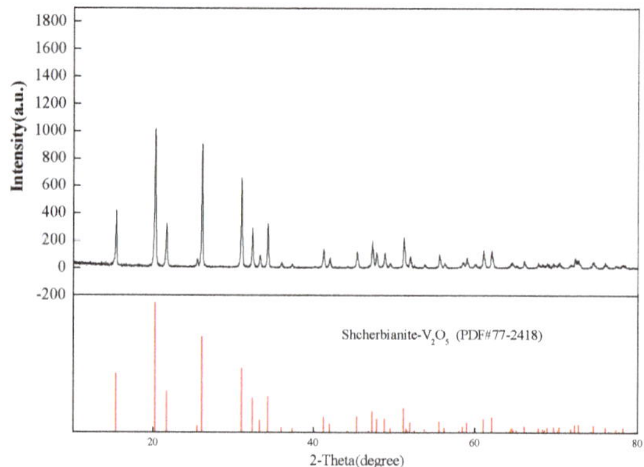

Figure 14. XRD patterns of the product obtained after calcination in air.

By comparing the XRD patterns of the experimental products with the standard patterns for VO_2 and V_2O_5, it was found that the XRD patterns of the products were identical to those of the pure VO_2 and V_2O_5. This verified that the desired products had been successfully prepared by calcining the $VO(OH)_2$ in argon and air. The compositions of the products (Table 3) were evaluated using ICP analysis, after dissolving the material in a suitable solvent. The results indicated that relatively pure products had been prepared using the proposed process flowchart.

Table 3. Chemical composition of VO_2 and V_2O_5 products.

Element	VO_2	V_2O_5	Fe	Al	S	P	Si	$Na_2O + K_2O$
VO_2 (wt.%)	98.82	-	0.26	0.005	0.003	0.03	0.001	1.1
V_2O_5 (wt.%)	-	98.70	0.25	0.006	0.003	0.03	0.002	1.2

5. Conclusions

In this reported study, a new method for the preparation of VO_2 and V_2O_5 from a typical V(IV) strip liquor, which avoids using ammonium salts, was investigated. Thermodynamic analysis of the resulting experimental products showed that a vanadium precursor, $VO(OH)_2$ could be obtained from the liquor, over a suitable pH range. It was found that adjusting the pH of the liquor to ~5.6, at room temperature, caused the vanadium precipitation yield to quickly reach its maximum. After obtaining the VO(OH)$_2$ product, washing it with dilute acid ensured a low impurity-content. VO$_2$ and V$_2$O$_5$ were then prepared from the VO(OH)$_2$, using a tube furnace, with sequential atmospheres of air and argon. Characterization of products confirmed their structure and purity, which demonstrated the feasibility and performance of this process. According to the product standards [37], the products prepared by this method are suitable for applications in metallurgy, like steel additive. Further purification is necessary for a high-purity product. Nevertheless, compared to the traditional precipitation method, this proposed eco-friendly method employs simple equipment, together with a low reagent and energy consumption. Therefore, it is a feasible process for promoting cleaner production of vanadium oxide, in the vanadium chemical industry.

Author Contributions: Y.M performed the experiments and wrote the paper; X.W, M.W. and S.S. contributed the reagents/materials/analysis tools; Y.M., D.K., H.W. and B.F. analyzed the data.

Funding: This research received no external funding.

Acknowledgments: One of the authors (Y.M.) is grateful to the Chinese Government for providing a scholarship.

Conflicts of Interest: The authors declare no conflict of interest.

References

1. Perles, T. Vanadium Market Fundamentals and Implications. In Proceedings of the Metal Bulletin 28th International Ferroalloys Conference, Berlin, Germany, 13 November 2012.
2. Skyllaskazacos, M.; Cao, L.Y.; Kazacos, M.; Kausar, N.; Mousa, A. Vanadium electrolyte studies for the vanadium redox battery—A review. *Cheminform* **2016**, *47*, 1521–1543. [CrossRef]
3. Zheng, Q.; Zhang, Y.; Huang, J.; Liu, T.; Xue, N.; Shi, Q. Optimal location of vanadium in muscovite and its geometrical and electronic properties by DFT calculation. *Minerals* **2017**, *7*, 32. [CrossRef]
4. Kear, G.; Shah, A.; Walsh, F. Development of the all-vanadium redox flow battery for energy storage: A review of technological, financial and policy aspects. *Int. J. Energy Research* **2012**, *36*, 1105–1120. [CrossRef]
5. Yang, X.; Zhang, Y.; Bao, S. Preparation high purity V$_2$O$_5$ from a typical low-grade refractory stone coal using a pyro-hydrometallurgical process. *Minerals* **2016**, *6*, 69. [CrossRef]
6. Zhang, Y.; Bao, S.; Liu, T.; Chen, T.; Huang, J. The technology of extracting vanadium from stone coal in China: History, current status and future prospects. *Hydrometallurgy* **2011**, *109*, 116–124. [CrossRef]
7. Moskalyk, R.R.; Alfantazi, A.M. Processing of vanadium: A review. *Miner. Eng.* **2003**, *16*, 793–805. [CrossRef]
8. Raja, B. Vanadium market in the world. *Steelworld* **2007**, *13*, 19–22.
9. Wu, Y.; Zhang, G.; Chou, K. Preparation of high quality ferrovanadium nitride by carbothermal reduction nitridation process. *J. Min. Metall. B Metall.* **2017**, *53*, 383–390. [CrossRef]
10. Sikong, L.; Kumbour, P. A new route for thermochromic vanadium dioxide synthesis. *Digest J. Nanomaterials Biostructures* **2015**, *10*, 135–140.
11. Osmolovskaya, O.; Murin, I.; Smirnov, V.; Osmolovsky, M. Synthesis of vanadium dioxide thin films and nanopowders a brief review. *Rev. Adv. Mater. Sci.* **2014**, *36*, 70–74.
12. Lu, Y.; Zhou, X. Synthesis and characterization of nanorod-structured vanadium oxides. *Thin Solid Films* **2018**, *660*, 180–185. [CrossRef]
13. Abbasi, M.H.; Safarnoorallah, M. extraction of vanadium oxide from boiler fuel ash. *J. Adv. Mater. Eng.* **2014**, *17*, 165–175.
14. Chen, F.; Zhang, Y.; Huang, J.; Liu, T.; Xue, N. Mechanism of enhancing extraction of vanadium from stone coal by roasting with MgO. *Minerals* **2017**, *7*, 33. [CrossRef]
15. Li, X.; Xie, B. Extraction of vanadium from high calcium vanadium slag using direct roasting and soda leaching. *Int. J. Miner. Metall. Mater.* **2012**, *19*, 595–601. [CrossRef]
16. Dash, H.; Rout, D.; Kar, B. Extraction of vanadium from vanadium bearing spent catalyst through heat treatment method. *Int. J. Innovative Res. Sci. Eng. Technol.* **2013**, *2*, 3201–3203.
17. Jena, B.; Dresler, W.; Reilly, I. Extraction of titanium, vanadium and iron from titanomagnetite deposits at pipestone lake, Manitoba, Canada. *Miner. Eng.* **1995**, *8*, 159–168. [CrossRef]

18. Deng, Z.; Wei, C.; Fan, G.; Li, M.; Li, C.; Li, X. Extracting vanadium from stone-coal by oxygen pressure acid leaching and solvent extraction. *Trans. Non-Ferrous Metall. Soc. China* **2010**, *20*, s118–s122. [CrossRef]

19. Ning, P.; Lin, X.; Wang, X.; Cao, H. High-efficient extraction of vanadium and its application in the utilization of the chromium-bearing vanadium slag. *Chem. Eng. J.* **2016**, *301*, 132–138. [CrossRef]

20. Wang, Y.; Li, D.; Zhang, H. Main factors affecting precipitation of vanadium with acidic ammonium salt and its countermeasures. *Iron Alloy* **2012**, *4*, 12–16. (In Chinese)

21. Wu, X.; Ming, X. Experimental study of V_2O_3 production with V_2O_5 as raw material. *Rare Met. Ceme. Carbides* **2015**, *43*, 15–17.

22. Xu, C.; Pang, M. Preparation of VO_2 powder by deoxidizing V_2O_5. *J. Mater. Sci. Eng.* **2006**, *24*, 252–254.

23. Surmacz-Górska, J.; Cichon, A.; Miksch, K. Nitrogen removal from wastewater with high ammonia nitrogen concentration via shorter nitrification and denitrification. *Water Sci. Tech.* **1997**, *36*, 73–78. [CrossRef]

24. Lin, L.; Yuan, S.; Chen, J.; Xu, Z.; Lu, X. Removal of ammonia nitrogen in wastewater by microwave radiation. *J. Hazard. Mater.* **2009**, *161*, 1063–1068. [CrossRef] [PubMed]

25. Li, H.; Feng, Y.; Zou, X.; Luo, X. Study on microbial reduction of vanadium metallurgical waste water. *Hydrometallurgy* **2009**, *99*, 13–17. [CrossRef]

26. Huang, W.; Zhang, Y.; Bao, S. Vacuum membrane distillation treatment of high concentration ammonia wastewater produced in vanadium extraction from stone coal. *Min. Metall. Eng.* **2012**, *32*, 103–106.

27. Liu, B.; Zheng, S.; Wang, S.; Zhang, Y.; Du, H. The electrowinning of vanadium oxide from alkaline solution. *Hydrometallurgy* **2016**, *165*, 244–250. [CrossRef]

28. Zhang, G.; Zhang, Y.; Bao, S.; Huang, J.; Zhang, L. A novel eco-friendly vanadium precipitation method by hydrothermal hydrogen reduction technology. *Minerals* **2017**, *7*, 182. [CrossRef]

29. Wang, M.; Huang, S.; Chen, B.; Wang, X. A review of processing technologies for vanadium extraction from stone coal. *Trans. Inst. Min. Metall.* **2018**. [CrossRef]

30. Liu, Y.; Yang, C.; Li, P.; Li, S. A new process of extracting vanadium from stone coal. *Int. J. Miner. Metall. Mater.* **2010**, *17*, 381–388. [CrossRef]

31. Li, X.; Wei, C.; Deng, Z.; Li, M.; Li, C.; Fan, G. Selective solvent extraction of vanadium over iron from a stone coal/black shale acid leach solution by D2EHPA/TBP. *Hydrometallurgy* **2011**, *105*, 359–363. [CrossRef]

32. Yang, X.; Zhang, Y.; Bao, S.; Shen, C. Separation and recovery of vanadium from a sulfuric-acid leaching solution of stone coal by solvent extraction using trialkylamine. *Sep. Purif. Technol.* **2016**, *164*, 49–55. [CrossRef]

33. Ma, Y.; Wang, X.; Wang, M.; Jiang, C.; Xiang, X.; Zhang, X. Separation of V(IV) and Fe(III) from the acid leach solution of stone coal by D2EHPA/TBP. *Hydrometallurgy* **2015**, *153*, 38–45. [CrossRef]

34. Zhou, X.; Wei, C.; Li, M.; Qiu, S.; Li, X. Thermodynamics of vanadium-sulfur-water systems at 298 K. *Hydrometallurgy* **2011**, *106*, 104–112. [CrossRef]

35. Dean, J. *Langes's Handbook of Chemistry*, 13th ed.; McGraw-Hill, Inc.: New York, NY, USA, 1985.

36. Wang, M.; Wang, X.; Li, B. A Novel Technology of Vanadium Extraction from Stone Coal. In *Rare Metal Technology 2015*; John Wiley & Sons, Inc.: Hoboken, NJ, USA, 2015; pp. 187–192.

37. *National Standard of People's Republic of China, Vanadium pentoxide*; YB/T 5304-2011; Standards Press of China: Beijing, China, 2005. (In Chinese)

 *metals*

Article

Synthesis of Magnesium Carbonate via Carbonation under High Pressure in an Autoclave

Srecko Stopic [1,*], Christian Dertmann [1], Giuseppe Modolo [2], Philip Kegler [2], Stefan Neumeier [2], Dario Kremer [3], Hermann Wotruba [3], Simon Etzold [4], Rainer Telle [4], Diego Rosani [5], Pol Knops [6] and Bernd Friedrich [1]

[1] IME Process Metallurgy and Metal Recycling, RWTH Aachen University, Intzestrasse 3, 52056 Aachen, Germany; cdertmann@metallurgie.rwth-aachen.de (C.D.); bfriedrich@ime-aachen.de (B.F.)

[2] Forschungszentrum Jülich GmbH, Wilhelm Johnen Strasse, 52428 Jülich, Germany; g.modolo@fz-juelich.de (G.M.); p.kegler@fz-juelich.de (P.K.); s.neumeier@fz-juelich.de (S.N.)

[3] AMR Unit of Mineral Processing, RWTH Aachen University, Lochnerstrasse 4-20, 52064 Aachen, Germany; kremer@amr.rwth-aachen.de (D.K.); wotruba@amr.rwth-aachen.de (H.W.)

[4] Department of Ceramics and Refractory Materials, GHI-Institute of Mineral Engineering, Mauerstrasse 5, 52064 Aachen, Germany; etzold@ghi.rwth-aachen.de (S.E.); Telle@ghi.rwth-aachen.de (R.T.)

[5] HeidelbergCement Technology Center–Global R&D, Oberklamweg 2-4, 69181 Leimen, Germany; Diego.Rosani@heidelbergcement.com

[6] Green Minerals, Rijksstraatweg 128, NL 7391 MG Twello, The Netherlands; planbco2@gmail.com

[*] Correspondence: sstopic@ime-aachen.de; Tel.: +49-241-80-95-860

Received: 5 November 2018; Accepted: 23 November 2018; Published: 27 November 2018

Abstract: Magnesium carbonate powders are essential in the manufacture of basic refractories capable of withstanding extremely high temperatures and for special types of cement and powders used in the paper, rubber, and pharmaceutical industries. A novel synthesis route is based on CO_2 absorption/sequestration by minerals. This combines the global challenge of climate change with materials development. Carbon dioxide has the fourth highest composition in earth's atmosphere next to nitrogen, oxygen and argon and plays a big role in global warming due to the greenhouse effect. Because of the significant increase of CO_2 emissions, mineral carbonation is a promising process in which carbon oxide reacts with materials with high metal oxide composition to form chemically stable and insoluble metal carbonate. The formed carbonate has long-term stability and does not influence the earth's atmosphere. Therefore, it is a feasible and safe method to bind carbon dioxide in carbonate compounds such as magnesite. The subject of this work is the carbonation of an olivine (Mg_2SiO_4) and synthetic magnesia sample (>97 wt% MgO) under high pressure and temperature in an autoclave. Early experiments have studied the influence of some additives such as sodium bicarbonate, oxalic acid and ascorbic acid, solid/liquid ratio, and particle size on the carbonation efficiency. The obtained results for carbonation of olivine have confirmed the formation of magnesium carbonate in the presence of additives and complete carbonation of the MgO sample in the absence of additives.

Keywords: $MgCO_3$-powder; synthesis; CO_2- absorption; olivine carbonation; autoclave; thermal decomposition; CO_2 utilization

---

1. Introduction

The significance of the results coming from greenhouse gas (GHG) emissions to both the atmosphere and our lives has already been urged and is nowadays well-known. Because of the continuous increase of CO_2 concentration in the atmosphere since the industrial revolution, various techniques are proposed. Carbon capture and utilization (CCU) is considered as the most promising technique in order to use the product in cement, transforming it into insoluble carbonate (mainly calcite and magnesite), that is able to remain stable in a geological timeframe [1].

In order to accelerate mineral carbonation, some pretreatment processes are required (microwave heating, grinding, sieving, separation, thermal decomposition, and chemical treatment). The main goal of pre-treatment processes is to increase the carbonation rate and improve the process kinetics. Typical pre-treatment methods are particle size reduction, magnetic separation and thermal treatment. Particle size reduction incorporates various grinding methods for an increase of specific surface area. In magnetic separation, undesired ferrous particles are separated from the rest of the feedstock. Thermal treatment is necessary for hydrated minerals, such as serpentine that contains H_2O molecules in the mineral structure. Pre-treatment is usually required in direct carbonation processes [2]. However, one must consider the balance between increase of reaction rate and additional energy costs, possible CO_2-production related to energy supply and the influence on the beneficial utilization of the final products. The major problem of pre-treatment is its high energy input, i.e., thermal treatment should be avoided due to the high energy demand and CO_2-emissions (depending on the energy source). Overall, the most potentially economical pre-treatment proved to be size reduction [3]. Although there are large resources, it is not a feasible feedstock material because of its crystallographic stability and thus the necessary step of thermal pre-treatment.

Industrially produced by-products containing alkaline metals are also feasible for mineral carbonation, such as numerous types of slags, scraps, red gypsum, combustion residues, fly ashes and other forms of metal oxide materials such as red mud [4]. Unlike natural feasible materials, industrial by-products usually do not require mining and pretreatment processes for utilization because they already have high alkaline metal contents which are sufficient for mineral carbonation [5]. Because of high availability in minerals and secondary materials among all of the possible materials selections, calcium oxide and magnesium oxide are the most favorable options, as shown in Equations (1) and (2) [6,7].

$$CaO_{(s)} + CO_{2(g)} = CaCO_{3(S)} + 179 kJ/mol \tag{1}$$

$$MgO_{(s)} + CO_{2(g)} = MgCO_{3(s)} + 118 kJ/mol \tag{2}$$

Although the carbonation process is an exothermic reaction, it requires an additional heat for better dissolution of carbon dioxide in water in order to form magnesium carbonate. The benefits of exothermic mineral carbonation may result in a positive net energy balance, which improves the net efficiency of a combined cycle power plant [8]. Calculating energy balances of the process is essential for the determination of the process' profitability which might be influenced by high energy costs. Furthermore, the overall reduction of carbon dioxide emissions has to be considered due to further emissions in the mineral carbonation process resulting from, e.g., transportation, grinding and processing of by-products. Furthermore, the potential use of the formed products should also be taken into consideration.

There are abundant calcium and magnesium rich minerals available in the earth's crust. Although MgO and CaO are the most abundant alkali and alkaline metal oxides, they cannot be found as binary oxides in nature. Usually, they exist as hydroxides or silicate minerals. In a mineral carbonation process, these can also be used as feedstock to form carbonates that are chemically stable in a geological timeframe. Silicate minerals usually are richer in alkaline metal content such as magnesium, sodium, and calcium. Common silicate minerals suitable for carbonation are forsterite (Mg_2SiO_4), antigorite ($Mg_3Si_2O_5(OH)_4$) and wollastonite ($CaSiO_3$) and their overall reaction conversions are given in Equations (3) to (5).

$$Mg_2SiO_4(s) + 2CO_2(g) + H_2O(l) = 2MgCO_3(s) + H_4SiO_4(aq) + 89 kJ/mol \tag{3}$$

$$Mg_3Si_2O_5(OH)_4(s) + 3CO_2(g) + 2H_2O(l) = 3MgCO_3(s) + 2H_4SiO_4(aq) + 64 kJ/mol \tag{4}$$

$$CaSiO_3(s) + CO_2(g) + 2H_2O(l) = CaCO_3(s) + H_4SiO_4(aq) + 90 kJ/mol \tag{5}$$

The aforementioned mineral carbonation using slags as reactant also have the chance to be profitable if they are built closely to the steel production site in order to reduce production costs [9]. Furthermore, heat integration with the steel production plant can reduce overall energy costs significantly [9]. Carbonation of different types of slags are widely studied, one study by Georgakopoulos [4] suggests that blended hydraulic cement BHC slag has the highest conversion rate which is 68.3%.

Red gypsum usually exists in the form of calcium sulfate dihydrate ($CaSO_4 \cdot 2H_2O$). Typically, red gypsum has a purity of 95% and has large resources in Malaysia [10]. Due to the large calcium content its carbonation is also a promising CCS option, i.e., one ton of red gypsum can stably bind 0.26 tons of gaseous CO_2. A big advantage of red gypsum as carbonation resource is no mining cost is required since it exists in the form of fine powder which favors the carbonation reaction.

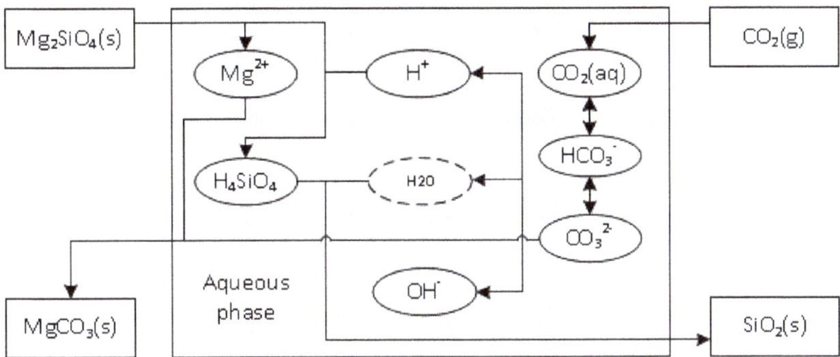

Figure 1. Reaction path of direct forsterite carbonation in aqueous solution.

Generally, the reaction path for the indirect carbonation of forsterite in an aqueous solution can be described by Equations (6) to (12) which is also illustrated in Figure 1 [11,12]. For simplicity, olivine consists only of Mg_2SiO_4, namely forsterite. First, gaseous carbon dioxide dissolves in the aqueous solution at a certain mass transfer rate as in Equation (6). Simultaneously, forsterite is dissolved in the aqueous solution (Equation (10)). In the aqueous solution, all species are assumed to be at equilibrium: Aqueous CO_2 dissociates into bicarbonate, which further dissociates into carbonate (Equations (7) and (8)). Self-ionization of water is given by Equation (9). Aqueous silicic acid then precipitates as amorphous silica, which is a by-product, and lastly magnesium ions and carbonate form magnesite (Equations (11) and (12)).

$$CO_2(g) \xrightarrow{r_{CO_2}} CO_2(aq) \tag{6}$$

$$CO_2(aq) + H_2O(l) \overset{K_{C1}}{\rightleftharpoons} HCO_3^-(aq) + H^+(aq) \tag{7}$$

$$HCO_3^-(aq) \overset{K_{C2}}{\rightleftharpoons} CO_3^{2-}(aq) + H^+(aq) \tag{8}$$

$$H_2O(l) \overset{K_W}{\rightleftharpoons} OH^-(aq) + H^+(aq) \tag{9}$$

$$Mg_2SiO_4(s) + 4H^+(aq) \xrightarrow{r_{Mg_2SiO_4}} 2Mg^{2+}(aq) + H_4SiO_4(aq) \tag{10}$$

$$H_4SiO_4(aq) \xrightarrow{r_{SiO_2}} SiO_2(s) + 2H_2O(l) \tag{11}$$

$$Mg^{2+}(aq) + CO_3^{2-}(aq) \xrightarrow{r_{MgCO_3}} MgCO_3(s) \tag{12}$$

The particular process is characterized by several equilibrium and non-equilibrium reactions. The determination of process parameters such as temperature, pressure and pH for maximum overall conversion rates is elementary. Direct CO_2 sequestration at high pressure with olivine as a feedstock has already been performed in numerous studies at different temperatures and pressures with or without the use of additives such as carboxylic acid, and sodium hydroxide [11,12]. It is reported that optimal reaction conditions are in the temperature range of 150–185 °C and in the pressure range of 135–150 bar [10]. Additives are reported to have a positive influence on carbonation rate. Optimal addition of additives are reported by Bearat et al. [13] in studies about the mechanism that limits aqueous olivine carbonation reactivity under the optimum sequestration reaction conditions observed as follows: 1 M NaCl + 0.64 M $NaHCO_3$, at 185 °C and P (CO_2) about 135 bar. A reaction limiting silica-rich passivating layer forms on the feedstocks grains, slowing down carbonate formation and raising process costs. Eikeland [14] reported that NaCl does not have significant influence on carbonation conversion. The presented results show a conversion rate of more than 90% using a $NaHCO_3$ concentration of 0.5 M, without adding of NaCl. Ideally, the solid phases exist as pure phases without growing together. In reality, different observations are made on the behavior of solid phases. Daval et al. [15] reported about high influence of amorphous silica layer formation on the dissolution rate of olivine at 90 °C and elevated pressure of carbon dioxide. This passivating layer may either built up from non-stoichiometric dissolution, precipitation of amorphous silica on forsterite particles or a combination of both. In contrast to that, Oelkers et al. [16] and Hänchen [17] observed stoichiometric dissolution and no build-up of a passivating layer except during start-up of experiments. Additionally, magnesite may precipitate on undissolved forsterite particles leading to a surface area reduction and therefore a reduction on forsterite dissolution rate, which was reported by Turri et al. [18]. Besides this undesired intermixing of solids, they observed pure particles of magnesite to be predominant in the smallest particle class, amorphous silica particles to be mainly present in the intermediate particle class and unreacted olivine particles to be predominant in the largest particle class. This knowledge may be of value for subsequent separation of products.

CO_2 sequestration with olivine as a feedstock was performed in a rocking batch autoclave at 175 °C and 100 bars in an aqueous solution and a CO_2-rich gas phase from 0.5 to 12 h. Turri showed maintainable recovery of separate fractions of silica, carbonates and unreacted olivine. Characterization of the recovered solids revealed that carbonates predominate in particle size range below 40 μm. The larger, residue fraction of final product after carbonation consisted mainly of unreacted olivine, while silica is more present in the form of very fine particles. An addition of sodium hydrogen carbonate at 0.64 M, oxalic acid at 0.5 M and ascorbic acid at 0.01 M was successfully applied in order to obtain maximal carbonation. The positive influence of the above-mentioned additives on the carbonation efficiency was reported by Olajire [19]. They studied the technology of CO_2 sequestration by mineral carbonation with current status and future prospects, but the positive influence of additives was not explained in detail.

Formation of submicron magnesite during reaction of natural forsterite in H_2O-saturated supercritical CO_2 was studied between 35 and 80 °C and at pressure of 90 bars [20]. The magnesite particles formed under below-mentioned conditions exhibited an extremely uniform submicron grain-size and nearly identical rhombohedral morphologies at all temperatures. Then an evidence for carbonate surface complexation during forsterite carbonation in wet supercritical carbon dioxide was also considered. The effect of Fe on the measured rates of olivine carbonation and its role in the formation of Si-rich surface layers, which can significantly inhibit olivine dissolution and limit the extent of the carbonation reaction was considered by Saldi et al. [21]. A series of batch and flow-through reactor experiments was conducted in pure water at 90 and 150 °C and under a CO_2 partial pressure of 100 and 200 bar, using both a natural sample of Fe-bearing olivine and a synthetic sample of pure forsterite. Experimental results show that Fe plays an ambivalent role in the carbonation.

The preparation of a magnesium hydroxy carbonate from magnesium hydroxide and carbon dioxide includes the formation of a magnesium hydroxide slurry and sparging CO_2 gas through it. Various experimental conditions are evaluated in order to obtain the conditions that result in the formation of the magnesium hydroxy carbonate [22].

Our paper deals with the formation of magnesium carbonate using an Italian olivine (35.57 wt% MgO) and a synthetic reference material, mainly consisting of magnesia (97.56 wt% MgO) with special attention on the influence of the additives and different solid/liquid ratio. After carbonation, the settled solid fraction contained mainly carbonation products, which was studied by structure and composition analysis (X-Ray Diffraction XRD) and reactivity (Thermogravimetric analysis TGA and Differential Scanning Calorimetry DSC). The water solution was analyzed by Inductively Coupled Plasma Optical Emission Spectroscopy ICP-OES in order to determine the concentration of nickel, iron, magnesium and cobalt.

2. Experimental Section

The samples used were Italian olivine (Figure 2a) and a high-grade synthetic dead burned magnesia (Figure 2b). From its chemical composition apart from the chromite present at approx. 0.45 wt% and other inert minerals at trace levels, the olivine was considered as a mixed Mg-Ni and Fe silicate.

The olivine has been delivered with a particle size of below 200 μm, the used magnesia has a grain size between 10 and 30 mm. Particle Size Distribution PSD Analysis was performed used Mastersizer Hydro 2000G (Malvern PANalytical GmbH, Kassel, Germany)

The carbonation tests were planned for the three different particle size fractions <20 μm, 20–63 μm and 100–200 μm in order to evaluate the optimal process parameters and the use of additives.

(a)　　　　　　　　　　　　　(b)

Figure 2. (**a**) Photos of Italian Olivine; (**b**) Reference material after grinding.

The olivine with $d_{90} = 100$ μm (90% below 100 microns) as it is presented in the sieve analysis in Figure 3 has been sieved wet to produce the three grain size fractions. The magnesia sample has been crushed and milled in a lab-scale jaw crusher and has also been sieved wet into the required grain size fractions.

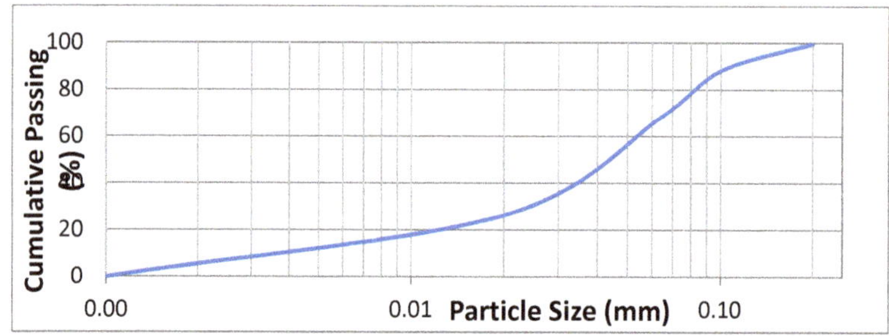

Figure 3. Sieve Analysis (Particle Size Distribution PSD) of olivine <200 μm.

The chemical composition of olivine and magnesia was analyzed by X-ray fluorescence XRF using Device PW2404 (Malvern Panalytical B.V., Eindhoven, Netherlands), such is presented in Table 1.

Table 1. Chemical composition of the investigated olivine and magnesia in wt%.

Components	Olivine	Magnesia
SiO_2	46.43	0.32
Al_2O_3	2.55	0.20
Fe_2O_3	10.88	0.58
TiO_2	0.11	0.05
CaO	2.16	0.75
MgO	35.57	97.56
K_2O	0.39	0.02
Na_2O	0.17	0.10
MnO	0.17	0.24
Cr_2O_3	0.45	0.00
P_2O_5	0.00	0.00
ZrO_2	0.02	0.03
SO_3	0.00	0.00
BaO	0.00	0.00
ZnO	0.08	0.08
NiO	0.89	0.09
Co_3O_4	0.08	0.00
CuO	0.06	0.00
Total	100.00	100.00

The planned experiments are shown in Table 2.

Table 2. Experimental plan (T = 175 °C, pCO_2: 117 bar, 300 rpm, 4 h).

Exp. No	S/L (g/mL H_2O)	Fraction Size (μm)	Concentration of Additives in Water, (mol/L)	Material
1	10/150 (0.066)	100–200	No	Olivine, Italy (35.57 wt% MgO)
2	10/150 (0.066)	<20	No	Olivine, Italy (35.57 wt% MgO)
3	30/150 (0.2)	20–63	0.64 $NaHCO_3$ 0.06 $H_2C_2O_4$ 0.003 $C_6H_8O_6$,	Olivine, Italy (35.57 wt% MgO)
4	30/150 (0.2)	20–63	No	Synthetic magnesia (97.56 wt% MgO)

Carbonation tests have been carried out in the 250 mL autoclave from Parr Instrument Company (Moline, IL, USA), USA as shown in Figure 4 at 175 °C and 117 bars with pure grade CO_2. An amount ranging from 10 to 30 g olivine has been added to 150 mL solution in different experiments.

After reaction, the liquid had very low contents of metal cations, therefore characterization of the reaction products was restricted to the solid phase by TGA/DSC using Instrument STA 449F3 with Proteus Software (NETZSCH, Selb, Germany) and XRD-Analysis using Bruker D8 Advance with LynxEye detector (Bruker AXS, Karlsruhe, Germany). X-ray powder diffraction patterns were collected on a Bruker-AXS D4 Endeavor diffractometer in Bragg–Brentano geometry, equipped with a copper tube and a primary nickel filter providing Cu Kα1,2 radiation (λ = 1.54187 Å).

Figure 4. Parr Autoclave with maximum pressure of 200 bar and maximum temperature of 250 °C.

3. Results and Discussion

The characterization of products was performed using TGA, DSC, XRD and ICP-OES analysis (SPECTRO ARCOS, SPECTRO Analytical Instruments GmbH, Kleve, Germany) in order to confirm the formation of $MgCO_3$. Additionally, the influence of additives on carbonation was discussed.

3.1. Product Characterization–XRD Analysis of Product after Carbonation

To evaluate the overall capability of the carbonation process, an experiment was performed on a synthetic reference material (>97 wt% MgO) considering the present mineralogical phases detected via XRD before and after the carbonation (Figures 5 and 6).

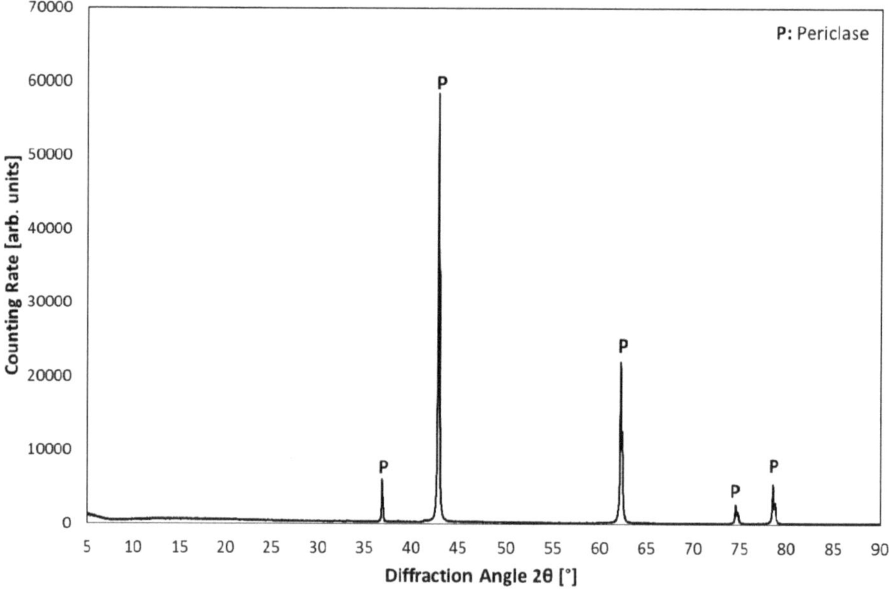

Figure 5. XRD analysis of reference material.

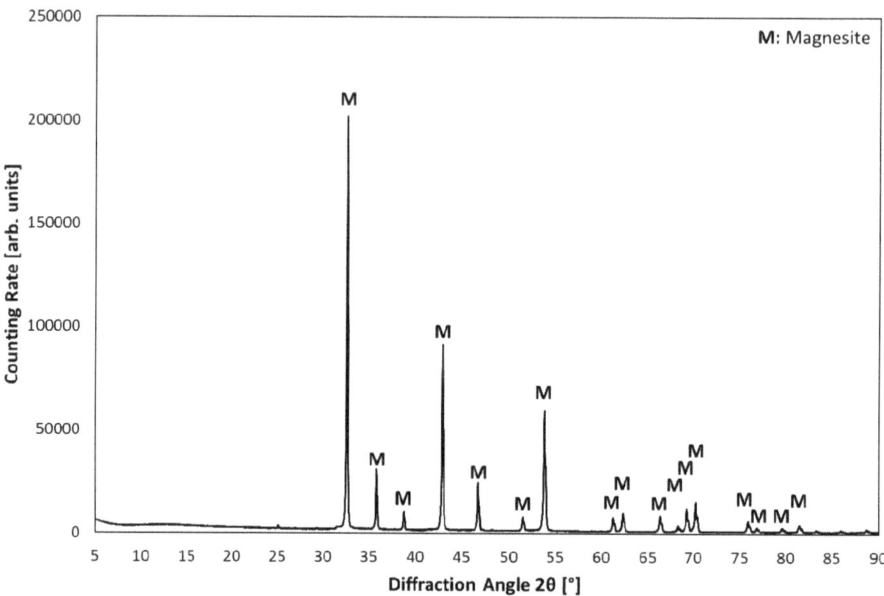

Figure 6. XRD analysis of reference sample after carbonation process.

The results prove the formation of magnesite ($MgCO_3$) out of periclase (MgO) in the absence of any additives. Both XRD patterns show the existence of a single phase, which underlines the capability of both the reference material and the carbonation process. In the next step of the present study, the carbonation process was applied to an Italian olivine sample as an exemplary natural raw material aiming at a comparable $MgCO_3$ formation as observed utilizing a synthetic reference material. XRD analysis of the initial olivine sample confirms the presence of forsterite, enstatite, clinoenstatite, lizardite, spinel and tremolite, based on hydroxide and silicate of magnesium, calcium and iron (Figure 7).

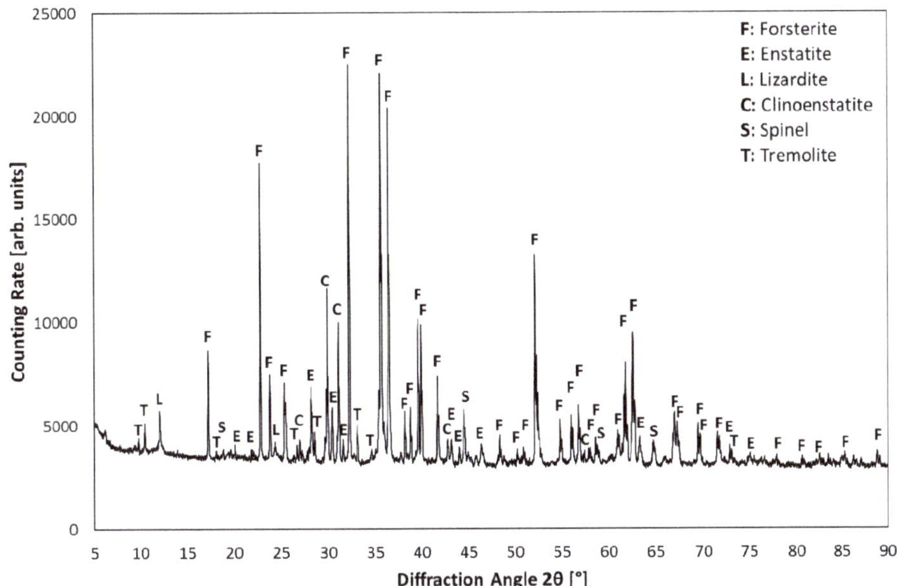

Figure 7. XRD analysis of initial olivine sample.

Unfortunately, the chosen parameters (investigated fraction of 100–200 µm, solid/liquid ratio of 0.066 at 175 °C and 117 bar) did not contribute to the formation of magnesite as analyzed by XRD. Applying the same experimental conditions on an olivine sample ground to a particle size below 20 µm did not yield any magnesite formation as well. However, followed by the addition of $NaHCO_3$, $H_2C_2O_4$ and $C_6H_8O_6$ the formation of $MgCO_3$ can be proved via XRD (Figure 8) when using a fraction size of 20–63 µm.

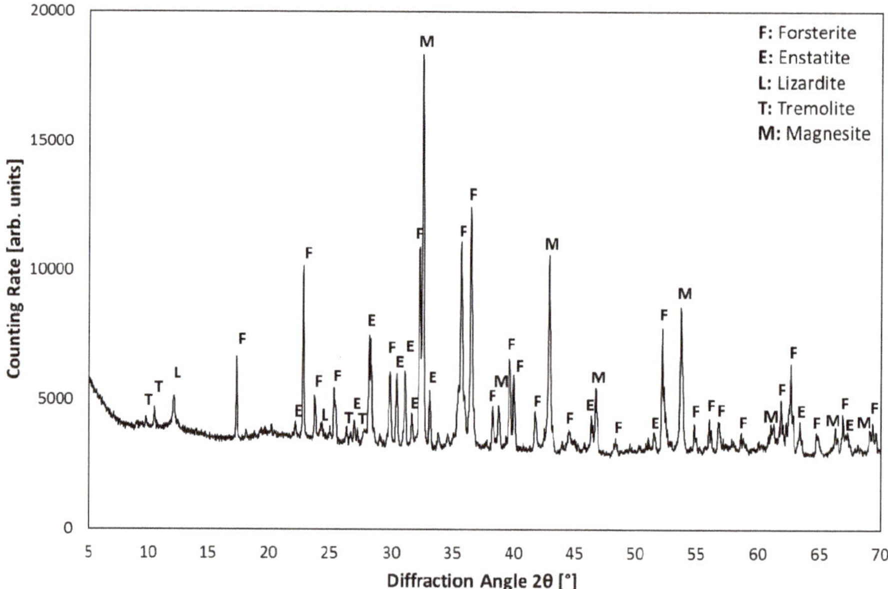

Figure 8. XRD analysis of olivine sample after carbonation.

3.2. Analysis of Water Solution after Carbonation of an Olivine

The ICP-OES analysis of water solution after carbonation, as shown in Table 3 confirms very small dissolution of valuable elements such as cobalt. The most dominant species are magnesium and silicon in the order of magnitude of mg/L what corresponds to few percent of leaching efficiency. The pH value confirmed that the solution is a neutral medium.

Table 3. Chemical analysis of Si, Mg, Fe, Ni, Cr, and Al in solution (mg/L) after carbonation.

Exp. No.	Si	Mg	Fe	Ni	Cr	Co	Al
2 (pH-6.7)	166	229	0.11	0.73	<1	<1	<1
3 (pH-7.3)	195	705	67	10	<1	<1	<1
4 (pH-2.32)	<1	211	<1	<1	<1	<1	<1

3.3. Carbonation Extent

Thermogravimetric analysis measurements were performed in order to establish the carbonation effect. The calculated carbonation was about 45% in the presence of additives, as shown in Figure 9.

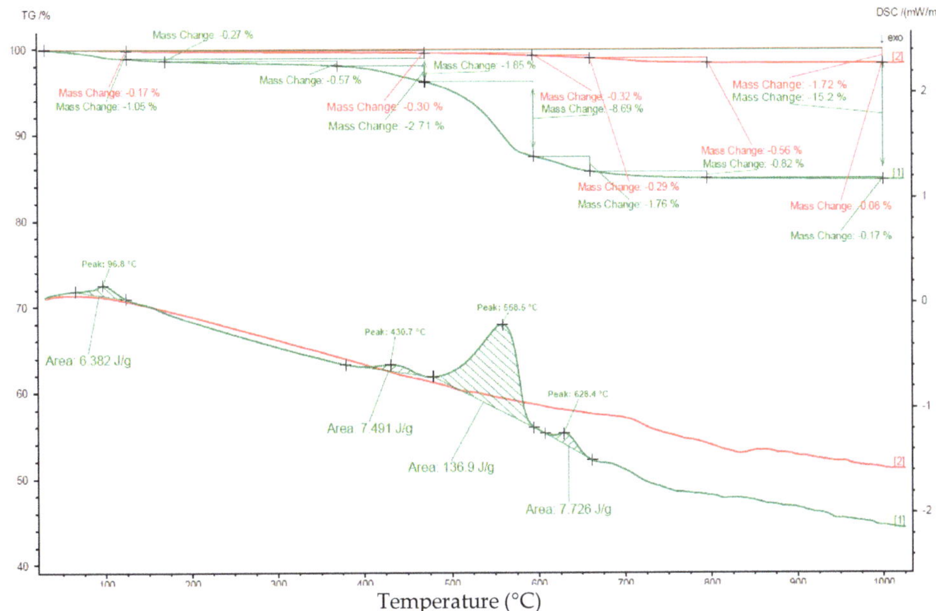

Figure 9. TGA/DSC Analysis in Exp. 3 in the presence of additives.

The total analysis of thermal decomposition of formed $(Mg,Fe)CO_3$ is shown in Table 4.

Table 4. Thermal Decomposition of samples before carbonation and after carbonation.

Interval	Before Carbonation		After Carbonation	
Interval [°C]	Loss of weight [%]	Peak area [J/g]	Loss of weight [%]	Peak area [J/g]
25–125	0.17		1.05	6.38
125–385	0.22		0.93	No peak
385–470	0.08	no peak	1.77	8.36
470–595	0.32		8.69	137
595–660	0.29		0.75	7.73
660–725	0.44	4.40	0.62	4.10
725–1000	0.20	No peak	0.36	No peak
25–1000	1.7		15.2	

As shown in Table 4, the total weight loss for an initial sample of the Italian olivine in the interval between 25 and 1000 °C amounts 1.7%, what is the amount of bound water in the used sample. The total weight loss for the initial olivine material in the interval between 25 and 470 °C amounts 0.47% in comparison to 3.75% for this sample after carbonation. In difference for the sample without carbonation, in the temperature interval between 470 °C and 595 °C, the weight loss amounts 8.69%, with a thermal effect of 137 J/g, what confirms that this temperature range is most important for thermal decomposition of $(Mg,Fe)CO_3$. Above 775 °C the change of weight is not significant, and thermal decomposition of sample is minimal, what means that the thermal decomposition is finished. In contrast to this experiment in the presence of additives in Exp. 3 with loss of weight of 15.2%, the overall weight losses of experiment 1 is very small (few percent) what confirms a low carbonation degree (as confirmed with XRD analysis in Figure 6). The weight loss for experiment 2 is about 3.5% (red line at Figure 10) which confirms very small carbonation rate in the absence of additives in comparison to the experiments with additives.

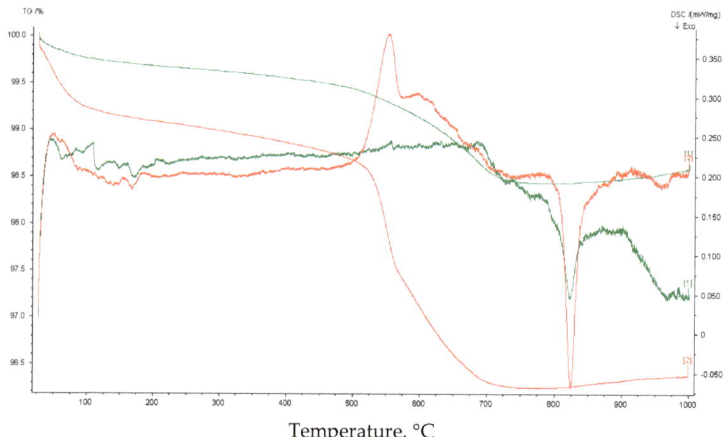

Figure 10. TGA and DSC Analysis in the experiments 1 and 2 (TGA and DSC for experiment 1- green color; TGA and DSC for experiment 2- red color).

This positive effect may be due to "reaction-driven cracking" in the presence of NaHCO$_3$, formation of etch pits, and/or other processes that continually renew the reactive surface area of Mg$_2$SiO$_4$.

$$NaHCO_3 \text{ (aq)} \rightarrow Na^+ + H^+ + CO_3^{2-} \tag{13}$$

An addition of oxalic acid leads to formation of Mg-ions in solution, which react with carbonate ions forming magnesium carbonate.

$$Mg_2SiO_4 + 2\,H_2C_2O_4 \rightarrow 2Mg^{2+} + C_2O_4^{2-} + H_4SiO_4 \tag{14}$$

$$Mg^{2+} + CO_3^{2-} \rightarrow MgCO_3 \text{ (s)} \tag{15}$$

The two analyses of thermal decomposition of products after carbonation of magnesia (97 wt%) confirmed total decomposition of the produced MgCO$_3$.

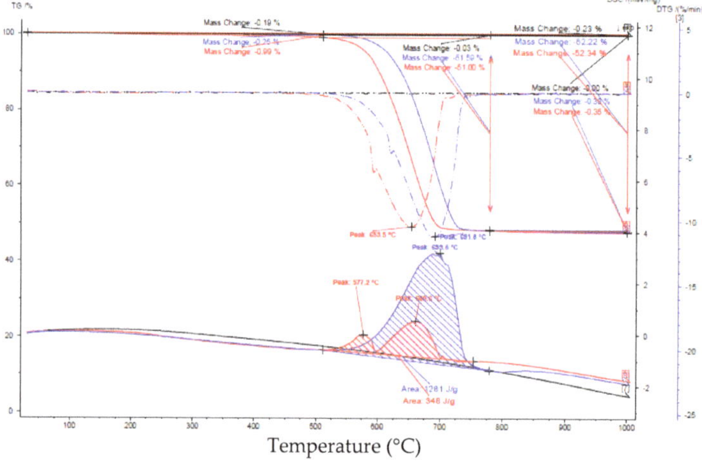

Figure 11. TGA and DSC Analysis of formed product after carbonation of synthetic magnesia (20 °C/min, nitrogen, Exp. 4).

As shown at Figure 11, the weight loss of 52.34 % (TGA-red line) and 52.22 (TGA-blue line) with maximal DSC effect (1281 J/g; blue area) at T_{max} = 699.8 °C confirm the formation of magnesium carbonate, what was compared with theoretical value, according to the Equation (16):

$$MgCO_3(s) \xrightarrow{T} MgO + CO_2(aq) \tag{16}$$

where: M (Mg) = 24.30 g/mol; M (MgCO$_3$) = 84.30 g/mol, M (MgO)= 40.30 g/mol, M (CO$_2$) = 44.0 g/mol.

Using a ratio between molar mass of magnesium carbonate and magnesium oxide, theoretical calculated loss of carbon dioxide amounts 52.55%, what is in good accordance with an experimental determined weight loss. Finally, it confirms the completed carbonation of synthetic magnesia under the chosen parameters and formation of magnesium carbonate, which particle size distribution was shown at Figure 12.

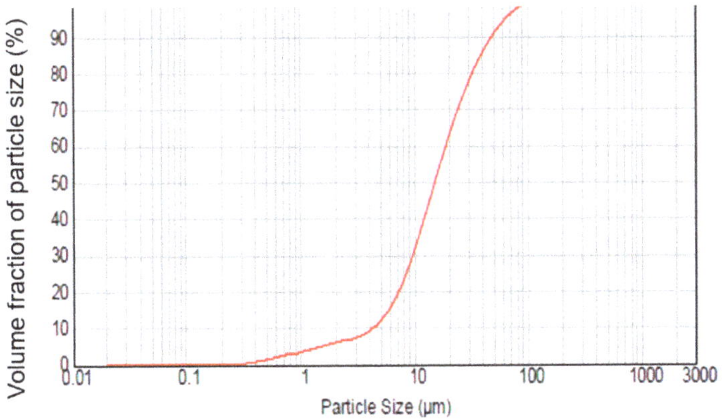

Figure 12. Particle size distribution of MgCO$_3$ after carbonation of synthetic magnesia.

The measured values of produced magnesium carbonate d$_{50}$ and d$_{84}$ amount 15.066 and 37.066 µm, respectively.

4. Conclusion

Synthesis of magnesium carbonate was studied via carbonation of olivine using different size fractions (under 20 µm, between 20 and 63 µm, and between 100 and 200 µm) with different solid/liquid ratios of 1:15 and 1:5; at 175 °C and pressure of CO$_2$ (117 bar) in an autoclave in the presence and in the absence of additives. The characterization of products using XRD, TGA, PSD and DSC analysis has confirmed the formation of MgCO$_3$. In contrast to carbonation of olivine in the absence of additives the formation of magnesium carbonate is possible at high pressure and temperature with olivine (35.57% MgO) from Italy in the presence of sodium hydrogen carbonate, oxalic acid, and ascorbic acid at 175 °C, 117 bar in 4 h (Exp. 3). The maximum carbonation (more than 95%) was obtained at the same conditions for synthetic magnesia (97.56 wt% MgO) in the absence of additives. In order to validate the first results in 0.25 L autoclave, new scale up experiments will be performed in 1.0 and 10.0 L autoclaves. Especially, the influence of rotating speed, pH-values and different initial secondary materials such as slag and red mud shall be analyzed in our future work. The analysis of the obtained solution after carbonation revealed very small content of cobalt and chromium, but it will be also considered in our future work in the presence of pH buffering agents in order to increase an extraction efficiency. Especially, a life-cycle-assessment of the carbonation process in the presence of additives will be performed in our future work.

Author Contributions: S.S. conceptualized and managed the research, and co-wrote the paper. D.K. performed the preparation of the olivine materials (grinding, sieving) and co-wrote the paper. H.W. co- wrote the paper. S.N. has performed experiments in an autoclave. P.K. has performed of XRD and TGA-analysis of initial sample and obtained products. G.M. helped in discussion of TGA and XRD-analysis. C.D. analyzed the data and co-wrote the paper. S.E. supervised the XRF- and XRD-analyses and co-wrote the paper with R.T., P.K. conceptualized the research and provided industrial advice. D.R. managed the laboratory facilities for TGA and DSC analysis in Heidelberg Cement Technology Center. B.F. supervised the personnel, provided funding and co-wrote the paper.

Acknowledgments: We would like to thank the BMBF (Federal Ministry of Education and Research) in Berlin for the financial support for the CO2MIN Project (No. 033RCO14B) in period from 01.06.2017 to 31.05.2020. For a continuous support and cooperation we would like to thank Andreas Bremen, AVT and Hesam Ostovari, LTT, RWTH Aachen University.

Conflicts of Interest: The authors declare no conflict of interest.

References

1. Nduagu, E.; Björklöf, T.; Fagerlund, J.; Mäkilä, E.; Salonen, J.; Geerlings, H.; Zevenhoven, R. Production of magnesium hydroxide from magnesium silicate for the purpose of CO_2 mineralization–Part 2: Mg extraction modeling and application to different Mg silicate rocks. *Miner. Eng.* **2012**, *30*, 87–94. [CrossRef]

2. Huijgen, W.J.J.; Comans, R.N.J. *Carbon Dioxide Sequestration by Mineral Carbonation. Literature Review 2003–2004*; Technol Report; Energy Research Centre of the Netherlands: Petten, The Netherlands, December 2005.

3. Santos, R.; Verbeeck, W.; Knops, P.; Rijnsburger, K.; Pontikes, Y.; Gerven, T. Integrated mineral carbonation reactor technology for sustainable carbon dioxide sequestration: "CO_2 Energy Reactor". *Energy Procedia* **2013**, *37*, 5884–5891. [CrossRef]

4. Georgakopoulos, E.D. Iron and steel slag valorization through carbonation and supplementary processes. Ph.D. Thesis, Cranfield University, Cranfield, UK, December 2016.

5. Santos, R.; Audenaerde, A.; Chiang, Y.; Iacobescu, R.; Knops, P.; Gerven, T. Nickel Extraction fom olivine: Effect of carbonation pretreatment. *Metals* **2015**, *5*, 1620–1644. [CrossRef]

6. Chang, E.; Pan, S.; Chen, Y.; Tan, C.; Chiang, P. Accelerated carbonation of steelmaking slags in a high-gravity rotating packed bed. *J. Hazard. Mater.* **2012**, *227*, 97–106. [CrossRef]

7. Aminu, M.; Nabavi, S.; Rochelle, C.; Manovic, V. A review of developments in carbon dioxide storage. *Appl. Energy* **2017**, *208*, 1389–1419. [CrossRef]

8. Azdarpour, A.; Asadullah, M.; Mohammadian, E.; Hamidi, H.; Junin, R.; Karaei, M. A review on carbon dioxide mineral carbonation through pH-swing process. *Chem. Eng. J.* **2015**, *279*, 615–630. [CrossRef]

9. Chang, E.; Pan, S.; Chen, Y.; Chu, H.; Wang, C.; Chiang, P. CO_2 sequestration by carbonation of steelmaking slags in an autoclave reactor. *J. Hazard. Mater.* **2011**, *195*, 107–114. [CrossRef] [PubMed]

10. Rahmani, O.; Junin, R.; Tyrer, M.; Mohsin, R. Mineral carbonation of red gypsum for CO_2 sequestration. *Energy Fuels* **2014**, *28*, 5953–5958. [CrossRef]

11. Bremen, A.M.; Mhamdi, A.; Mitsos, A. Mineral carbonation modeling and investigation on relevant parameters. In Proceedings of the Internal Project Meeting, Aachen, Germany, 13 November 2017.

12. Bremen, A.M.; Mhamdi, A.; Mitsos, A. Mineral carbonation model status update. In Proceedings of the Internal Project Meeting, Aachen, Germany, 7 September 2018.

13. Béarat, H.; McKelvy, M.; Chizmeshya, A.; Gormley, D.; Nunez, R.; Carpenter, R.; Squires, K.; Wolf, G. Carbon sequestration via aqueous olivine mineral carbonation: Role of Passivating layer formation. *Environ. Sci. Technol.* **2006**, *40*, 4802–4808. [CrossRef] [PubMed]

14. Eikeland, E.; Bank, A.; Tyrsted, C.; Jensen, A.; Iversen, B. Optimized carbonation of magnesium silicate mineral for CO_2 storage. *Appl. Mater. Interfaces* **2015**, *7*, 5258–5264. [CrossRef] [PubMed]

15. Daval, D.; Sissmann, O.; Menguy, N.; Salsi, G.; Gyot, F.; Martinez, I.; Corvisier, J.; Machouk, I.; Knauss, K.; Hellmann, R. Influence of anorphous silica lyer formation on the dissolution rate of olivine at 90 °C and elevated pCO_2. *Chem. Geol.* **2011**, *284*, 193–209. [CrossRef]

16. Oelkers, E. An experimental study of forsterite dissolution rates as afunction of temperature and aqueous Mg and Si concentrations. *Chem. Geol.* **2001**, *175*, 485–494. [CrossRef]

17. Hänchen, M.; Prigiobbe, V.; Storti, G.; Seward, T.M.; Mazzotti, M. Dissolution kinetics of forsteritic olivine at 90–150 °C including effect of the presence of CO_2. *Geochim. Cosmochim. Acta* **2006**, *70*, 4403–4416. [CrossRef]

18. Turri, L.; Muhr, H.; Rijnsburger, K.; Knops, P.; Lapicque, F. CO_2 sequestration by high pressure reaction with olivine in a rocking batch autoclave. *Chem. Eng. Sci.* **2017**, *171*, 27–31. [CrossRef]

19. Olajire, A. A review of mineral carbonation technology in sequestration of CO_2. *J. Pet. Sci. Eng.* **2013**, *109*, 364–392. [CrossRef]

20. Qafoku, O.; Hu, J.; Hess, N.; Hu, M.; Ilton, E.; Feng, J.; Arey, B.; Felmy, A. Formation of submicron magnesite during reaction of natural forsterite in H_2O-saturated supercritical CO_2. *Geochim. Cosmochim. Acta* **2014**, *134*, 197–209. [CrossRef]

21. Saldi, G.; Daval, D.; Morvan, G.; Knauss, K. The role of Fe and redox conditions in olivine carbonation rates: An experimental study of the rate limiting reactions at 90 and 150 °C in open and closed systems. *Geochim. Cosmochim. Acta* **2013**, *118*, 157–183. [CrossRef]

22. Botha, A.; Strydom, C. Preparation of a magnesium hydroxy carbonate from magnesium hydroxide. *Hydrometallurgy* **2001**, *62*, 175–183. [CrossRef]

Article

Preparation of Spherical Mo₅Si₃ Powder by Inductively Coupled Thermal Plasma Treatment

Jang-Won Kang [1,2], Jong Min Park [1], Byung Hak Choe [2], Seong Lee [3], Jung Hyo Park [3], Ki Beom Park [1], Hyo Kyu Kim [1], Tae-Wook Na [1], Bosung Seo [1] and Hyung-Ki Park [1,*]

[1] Gangwon Regional Division, Korea Institute of Industrial Technology, Gangneung 25440, Korea; jwkang@kitech.re.kr (J.-W.K.); pjm2639@kitech.re.kr (J.M.P.); hope92430@kitech.re.kr (K.B.P.); khk4529@kitech.re.kr (H.K.K.); arkasa86@kitech.re.kr (T.-W.N.); bs3863@kitech.re.kr (B.S.)

[2] Department of Advanced Metal and Materials Engineering, Gangneung-Wonju National University, Gangneung 25457, Korea; cbh@gwnu.ac.kr

[3] Agency for Defense Development, Daejeon 34186, Korea; myhome1992@hanmail.net (S.L.); melpjh@hanmail.net (J.H.P.)

* Correspondence: mse03@kitech.re.kr; Tel.: +82-33-649-4016

Received: 28 June 2018; Accepted: 1 August 2018; Published: 3 August 2018

Abstract: A method was developed to fabricate spherical Mo₅Si₃ powder by milling and spheroidizing using inductively coupled thermal plasma. A Mo₅Si₃ alloy ingot was fabricated by vacuum arc melting, after which it was easily pulverized into powder by milling due to its brittle nature. The milled powders had an irregular shape, but after being spheroidized by the thermal plasma treatment, they had a spherical shape. Sphericity was increased with increasing plasma power. After plasma treatment, the percentage of the Mo₃Si phase had increased due to Si evaporation. The possibility of Si evaporation was thermodynamically analyzed based on the vapor pressure of Mo and Si in the Mo₅Si₃ liquid mixture. By this process, spherical Mo silicide powders with high purity could be fabricated successfully.

Keywords: Mo silicide; Mo₅Si₃; spheroidizing; powder; inductively coupled thermal plasma

---

1. Introduction

Refractory metal-based silicide alloys, which are also referred to as refractory metal in situ composites, currently receive a lot of attention as structural materials for ultrahigh temperature applications [1–3]. Among these materials, Mo silicide-based alloy [4,5] and Nb silicide-based alloy [6,7] have been intensively studied due to their excellent strength, creep resistance and oxidation resistance at ultrahigh temperature. Furthermore, Mo and Nb have a relative low density compared to other refractory metals, such as Ta and W [8].

Mo silicide-based alloys are composed of α-Mo and Mo silicide. Mo silicides are formed of three main phases: Mo₅Si₃, Mo₃Si, and MoSi₂ [9]. Of these, Mo₅Si₃ has the highest melting temperature of 2180 °C [10]. Therefore, many studies have evaluated the high-temperature creep and oxidation resistance of Mo₅Si₃.

Unfortunately, Mo silicide-based alloys have low fracture toughness at ambient temperature [11], along with low machinability due to the low thermal conductivity and brittle nature of Mo silicides [12,13]. Therefore, it is difficult to fabricate components of Mo silicide-based alloys by conventional casting and machining methods. In addition, since in the Mo–Si binary system the Mo₃Si phase is in between the Mo solid solution and the Mo₅Si₃ phases, Mo silicide-based alloys composed of Mo and Mo₅Si₃ cannot be fabricated by casting. However, based on the powder metallurgy process, Mo-based silicide alloys, where the microstructures consist of Mo5Si3–Mo3Si and Mo–Mo5Si3–Mo3Si, could be fabricated.

Hence, powder metallurgy processes are an attractive way to fabricate components of Mo silicide-based alloys. Previous studies have attempted to fabricate Mo silicide-based alloys [14,15] and Mo silicide powders [16] by mechanical alloying. However, powders fabricated by mechanical alloying suffer from low productivity and oxygen contamination, as well as an irregular morphology. With regard to sintering, spherical powders are much more favorable than those with irregular shapes, as they offer higher packing density and fluidity [17].

It is therefore necessary to develop a method to fabricate high-purity spherical Mo silicide powders. To our knowledge, there is no previous work on the preparation of Mo silicide powders or pre-alloyed Mo silicide-based alloy powders by inductively coupled thermal plasma processing. Thus, in this study, we fabricated Mo_5Si_3 powders by pulverizing a Mo_5Si_3 ingot, utilizing its brittle nature. To improve the sphericity of the powders, they were spheroidized by an inductively coupled thermal plasma treatment. The effect of the plasma power on the morphology and phase balance of the powders was examined, and the evaporation behavior of Si during plasma treatment was analyzed thermodynamically.

2. Experimental Procedures

Mo_5Si_3 ingots with a chemical composition of 85.06Mo–14.94Si in wt% (62.5Mo–37.5Si in atom%) were fabricated by vacuum arc melting. For vacuum arc melting, the chamber was evacuated to a high vacuum (10^{-5} torr) by oil diffusion pump and then high-purity argon gas was injected into the chamber until the pressure reached 400 torr. The ingot, which was 150 mm long × 75 mm wide × 10 mm high, was cast in a quadrangle-shaped cold copper crucible and its weight was 1100 g. To homogenize the composition, the ingot was remelted five times.

To analyze the chemical composition and oxygen concentration, the center of the ingot was cut to a cylinder 3 mm in diameter and 5 mm in height, and measurements were carried out five times in each sample. The chemical composition of the ingot and powders, as analyzed by inductively coupled plasma mass spectrometry (ICP-MS) (iCAP Q, Thermo Fisher Scientific, Waltham, MA, USA), is given in Table 1. To measure the concentration of Mo and Si in Mo silicide, ICP-MS was carried out following the procedure in [18].

Table 1. Mo and Si concentrations (in wt%) of the ingot and powders after spheroidizing at plasma powers 3–7 kW. The values given in parenthesis refer to standard deviation.

Sample	Power (kW)	Mo (wt%)	Si (wt%)
Ingot	-	85.17 (0.06)	14.83 (0.04)
	3	85.70 (0.05)	14.30 (0.03)
	4	85.91 (0.04)	14.09 (0.02)
Powders after Spheroidizing	5	86.54 (0.05)	13.46 (0.03)
	6	87.45 (0.06)	12.55 (0.04)
	7	88.85 (0.05)	11.15 (0.03)

The oxygen concentration, as analyzed by an inert gas fusion infrared absorption method (LECO, 736 series), is given in Table 2.

Table 2. Oxygen concentrations (in wt%) of the ingot, powders after milling, and powders after spheroidizing at a plasma power of 6 kW. The values given in parenthesis refer to standard deviation.

Samples	Oxygen (wt%)
Ingot	0.003 (0.001)
Powders after Milling	0.172 (0.002)
Powders after Spheroidizing (6 kW)	0.016 (0.001)

For pulverization, two Mo_5Si_3 ingots with a weight of 2200 g were first crushed using a jaw crusher into particles with a size of less than 3 mm. Then, the particles were ball-milled in a stainless-steel container, using tungsten carbide balls with a diameter of 5 mm as milling media. The ball-to-powder ratio was 5:1 by weight and the steel container was purged with high-purity argon to prevent oxygen contamination during ball milling. The milling was performed for 5 h at a rotational speed of 200 rpm. After milling, the powders were sieved to the range of 38–75 μm, because the powders larger than 75 μm were not perfectly spheroidized by the thermal plasma system used in this study.

The sieved powders were spheroidized using an inductively coupled thermal plasma treatment (RFP-10, PLASNIX, Incheon, Korea). To investigate the spheroidizing behavior with respect to plasma power, plasma powers of 3–7 kW were examined. The other parameters were fixed as follows. The plasma oscillation frequency was 13.56 MHz, and the chamber pressure was 80 KPa. The powder carrier and center gas were high-purity argon, with flow rates of 5 and 2 slm (standard liter per minute), respectively. Since the maximum temperature in the chamber almost reaches 10,000 K during plasma processing, the sheath gas was injected along the chamber wall for cooling. The sheath gas was a mixture of argon and helium (argon/helium = 4:1) with a flow rate of 75 slm. To feed the powders into the plasma chamber, a vibrating feeder was used with a feeding rate of 300 g/h.

The morphology of the powders after sieving and spheroidizing was observed by field emission-scanning electron microscopy (FE-SEM, FEI, QUANTA FEG 250, Thermo Fisher Scientific, Waltham, MA, USA) along with energy-dispersive spectrometry (EDS, Octane Elite EDS, EDAX, Mahwah, NJ, USA). To investigate the phase balance, the powders were analyzed using X-ray diffractometry (XRD, Empyrean, PANalytical, Almelo, The Netherlands) with Cu Kα radiation, in the 2θ range of 25–75°. For the XRD analysis, the samples were selected at random. The powder size distribution was investigated using a powder size analyzer (Mastersizer 3000, Malvern Panalytical, Malvern, UK). To understand the Si evaporation behavior during plasma treatment, thermodynamic parameters such as the standard Gibbs free energy change and activity coefficient were calculated by Thermo-Calc using the SSOL database.

3. Results and Discussion

Figure 1 is the microstructure of the cast Mo_5Si_3 ingot observed by back-scattered electron imaging. The ingot was composed of two phases. By EDS analyses, the Si concentrations of the dark area (Point 1) and bright area (Point 2) are 14.28 and 9.15 wt%, respectively, which correspond with the Si concentrations of Mo_5Si_3 and Mo_3Si (14.94 and 8.89 wt%). This result indicates that the dark and bright areas were the Mo_5Si_3 and Mo_3Si phases, respectively.

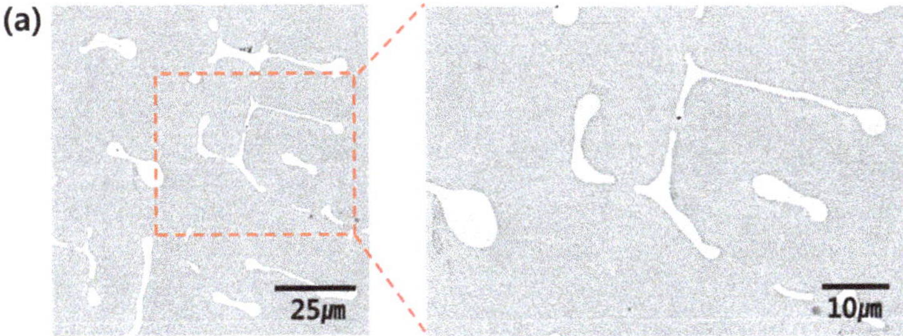

Figure 1. *Cont.*

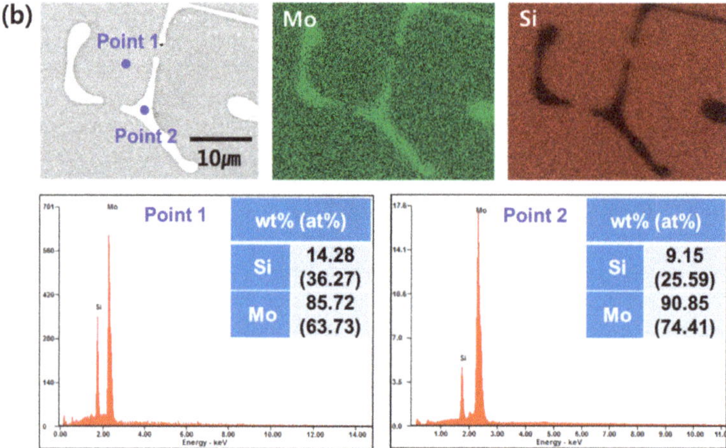

Figure 1. (**a**) The microstructure of the cast Mo_5Si_3 ingot observed by back-scattered electron imaging, and (**b**) EDS area and point analyses result. Points 1 and 2 were determined as Mo_5Si_3 and Mo_3Si, respectively.

Figure 2 shows the morphology of the powders before and after spheroidizing by inductively coupled thermal plasma treatment. As shown in Figure 2a, the powders after milling and sieving had an irregular shape. The powder size distribution of the powder after milling was examined by a powder analyzer; the d_{10}, d_{50}, and d_{90} values were 47.9, 75.4, and 117.0 μm, respectively. Mo silicides, including Mo_5Si_3, have a brittle nature; grain boundary cracking occurs easily at room temperature [12]. Chu et al. [19] researched this by fabricating single crystal Mo_5Si_3 by the Czochralski method and evaluating the mechanical properties with respect to crystal orientation. The room temperature fracture toughness, which ranged from 2 to 2.5 MPa$\sqrt{m}$, was not severely affected by the crystal orientation. Therefore, the Mo_5Si_3 ingot was easily pulverized by jaw crushing and ball milling.

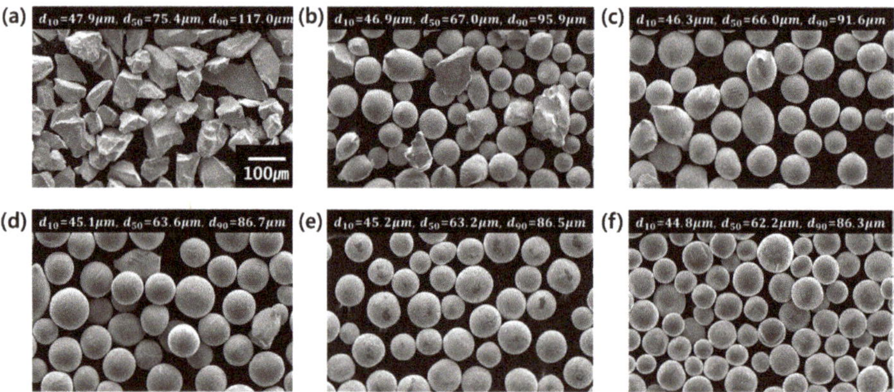

Figure 2. Morphology of the powders (**a**) after ball-milling and sieving, and (**b**–**f**) after spheroidizing by inductively coupled thermal plasma, with plasma powers of (**b**) 3 kW, (**c**) 4 kW, (**d**) 5 kW, (**e**) 6 kW, and (**f**) 7 kW.

After spheroidizing, the shape of the powders had changed from irregular to spherical. The sphericity of the powders was higher with increased plasma power. The powder size tended to become smaller with increasing plasma power, and the d_{10}, d_{50}, and d_{90} values of the powder spheroidized at 7 kW were 44.8, 62.2, and 86.3 µm, respectively.

Previously, the temperature profile during inductively coupled thermal plasma treatment was simulated by COMSOL Multiphysics. The simulation indicated that the maximum temperature of the chamber was almost 10,000 K during plasma processing [20,21] and the powder reached maximum temperature within 10 ms after injecting in the plasma chamber [22]. Therefore, when powders with an irregular shape are injected into the plasma chamber, they are fully melted and spheroidized to reduce the surface area.

To analyze the phases present, the powders were examined by XRD. Figure 3a shows the XRD patterns for 2θ = 25–75° of the powders before and after spheroidizing. The powders were composed of two phases, Mo_5Si_3 and Mo_3Si. To investigate the percentage of Mo_5Si_3 and Mo_3Si in the powders, the XRD patterns were analyzed by the Rietveld method, and the result is shown in Figure 3b. The percentages of Mo_5Si_3 and Mo_3Si in the powder before spheroidizing were 92.5% and 7.5%, respectively. As the plasma power increased, the percentage of Mo_5Si_3 was gradually decreased in the spheroidized powders, reaching a minimum of 63.1% with the power of 7 kW.

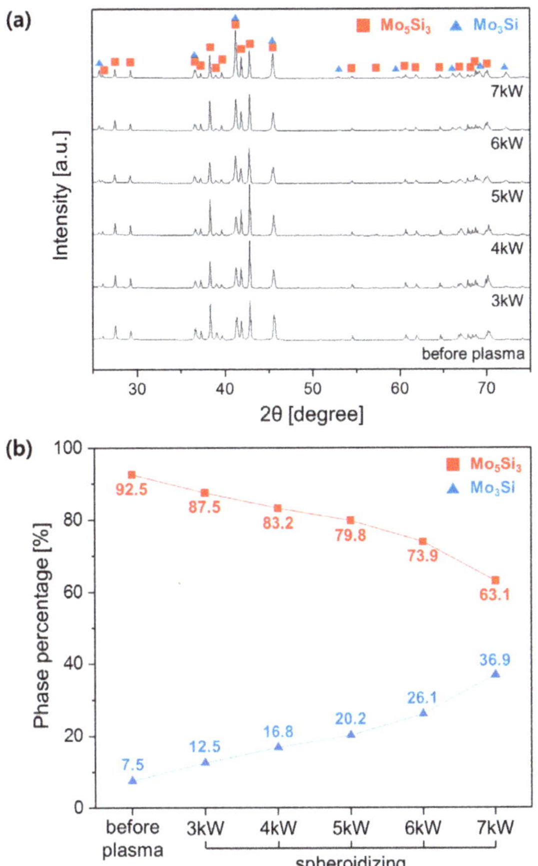

Figure 3. (a) XRD patterns of the powders before and after spheroidizing by inductively coupled thermal plasma treatment. (b) The percentage of each phase in the powders was analyzed by the Rietveld method based on the XRD data in Figure 3a.

By the chemical stoichiometry, the concentration of Si is greater in Mo_5Si_3 than in Mo_3Si (14.94 wt% vs. 8.89 wt%, respectively). The fact that the amount of the Si-rich phase reduces with increased plasma power suggests that Si evaporation may occur during plasma treatment. Therefore, the Si concentration of the powders was analyzed to examine this possibility.

Table 1 shows the Si concentration of the ingot and the powders after spheroidizing with different plasma powers. The weight concentration of Si in the Mo_5Si_3 having perfect stoichiometry is 14.94 wt%. The Si concentration of the ingot was 14.83 wt%, which is almost same as that of the Mo_5Si_3 phase. As the plasma power increased, the Si concentration gradually decreased, reaching a minimum of 11.15 wt% with the power of 7 kW. During the plasma treatment, the evaporation rate of Si would much higher than that of Mo; therefore, the Si concentration was decreased after spheroidizing.

The evaporation behavior of the element is determined by the vapor pressure [23,24]. An element with a higher vapor pressure will have a higher evaporation rate. The vapor pressure of a pure element i (p_i^o) is calculated by the following Equation [25]: $p_i^o = exp\left(-\frac{\Delta G_i^o}{RT}\right)$, where R is the gas constant, T is the temperature, and ΔG_i^o is the standard Gibbs free energy change of element i during the phase transformation from liquid to gas. Figure 4a shows the vapor pressure of pure Mo and Si with respect to the temperature. Both vapor pressures increase with increasing temperature. Furthermore, p_{Si}^o is higher than p_{Mo}^o over the whole temperature range, which means that the evaporation rate of Si is higher than that of Mo.

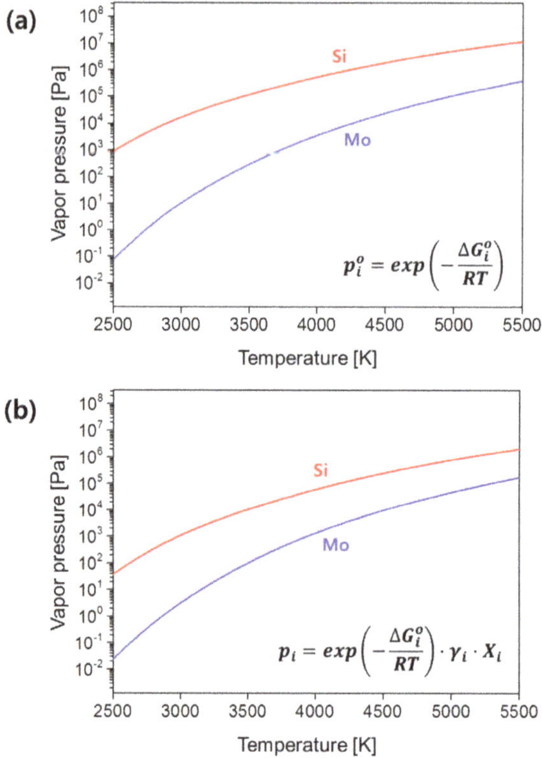

Figure 4. (**a**) Vapor pressure of pure Mo and Si with respect to temperature. (**b**) Vapor pressure of Mo and Si in molten Mo_5Si_3.

However, p_i^o is the vapor pressure for the pure element *i*. To more accurately analyze the evaporation behavior, the vapor pressures of Mo and Si in molten Mo_5Si_3 should be considered. The vapor pressure of element *i* in a liquid mixture (p_i) is calculated by the following equation: $p_i = exp\left(-\frac{\Delta G_i^o}{RT}\right) \cdot \gamma_i \cdot X_i$, where γ_i is the activity coefficient of element *i* in a liquid mixture and X_i is the molar fraction of element *i* [25]. To calculate the vapor pressure of Mo and Si in molten Mo_5Si_3 with respect to the temperature, activity coefficients of Mo and Si should be known. Therefore, the activity coefficients at 2500, 3000, 3500, 4000, 4500, 5000, and 5500 K of Mo and Si in the Mo–Si binary system were calculated by Thermo-Calc using the SSOL database. Then, the activity coefficients of Mo and Si in the composition of Mo_5Si_3 were extracted, and they were used to determine the p_{Mo} and p_{Si}.

Figure 4b shows the vapor pressures of Mo and Si in molten Mo_5Si_3 with respect to temperature. Since the evaporation rate in a liquid mixture is affected by the activity coefficient, the vapor pressures of Mo and Si in molten Mo_5Si_3 were different to that for pure Mo and Si. However, the vapor pressure of Si was still higher than that of Mo, confirming that the amount of Si evaporation was greater than that of Mo during plasma treatment. Therefore, the decreasing amount of the Si-rich phase (Mo_5Si_3) with increasing plasma power (Table 1) was caused by Si evaporation during spheroidizing. Even though Si was evaporated during spheroidizing, any condensation of Si was not observed. Si nanoparticles could be nucleated during plasma treatment; however, they would be filtered by the cyclone system due to their small size and low weight.

To determine the extent of oxygen contamination during the fabrication of the spherical Mo silicide powder, the oxygen concentrations of the ingot, the powders after milling, and the powders after spheroidizing at 6 kW were analyzed. The oxygen concentrations of the three samples are shown in Table 2. The oxygen concentration of the ingot was 0.003%, which increased to 0.171% after milling. This could be caused by the increase in specific surface area by pulverizing, as well as the contamination during milling, as argon purging may not perfectly eliminate oxygen from the milling container. After spheroidizing at 6 kW, the oxygen concentration was decreased to 0.016%. This would be due to the reducing environment during the plasma treatment, due to the extremely high temperature and low oxygen partial pressure in the chamber [21]. Thus, while the oxygen concentration was higher in the spheroidized powder than in the ingot, it was still considerably low.

To examine the internal microstructure of the spherical powder, the powder spheroidized at 6 kW was observed by FE-SEM. Figure 5a shows the cross-sectional microstructure of the powder observed by back-scattered electron imaging. As expected from the XRD results (Figure 3), the powders were composed of two phases. To identify the phases, EDS area and point analyses were performed, and the result is shown in Figure 5b. The Si concentrations of the dark area (Point 1) and bright area (Point 2) are 14.57 and 9.10 wt%, respectively. Therefore, the dark and bright areas can be defined as Mo_5Si_3 and Mo_3Si, respectively. The melting temperatures of Mo_5Si_3 and Mo_3Si are 2453 and 2298 K, respectively [26]. As the powders cooled during plasma treatment, the Mo_5Si_3 phase solidified first, after which Mo_3Si solidified. Therefore, the microstructure shown in Figure 5 would be formed by the first nucleation of Mo_5Si_3.

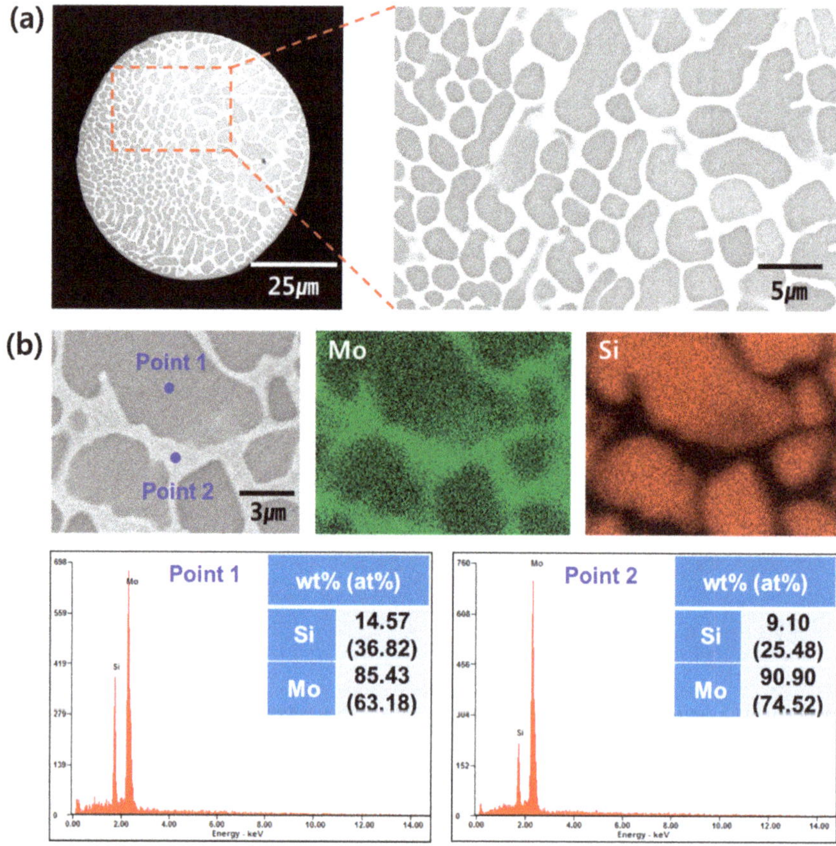

Figure 5. (**a**) Cross-sectional microstructure of the powder observed by back-scattered electron imaging, and (**b**) EDS area and point analyses result. Points 1 and 2 were determined as Mo_5Si_3 and Mo_3Si, respectively.

4. Conclusions

The Mo_5Si_3 ingot was prepared by vacuum arc melting, after which it was pulverized to a powder by jaw crushing and ball milling. As the milled powders had an irregular shape, they were spheroidized by thermal plasma treatment. As the plasma power increased, the sphericity of the powders increased. They were perfectly spheroidized when the plasma power was higher than 6 kW. After plasma treatment, the ratio of Mo_5Si_3 to Mo_3Si had decreased due to Si evaporation. Based on the thermodynamic analysis, Si has a higher vapor pressure than Mo in the Mo_5Si_3 liquid mixture. By this process, spherical Mo silicide powders with a low oxygen concentration of 0.016% could be fabricated successfully.

Author Contributions: Conceptualization, J.-W.K., J.M.P., B.-H.C. and S.L.; Data curation, K.B.P.; Formal analysis, J.M.P., J.H.P. and B.S.; Investigation, T.-W.N.; Methodology, H.K.K.; Writing original draft, H.-K. P.

Acknowledgments: This work was supported by a research fund of Korea Institute of Industrial Technology (KITECH EO-18-0012) and Agency for Defense Development (Project No. 811555-912515201).

References

1. Petrovic, J.J. MoSi$_2$-based high-temperature structural silicides. *MRS Bull.* **1993**, *18*, 35–41. [CrossRef]
2. Chu, F.; Thoma, D.J.; McClellan, K.J.; Peralta, P. Mo$_5$Si$_3$ single crystals: Physical properties and mechanical behavior. *Mater. Sci. Eng. A* **1999**, *261*, 44–52. [CrossRef]
3. Yeh, C.L.; Chou, C.C.; Hwang, P.W. Experimental and Numerical Studies on Self-Propagating High-Temperature Synthesis of Ta$_5$Si$_3$ Intermetallics. *Metals* **2015**, *5*, 1580–1590. [CrossRef]
4. Schneibel, J.H.; Liu, C.T.; Easton, D.S.; Carmichael, C.A. Microstructure and mechanical properties of Mo-Mo$_3$Si-Mo$_5$SiB$_2$ silicides. *Mater. Sci. Eng. A* **1999**, *261*, 78–83. [CrossRef]
5. Kwon, N.Y.; Kim, Y.D.; Suk, M.J.; Lee, S.; Oh, S.T. Synthesis of Mo-Si-B intermetallic compounds with continuous α-Mo matrix by pulverization of ingot and hydrogen reduction of MoO$_3$ powders. *Int. J. Refract. Met. Hard Mater.* **2017**, *65*, 25–28. [CrossRef]
6. Bewlay, B.P.; Jackson, M.R.; Zhao, J.C.; Subramanian, P.R. A review of very-high-temperature Nb-silicide-based composites. *Metall. Mater. Trans. A* **2003**, *34*, 2043–2052. [CrossRef]
7. Tang, Y.; Gua, X. High temperature deformation behavior of an optimized Nb-Si based ultrahigh temperature alloy. *Scr. Mater.* **2016**, *116*, 16–20. [CrossRef]
8. Byun, J.M.; Hwang, S.H.; Lee, S.; Kim, Y.D. Research trend of Mo based superalloys. *J. Kor. Powd. Met. Inst.* **2013**, *20*, 487–493. [CrossRef]
9. Kieffer, R.; Benesovsky, F. Recent developments in the field of silicides and borides of the high-melting-point transition metals. *Powder Metall.* **1958**, *1*, 145–171. [CrossRef]
10. Akinc, M.; Meyer, M.K.; Kramer, M.J.; Thom, A.J.; Huebsch, J.J.; Cook, B. Boron-doped molybdenum silicides for structural applications. *Mater. Sci. Eng. A* **1999**, *261*, 16–23. [CrossRef]
11. Choe, H.; Chen, D.; Schneibel, J.H.; Ritchie, R.O. Ambient to high temperature fracture toughness and fatigue-crack propagation behavior in a Mo–12Si–8.5B (at.%) intermetallic. *Intermetallics* **2001**, *9*, 319–329. [CrossRef]
12. Ström, E.; Zhang, J. Mechanical anisotropy and segregation in Mo$_5$Si$_3$ studied by EBSD. *Intermetallics* **2005**, *13*, 367–372. [CrossRef]
13. Mendiratta, M.G.; Dimiduk, D.M. Strength and toughness of a Nb/Nb$_5$Si$_3$ composite. *Metall. Mater. Trans. A* **1993**, *24*, 501–504. [CrossRef]
14. Zamani, S.; Bakhsheshi-Rad, H.R.; Shokuhfar, A.; Vaezi, M.R.; Kadir, M.R.A.; Shafiee, M.R.M. Synthesis and characterization of MoSi$_2$-Mo$_5$Si$_3$ nanocomposite by mechanical alloying and heat treatment. *Int. J. Refract. Met. Hard Mater.* **2012**, *31*, 236–241. [CrossRef]
15. Krüger, M.; Schmelzer, J.; Helmecke, M. Similarities and Differences in Mechanical Alloying Processes of V-Si-B and Mo-Si-B Powders. *Metals* **2016**, *6*, 241. [CrossRef]
16. Kermani, M.; Razavi, M.; Rahimipour, M.R.; Zakeri, M. The effect of mechanical alloying on microstructure and mechanical properties of MoSi$_2$ prepared by spark plasma sintering. *J. Alloy. Compd.* **2014**, *593*, 242–249. [CrossRef]
17. Fang, Z.Z.; Paramore, J.D.; Sun, P.; Chandran, K.S.R.; Zhang, Y.; Xia, Y.; Cao, F.; Koopman, M.; Free, M. Powder metallurgy of titanium—Past, present, and future. *Int. Mater. Rev.* **2017**. [CrossRef]
18. Danzaki, Y.; Wagatsuma, K.; Syoji, T.; Yoshimi, K. Dissolution of molybdenum-silicon (-boron) alloys using a mixture of sulfuric, nitric and hydrofluoric acids and a sequential correction method for ICP-AES analysis. *Fresen J. Anal. Chem.* **2001**, *369*, 184–186. [CrossRef]
19. Chu, F.; Thoma, D.J.; McClellan, K.; Peralta, P.; He, Y. Synthesis and properties of Mo$_5$Si$_3$ single crystals. *Intermetallics* **1999**, *7*, 611–620. [CrossRef]
20. Boulos, M.I. The role of transport phenomena and modeling in the development of thermal plasma technology. *Plasma Chem. Plasma Process.* **2016**, *36*, 3–28. [CrossRef]
21. Leparoux, M.; Loher, M.; Schreuders, C.; Siegmann, S. Neural network modelling of the inductively coupled RF plasma synthesis of silicon nanoparticles. *Powder Technol.* **2008**, *185*, 109–115. [CrossRef]
22. Tong, J.B.; Lu, X.; Liu, C.C.; Pi, Z.Q.; Zhang, R.J.; Qu, X.H. Numerical simulation and prediction of radio frequency inductively coupled plasma spheroidization. *Appl. Therm. Eng.* **2016**, *100*, 1198–1206. [CrossRef]
23. Chakrabarti, O.; Weisensel, L.; Sieber, H. Reactive melt infiltration processing of biomorphic Si–Mo–C ceramics from wood. *J. Am. Ceram. Soc.* **2005**, *88*, 1792–1798. [CrossRef]

24. Xu, L.; Huang, C.; Liu, H.; Zou, B.; Zhu, H.; Zhao, G.; Wang, J. Study on in-situ synthesis of ZrB_2 whiskers in ZrB_2–ZrC matrix powder for ceramic cutting tools. *Int. J. Refract. Met. Hard Mater.* **2013**, *37*, 98–105. [CrossRef]
25. Su, Y.; Guo, J.; Jia, J.; Liu, G.; L, Y. Composition control of a TiAl melt during the induction skull melting (ISM) process. *J. Alloy. Compd.* **2002**, *334*, 261–266.
26. Liu, Y.; Shao, G.; Tsakiropoulos, P. Thermodynamic reassessment of the Mo-Si and Al-Mo-Si system. *Intermetallics* **2000**, *8*, 953–962. [CrossRef]

Article

Tuning the Morphology of ZnO Nanostructures with the Ultrasonic Spray Pyrolysis Process

Elif Emil [1,2], Gözde Alkan [3,*], Sebahattin Gurmen [1], Rebeka Rudolf [4], Darja Jenko [5] and Bernd Friedrich [3]

[1] Department of Metallurgical & Materials Eng., Istanbul Technical University, 34469 Istanbul, Turkey; emil.elif@gmail.com (E.E.); gurmen@itu.edu.tr (S.G.)

[2] Department of Materials Science & Tech., Turkish-German University, 34820 Istanbul, Turkey

[3] IME Institute of Process Metallurgy and Metal Recycling, RWTH Aachen University, Intzestrasse 3, 52056 Aachen, Germany; bfriedrich@ime-aachen.de

[4] Faculty of Mechanical Engineering, University of Maribor, Smetanova ulica 17, 2000 Maribor, Slovenia; rebeka.rudolf@um.si

[5] Institute of Metals and Technology, Laboratory of Applied Surface Science, Lepi pot 11, SI-1000 Ljubljana, Slovenia; darja.jenko@imt.si

* Correspondence: galkan@ime-aachen.de; Tel.: +49-241-95873

Received: 6 June 2018; Accepted: 19 July 2018; Published: 24 July 2018

Abstract: Nanostructured zinc oxide (ZnO) particles were synthesized by the one step Ultrasonic Spray Pyrolysis (USP) process from nitrate salt solution ($Zn(NO_3)_2 \cdot 6H_2O$). Various influential parameters, from $Zn(NO_3)_2 \cdot 6H_2O$ concentrations (0.01875–0.0375 M) in the initial solution, carrier gas (N_2) flow rates (0.5–0.75 L/min) to reaction temperature (400–800 °C), were tested to investigate their role on the final ZnO particles' morphology. For this purpose, Scanning Electron Microscopy (SEM), High Resolution Transmission Electron Microscopy (HRTEM) and (Selected Area Electron Diffraction) SAED techniques were used to gain insight into how the ZnO morphology is dependent on the USP process. It was revealed that, by certain parameter selection, different ZnO morphology could be achieved, from spherical to sphere-like structures assembled by interwoven nanoplate and nanoplate ZnO particles. Further, a more detailed crystallographic investigation was performed by XRD and Williamson-Hall (W-H) analysis on the ZnO with unique and non-typical planar morphology that was not reported before by USP synthesis. Moreover, for the first time, a flexible USP formation model was proposed, ending up in various ZnO morphologies rather than only ideal spheres, which is highly promising to target a wide application area.

Keywords: ZnO; ultrasonic spray pyrolysis; influential parameters; formation mechanism; structure; morphologies; characterization; TEM; HRTEM

---

1. Introduction

ZnO in a nanosized form is an indispensable candidate for electronic, optical, and gas sensors, as well as in catalysis applications, owing to its band gap value of 3.37 eV, large exciton binding energy of 60 meV and high electron mobility [1–5]. In the last decade, ZnO with various morphologies, including flower-like [6,7], nanodisc [1], nanobelt [8], and nanotube [9], targeting various application areas, have been investigated using different synthesis methods, e.g., sol-gel [10], hydrothermal [11], the microwave-assisted method [12] and spray pyrolysis [13]. Kajbafvala et al. reported synthesis of spherical and flower-like ZnO via the microwave-assisted method for organic dye photo-degradation via UV lamp irradiation. It was revealed that the degradation efficiency of the spherical particles is better than that of those with flower-like morphology due to their higher surface area, which provides the absorption of oxygen molecules and OH-ions, resulting in an increment of H_2O_2 and OH^-

radicals' formation rate [14]. In another study, Sin et al. [15] reported synthesised self-assembled ZnO microsphere formed by nanoplanar crystal units to improve photocatalytic performance for various endocrine-disrupting chemicals under UV irradiation by the chemical solution route. The enhancement in the photocatalytic performance was attributed to the combined effects of the hierarchical surface structure and the large surface area, which suppresses the recombination of photo-generated electrons (e−) and holes (h+), and expedites the diffusion of electrons [15]. In another study by Li et al., carbon-doped and coated spherical ZnO particles were synthesized for use as anode material in high power Zn-Ni batteries. ZnO microspheres exhibit excellent cycling stability and superior high-rate performance [16].

These previous studies revealed that ZnO finds wide application areas, e.g., planar nano ZnO favors catalytic properties owing to its higher surface area, while granular nano ZnO is preferred in optical applications due to better absorption behavior, and battery applications due to structural stability and anti-corrosion capability [13]. Considering these morphology dependent utilizations of ZnO, adjusting a synthesis method to end up with varying morphologies in a controlled way would be advantageous. There have already been some studies reported, that control ZnO morphology by wet chemical methods is possible; however, these used reactant agents, such as polyethylene glycol (PEG) and cetyltrimethylammonium bromide (CTAB). In our previous study focusing on Ag/ZnO core shell nanostructured materials for photocatalytic applications, under some synthesis conditions, variation from the typical spherical morphology Ultrasonic Spray Pyrolysis (USP) into entangled plates was observed in ZnO particles without any additives [17]. These findings raised the question of whether it is possible to tune ZnO particles morphology by the USP process with simplicity, good process control, high flexibility, and good scale-up potential, without any additives, obtaining high-purity products [17]. For this purpose, varying influential USP process parameters like $Zn(NO_3)_2 \cdot 6H_2O$ concentration in the initial solution, carrier gas (N_2) flow rate and reaction temperature (400–800 °C) were examined. Based on this, the aim was to synthesize ZnO nanoparticles and to investigate the newly formatted ZnO morphology through the USP process. With detailed ZnO morphological investigations by High Resolution Transmission Electron Microscopy (HRTEM) and Scanning Electron Microscopy (SEM), the role of each USP parameter was determined. A more detailed crystallographic investigation was performed via XRD and W-H analysis only on the non-conventional nanostructured ZnO particles. Moreover, the reason behind the different morphologies of ZnO were explained, and a reaction progress model was proposed considering the USP process and thermodynamic conditions. There has been no study in the literature utilizing USP to synthesize ZnO nanoparticles with the aim of obtaining different morphologies.

2. Experimental Procedures

2.1. Synthesis of Zinc Oxide Particles

For the synthesizing of ZnO particles with USP, an aqueous solution of zinc nitrate hexahydrate ($Zn(NO_3)_2 \cdot 6H_2O$, purity > 99.9%) was used, purchased from MERCK Chemical GmbH (Darmstadt, Germany) Zinc nitrate hexahydrate, at the determined amount given in Table 1, was dissolved in distilled water and stirred for 30 min by a magnetic stirrer. This solution represents the so-called precursor, which was atomized by an ultrasonic nebulizer at 1.7 MHz. The formed aerosol was transferred into a one-step USP device to the pre-heated furnace (Nabertherm, R 50/250/12, Lilienthal, Germany) through a quartz tube (0.7 m length and 0.02 m diameter) using N_2 gas. The experiments were conducted with 0.5 L/min and 0.75 L/min N_2 flow rate at 400, 600 and 800 °C reaction temperature for 3 h. The thermal decomposition of $Zn(NO_3)_2 \cdot 6H_2O$ under air atmosphere during the spray pyrolysis has been reported previously [17,18]. The chemical balance equations were given in 3 reaction steps:

$$Zn(NO_3)_2 \cdot 6H_2O \rightarrow Zn(NO_3)_2 \cdot H_2O + 5H_2O\uparrow \tag{1}$$

$$3Zn(NO_3)_2 \cdot H_2O \rightarrow Zn(NO_3)_2 \cdot 2Zn(OH)_2 + H_2O\uparrow + 4NO_x\uparrow \tag{2}$$

$$Zn(NO_3)_2 \cdot 2Zn(OH)_2 \rightarrow 3ZnO + 5H_2O\uparrow + 2NO_x\uparrow \qquad (3)$$

ZnO is the final product of all intermediate product decomposition of zinc nitrate. Synthesized ZnO particles were collected in a washing bottle filled with ethanol. Details of the synthesis procedure can be found elsewhere [19,20]. To characterize nanostructured ZnO particles, ethanol was evaporated in a drying oven at 70 °C for 300 s. A summary of the process parameters is illustrated in Table 1.

Table 1. Process parameters of nanostructured ZnO particles synthesized with various solution concentrations, reaction temperature and N_2 gas flow rate.

Sample Name	$Zn(NO_3)_2 \cdot 6H_2O$ Concentration (mol/L)	Reaction Temperature (°C)	N_2 Gas Flow Rate (L/min)
S1	0.01875	800	0.5
S2	0.02875	800	0.5
S3	0.03750	800	0.5
S4	0.02875	600	0.5
S5	0.02875	400	0.75
S6	0.02875	600	0.75
S7	0.02875	800	0.75

2.2. Characterization of Zinc Oxide Particles

The morphology of the ZnO nanoparticles (size and shape) was examined by Field Emission Scanning Electron Microscopy (FE-SEM, JSM 700F, JEOL, Tokyo, Japan), operating at 5 kV. During SEM sample preparation, the SEM holder was grinded, the particles were dispersed in ethanol, and then the suspension was added dropwise onto the SEM holders, and, afterwards, a conductivity Pt coating was added to prevent charging of the particles by Sputter Coater (Polaron Range SC7620, Quorom Technologies, East Sussex, UK). A Transmission Electron Microscope (TEM, JEM-2100 HR, JEOL, Tokyo, Japan) with integrated Selected Area Electron Diffraction (SAED) pattern analysis, operating at 200 kV was used. Samples of ZnO nanoparticles in demineralized water for TEM analyses were drop cast onto a copper TEM grid covered with a carbon support film, dried, and then used for investigations.

The as-synthesized ZnO nanoparticles were analyzed using a X-ray diffractometer operating at 40 mA and 40 kV with Cu-Kα radiation ($\lambda = 0.154051$ nm). The diffraction spectra were recorded 2θ in the range between 10° and 90° in 2θ steps of 0.02 degrees. Phase composition was determined with a digital library of crystallographic cards JCPDS. Based on the Full Width at Half Maximum (FWHM), zinc oxide crystallite size was evaluated by Williamson–Hall (W-H) analysis and the Debye-Scherrer (DS) method.

3. Results and Discussion

3.1. Effect of Precursor Concentration

SEM micrographs of nanostructured ZnO particles are given in Figure 1, revealing the effect of precursor concentration on final morphology.

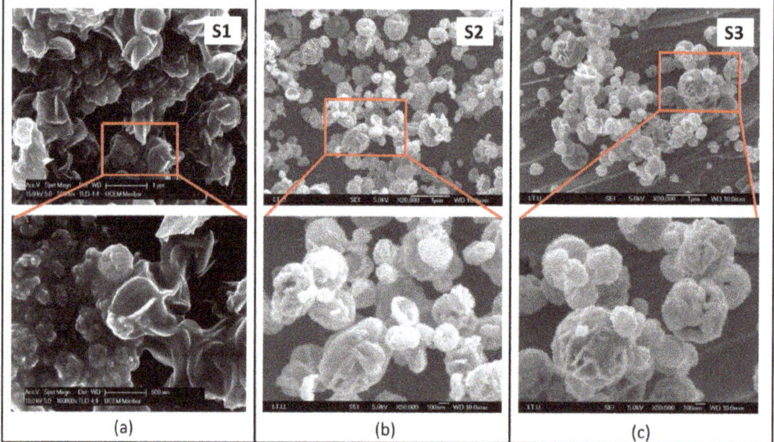

Figure 1. SEM micrographs of nanostructured ZnO particles (**a**) S1; (**b**) S2; and (**c**) S3, at 800 °C, 0.5 L/min of N_2 gas flow rate.

As can be observed, a relatively lower zinc nitrate concentration (0.01875 mol/L) exhibited dominant plates and accompanying sphere-like morphology, assembled by interwoven nanoplates, which is not commonly achieved via USP. A gradual change in the microstructure was revealed by increases in zinc nitrate concentrations. An increase to 0.02875 mol/L resulted in a sphere-like morphology assembled by nanoplates. A further increase to (0.03750 mol/L) resulted in dominant granular accompanied particles. Although there are many studies dealing with ZnO synthesis with USP, such a unique morphology has not been reported and explained previously. In order to better understand its crystallography and to see if this morphology difference was due to a different phase formation, a detailed crystallographic investigation was performed only on that sample. The corresponding XRD pattern is given in Figure 2.

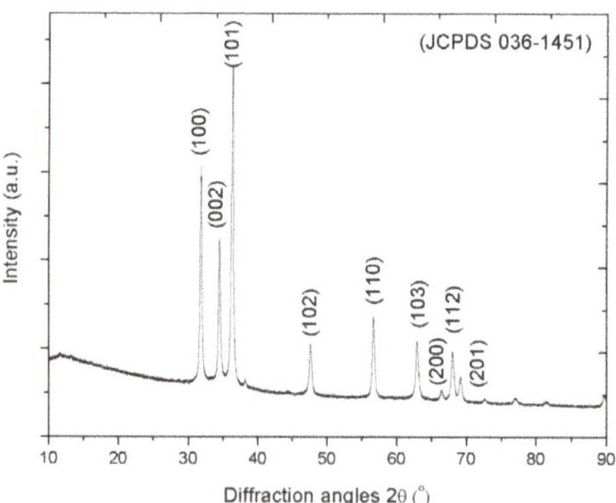

Figure 2. Indexed XRD pattern of sample S1.

As shown in Figure 2, reference sample S1 exhibited a phase pure zinc oxide hexagonal structure with a space group P63mc (186), unit cell of a = 3.2533 Å and c = 5.2073 Å corresponding to JCPDS Card 00-036-1451. The peaks at 2θ = 31.80°, 34.45°, 36.32°, 47.56°, 56.65°, 62.89°, 66.38°, 67.98° and 69.10° are assigned to (100), (002), (101), (102), (110), (103), (200), (112) and (201) diffraction planes, respectively. No characteristic peaks of zinc nitrate salt were detected in the diffraction pattern, implying termination of decomposition reaction. Based on the most intense peaks (101) diffraction planes, the crystallite size was calculated using the DS method, where the contribution of crystallite size and lattice strain to total peak broadening were not considered. The β parameter was corrected due to the effect of instrumental broadening on total peak broadening in DS, and the crystallite size of S1 was calculated as 24 nm by DS. However, the contribution of lattice strain on total peak broadening should be noted, since lattice strain is induced by a rapid heating/cooling rate and short residence time of the USP process, as reported previously [21]. This relation is expressed in Equation (4):

$$\beta_{hkl} = \beta_D + \beta s \tag{4}$$

Williamson Hall (W-H) analysis was preferred to evaluate the crystallite size and lattice strain. Details of the W-H analysis are explained elsewhere [22]. The intense peaks corresponding to (100), (002), (101), (102), (110), (103) and (112) planes were selected to conduct the W-H analysis. The strain-induced broadening was connected with crystal distortion, and is defined in Equation (5):

$$\varepsilon \approx \beta s / \tan \theta \tag{5}$$

DS relation, Equation (4), can be reformulated with Equation (5) which results in Equation (6):

$$\beta_{hkl}\cos \theta = (k\lambda/D) + (4\varepsilon\sin \theta) \tag{6}$$

where β_{hkl} is the total peak broadening, k is the shape factor, λ is the wavelength of Cu-Kα radiation ($\lambda = 0.154051$ nm), D is the crystallite size of the synthesized particles, ε is the lattice strain, β is FWHM of the peak, and θ are the Bragg angles. The individual contribution of crystallite size and lattice strain were evaluated by the W-H method integrated with Uniform Deformation Model (UDM). Graphics of (4sin θ) versus Bhklcosθ were drawn in Figure 3. Crystallite size and lattice strain respectively were estimated from the intercept on the y-axis and slope of linearly fitted data.

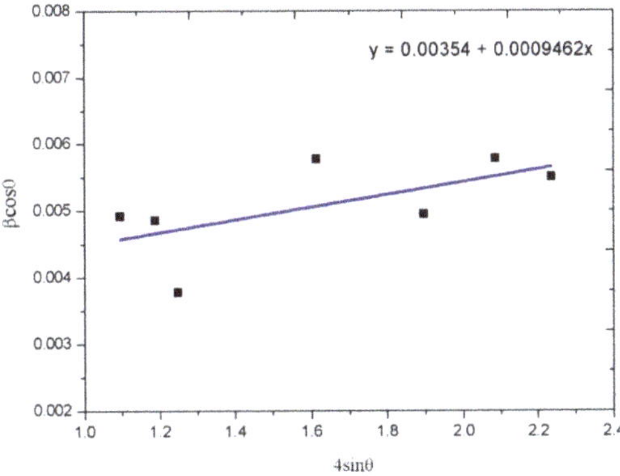

Figure 3. Williamson-Hall (W-H) analysis integrated with Uniform Deformation Model (UDM) model of S1.

Based on the broadening X-ray diffraction peak (FWHM), the crystallite sizes of the synthesized particles from 0.01875 mol/L were estimated at 38 nm and 24 nm using the W-H method integrated with UDM, and the DS method, respectively. This fine crystallite size and the XRD pattern given in Figure 2, which is consistent with the literature, imply the formation of secondary particles by coalescence of these primary crystals. Due to the neglection of the intrinsic strain at the lattice, the values of the crystallite sizes calculated using the W-H method differed from that evaluated by DS analysis, as expected. According to the plots of (4sin θ) vs. Bhklcos θ of particles, the intrinsic strain at the lattice for S1 was found to be 9.462×10^{-4}. It was proved that DS analysis calculates a smaller crystallite size than the size calculated using the W-H method in the presence of a tensile lattice strain. Moreover, these strain and crystallite size values were found to be similar to values reported previously for spherical ZnO particles [23,24]. Since the crystallographic findings do not differ dramatically from the previously reported data for spherical ZnO, the origin of this unique morphology was searched for in the USP formation mechanism.

In order to highlight microstructural features in detail, HRTEM was performed on the samples synthesized from varying precursor concentrations. Corresponding HRTEM images with electron diffraction patterns can be found in Figure 4.

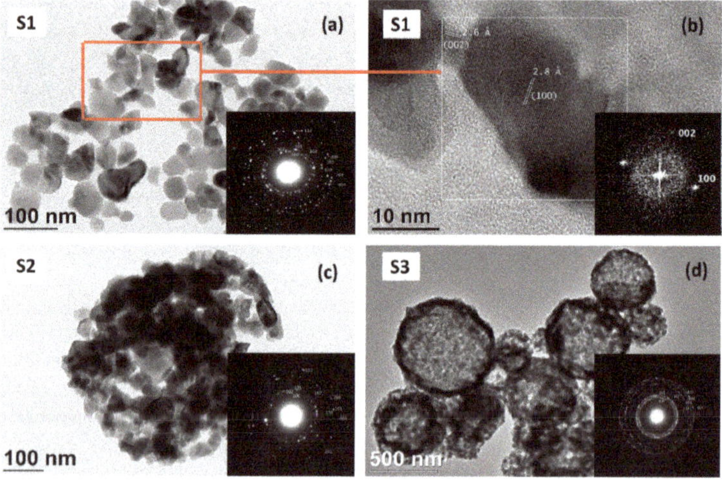

Figure 4. High Resolution Transmission Electron Microscopy (HRTEM) micrographs and electron diffraction ring pattern of S1 (**a**); HR-TEM micrographs and electron diffraction spot pattern S1 (**b**); HRTEM micrographs and electron diffraction ring pattern of S2 (**c**) and S3 (**d**).

The TEM Bright-Field (BF) image, and the corresponding Selected Area Electron Diffraction (SAED) pattern of S1 solution concentration, showed fine particle sizes in the range of 10–70 nm. The planar morphology was also revealed, and even some hexagonal particles were observed. The HR-TEM image (Figure 2b), with a 2-D Fast Fourier Transform (FFT), shows a lattice of two ZnO nanoparticles synthesized from 0.01875 mol/L. The bigger particle is oriented in the (100) direction, and the attached neighboring particle on the left of this particle is oriented in the (002) direction. The measured distance from the HR-TEM image between the lattice planes is 2.8 Å in the (100) direction and 2.6 Å in the (002) direction, which is in very good agreement with the cell parameters of ZnO (cell parameters a = 3.2498 Å, c = 5.2066 Å). In parallel with the XRD findings, with an increase in concentration, a gradual change in morphology was observed in the HRTEM images. Planar fine crystals were transformed into spherical ZnO nanostructured spheres (500–700 nm) with increasing concentrations. The SAED of all ZnO nanoparticles showed characteristic diffraction of a ring pattern, with some brighter and more distinct spots in the rings, which indicated the presence of some larger crystallites, though the rings were still relatively continuous, which meant that the crystallites were small, in the nm range, and in a random orientation. The electron diffraction spots can be described by a hexagonal crystalline-structured zinc oxide with a space group P63mc (JCPDS card No. 00-00-036-1451) with indices as shown in the inset of Figure 4a and in accordance with the XRD spectrum in Figure 2. Although the microstructure changed from planar to spherical with increasing solution concentrations, they exhibited the same crystallite structure [23]. In previous studies, it was reported that the decomposition of zinc nitrate hexahydrate into zinc oxide takes place via step-wise reaction with the formation of a $Zn(NO_3)_2 \cdot 2Zn(OH)_2$ intermediate compound. However, before the decomposition takes place, the initial salts melt and reaction takes place in the liquid phase [17,18]. Since particle formation in the melt phase is driven by nucleation and growth processes, the final morphology can be adjusted by controlling nucleation and growth rates via process parameters. The different morphologies achieved in this study at various temperatures, concentrations, and flow rates can be explained by this fact.

In the case of a higher precursor concentration, supersaturation induces higher nuclei rates, yielding more nuclei formation and less growth rate, and resulting in the formation of spherical-like ZnO particles. Similarly, when the temperature increased, the supersaturation degree increased again. This also increased the nucleation rate and, therefore, at 800 °C, almost all samples exhibited spherical-like morphologies. In the USP process, temperature change and flow rate also act parallel to concentration in terms of supersaturation degree. Increased temperatures and flow rate results in higher temperature gradients and increases the supersaturation rate. In order to demonstrate the change, the supersaturation degree plays a crucial role in the final morphology; the optimal concentration (0.02875 mol/L) was selected and the temperatures and flow rates were varied to observe their individual effects.

3.2. Effect of Reaction Temperature and N_2 Gas Flow Rate

USP reaction temperature and gas flow rate together determine the residence time of droplets/particles in the heating zone, and therefore they also play a significant role in the nucleation step and growth mechanism. In previous synthesis conditions, reaction temperature was fixed at 800 °C. Temperature was changed to 600 °C/400 °C while ensuring complete decomposition of zinc nitrate. SEM micrographs of the formatted ZnO particles can be found in Figure 5.

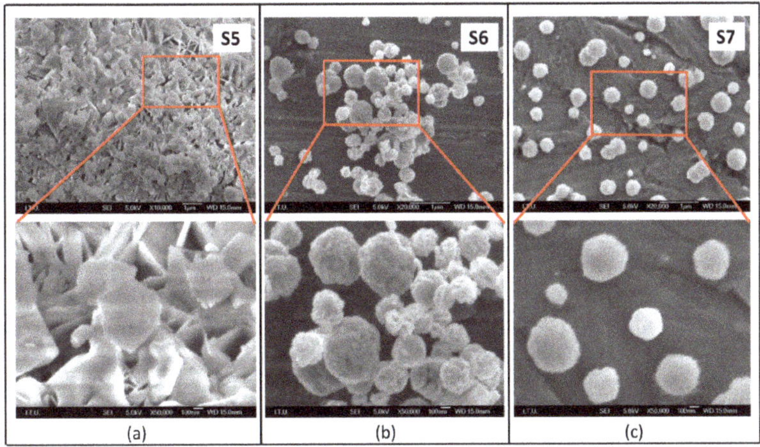

Figure 5. SEM micrographs of nanostructured ZnO particles synthesized at various reaction temperatures at S5 (**a**), S6 (**b**) and S7 (**c**), where concentration (0.02875 mol/L) and flow rate (0.5 L/min) were constant.

To begin with, it is worth emphasizing that all samples synthesized at different temperatures exhibited complete conversion to ZnO. With increasing synthesis temperature, similar to the change in concentration, a gradual morphological change was observed from plates to spheres. When the reaction temperature was high, the supersaturation rate and mobility of particles were expected to be high. Higher supersaturation degree yielded higher nucleation rates, while a higher mobility increased the collision rates and growth rates of individual particles. For low temperatures, the driving force of the reaction in the system and nucleation rates are low. Therefore, there can be an accumulation on already existing nuclei and directed growth may be observed, as shown in Figure 5, in parallel with low precursor concentrations (see Figure 1a). However, when a particle synthesized at the highest temperature was considered, there was no observed coarsening, which implies that the increased nucleation rate was more dominant.

In order to assess flow rate effect in morphology, a high temperature (800 °C) was used to ensure complete conversion and ZnO was synthesized at a relatively lower flow rate. The SEM micrographs presented in Figure 6 reveal the effect of flow rate.

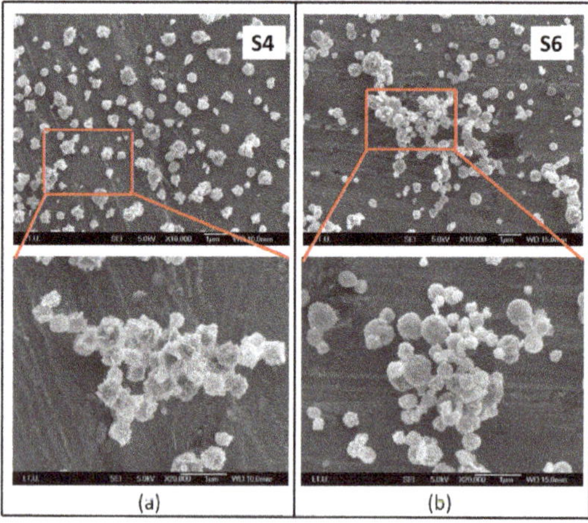

Figure 6. SEM micrographs of nanostructured ZnO particles synthesised from 0.02875 mol/L at 600 °C; (**a**) 0.75 L/min N_2 gas flow and (**b**) 0.5 L/min N_2 gas flow rate.

Figure 6 reveals that an obvious morphology change occured when the flow rate was varied. Although the particle size remained similar, when the flow rate decreased from 0.75 to 0.5 L/min, the granular morphology was replaced with plates and spheres assembled by these plates. In the case of higher flow rates, a higher gradual temperature was achieved within the droplet, which also increased the supersaturation and driving force of the reaction.

3.3. Formation Mechanism

In the USP process, the aerosol droplets undergo evaporation/drying, precipitation and thermolysis in a single-step process and under extreme synthesis conditions (high droplet/particle heating rate and high surface reaction), as presented in Figure 7. Within the short reaction time (2–3 s), intraparticle transport, solute nucleation and growth take place. The morphological findings presented in Figures 1, 5 and 6, revealing the effect of temperature, precursor concentration and flow rate, are consistent, implying that lower supersaturation degrees decrease the driving force of the reaction and, hence, nucleation rates. During the USP process, as summarized in Figure 7, in the case of lower nucleation rates, growth occurs from already existing nuclei in favored crystallographic directions, ending up with planar growth. In the case of a higher driving force, higher nucleation rates ensure homogenous nucleation, and precipitation takes place in the determined volume and shape of spherical droplets.

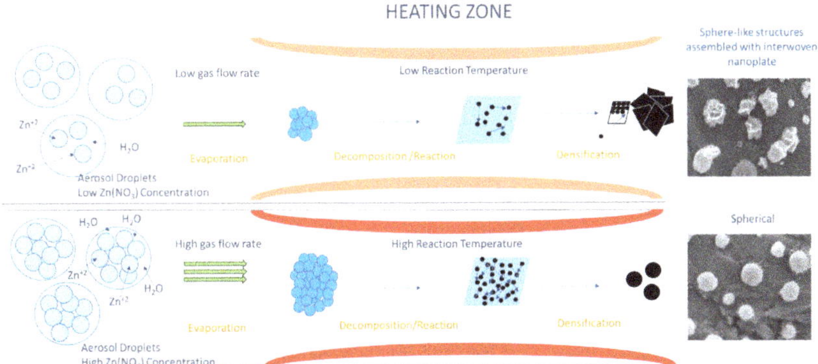

Figure 7. Schematic diagram of spherical and sphere-like structures assembled with interwoven nanoplate ZnO formation through Ultrasonic Spray Pyrolysis (USP).

Keeping in mind that morphological features of nanostructured ZnO particles are directly related to their functional properties, USP can be utilized as a suitable method targeting various application areas of ZnO. It has already been reported that higher surface areas of ZnO plates are favorable in photocatalysis applications [24]. For such applications, a synthesis strategy can be utilized dealing with low concentrations, temperatures and flow rates. On the other hand, the photocatalysis application necessitates the sintering of particles and UV-blockage properties, as spherical particles are more favorable owing to their good sinterability and good absorbance with less scattering of light [25,26]. For such applications, USP should be utilized at higher concentrations, temperatures and flow rates. A basic and empirical model can be found in Figure 7, which summarizes recipes to synthesize spherical or planar ZnO particles using the USP method.

4. Conclusions

The synthesis of pure and nanostructured ZnO particles was accomplished by USP. A lower concentration of $Zn(NO_3)_2 \cdot 6H_2O$ in the initial solution, and the reaction temperature and flow rate of N_2 resulted in lower Zn-saturation and, therefore, in lower nucleation rates. All these facts lead to the formation of nanostructured ZnO particles with a planar morphology, which is not typical for USP. Moreover, it was explained by the schematic representation of the formation mechanism, which showed that it is possible to control nanostructured ZnO particle morphologies, from spheres assembled by plates to plates via altering the USP parameters. This pioneering study aimed to explain the formation mechanisms of ZnO nanostructures and will contribute to targeting various applications in our next study.

Author Contributions: E.E., G.A., R.R., B.F. and designed the research, performed the experiments and analysed the data. D.J. provided the analysis tools and performed the characterisation. G.A., E.E. and R.R. wrote the article.

Acknowledgments: The authors would like to thank Gültekin GÖLLER and Technician Hüseyin SEZER for the SEM studies. The authors acknowledge the financial support from the Slovenian Research Agency (Research Core Funding No. P2-0120, P2-0132, BI-DE/17-19-12). The responsible proof reader for the English language is Shelagh Hedges, Faculty of Mechanical Engineering, University of Maribor, Slovenia.

Abbreviations

Ag	Silver
BF	Bright-Field
DS	Debye-Scherrer method
FE	SEM-Field Emission Scanning Electron Microscopy
HRTEM	High Resolution Transmission Electron Microscopy
N_2	nitrogen
OH^-	Hydroxide ion
SAED	Selected Area Electron Diffraction
SEM	Scanning Electron Microscopy
UDM	Uniform Deformation Model
UV	Ultraviolet
USP	Ultrasonic Spray Pyrolysis
ZnO	Nanostructured zinc oxide particles
XRD	X-ray diffraction analysis
W-H	Williamson-Hall analysis

References

1. Yang, F.; Liu, W.H.; Wang, X.W.; Zheng, J.; Shi, R.Y.; Zhao, H.; Yang, H.Q. Controllable low temperature vapor-solid growth and hexagonal disk enhanced field emission property of ZnO nanorod arrays and hexagonal nanodisk networks. *ACS Appl. Mater. Interfaces*, **2012**, *4*, 3852–3859. [CrossRef] [PubMed]
2. Ali, L.I.; El-Molla, S.A.; Ibrahim, M.M.; Mahmoud, H.R.; Naghmash, M.A. Effect of preparation methods and optical band gap of ZnO nanomaterials on photodegradation studies. *Opt. Mater.* **2016**, *58*, 484–490. [CrossRef]
3. Sun, Y.; Chen, L.; Bao, Y.; Zhang, Y.; Wang, J.; Fu, M.; Wu, J.; Ye, D. The Applications of Morphology Controlled ZnO in Catalysis. *Catalysts* **2016**, *6*, 188. [CrossRef]
4. Dedova, T.; Acik, I.O.; Krunks, M.; Mikli, V.; Volobujeva, O.; Mere, A. Effect of substrate morphology on the nucleation and growth of ZnO nanorods prepared by spray pyrolysis. *Thin Solid Films* **2012**, *520*, 4650–4653. [CrossRef]
5. Klubnuan, S.; Suwanboon, S.; Amornpitoksuk, P. Effects of optical band gap energy, band tail energy and particle shape on photocatalytic activities of different ZnO nanostructures prepared by a hydrothermal method. *Opt. Mater.* **2016**, *53*, 134–141. [CrossRef]
6. Gu, C.; Huang, J.; Wu, Y.; Zhai, M.; Sun, Y.; Liu, J. Preparation of porous flower-like ZnO nanostructures and their gas-sensing property. *J. Alloys Compd.* **2011**, *509*, 4499–4504. [CrossRef]
7. Singh, N.K.; Shrivastava, S.; Rath, S.; Annapoorni, S. Optical and room temperature sensing properties of highly oxygen deficient flower-like ZnO nanostructures. *Appl. Surf. Sci.* **2010**, *257*, 1544–1549. [CrossRef]
8. Kennedy, O.W.; Coke, M.L.; White, E.R.; Shaffer, M.S.P.; Warburton, P.A. MBE growth and morphology control of ZnO nanobelts with polar axis perpendicular to growth direction. *Mater. Lett.* **2018**, *212*, 51–53. [CrossRef]
9. Ding, J.; Fang, X.; Yang, R.; Kan, B.; Li, X.; Yuan, N. Transformation of ZnO polycrystalline sheets into hexagon-like mesocrystalline ZnO rods (tubes) under ultrasonic vibration. *Nanoscale Res. Lett.* **2014**, *9*, 1–6. [CrossRef] [PubMed]
10. Mahdavi, R.; Ashraf Talesh, S.S. The effect of ultrasonic irradiation on the structure, morphology and photocatalytic performance of ZnO nanoparticles by sol-gel method. *Ultrason. Sonochem.* **2017**, *39*, 504–510. [CrossRef] [PubMed]
11. Wang, H.; Xie, J.; Yan, K.; Duan, M. Growth Mechanism of Different Morphologies of ZnO Crystals Prepared by Hydrothermal Method. *J. Mater. Sci. Technol.* **2011**, *27*, 153–158. [CrossRef]
12. Huang, J.; Xia, C.; Cao, L.; Zeng, X. Facile microwave hydrothermal synthesis of zinc oxide one-dimensional nanostructure with three-dimensional morphology. *Mater. Sci. Eng. B Solid-State Mater. Adv. Tec.* **2008**, *150*, 187–193. [CrossRef]

13. Htay, M.T.; Hashimoto, Y.; Ito, K. Growth of ZnO submicron single-crystalline platelets, wires, and rods by ultrasonic spray pyrolysis. *Jpn. J. Appl. Phys. Part 1 Regul. Pap. and Short Notes and Rev. Pap.* **2007**, *46*, 440–448. [CrossRef]

14. Kajbafvala, A.; Ghorbani, H.; Paravar, A.; Samberg, J.P.; Kajbafvala, E.; Sadrnezhaad, S.K. Effects of morphology on photocatalytic performance of Zinc oxide nanostructures synthesized by rapid microwave irradiation methods. *Superlattices Microstruct.* **2012**, *51*, 512–522. [CrossRef]

15. Sin, J.C.; Lam, S.M.; Lee, K.T.; Mohamed, A.R. Self-assembly fabrication of ZnO hierarchical micro/nanospheres for enhanced photocatalytic degradation of endocrine-disrupting chemicals. *Mater. Sci. Semicond. Process.* **2013**, *16*, 1542–1550. [CrossRef]

16. Li, J.; Zhao, T.; Shangguan, E.; Li, Y.; Li, L.; Wang, D.; Wang, M.; Chang, Z.; Li, Q. Enhancing the rate and cycling performance of spherical ZnO anode material for advanced zinc-nickel secondary batteries by combined in-situ doping and coating with carbon. *Electrochim. Acta* **2017**, *236*, 180–189. [CrossRef]

17. Muñoz-Fernandez, L.; Alkan, G.; Milošević, O.; Rabanal, M.E.; Friedrich, B. Synthesis and characterisation of spherical core-shell Ag/ZnO nanocomposites using single and two-steps ultrasonic spray pyrolysis (USP). *Catal. Today* **2017**. [CrossRef]

18. Hidayat, D.; Ogi, T.; Iskandar, F.; Okuyama, K. Single crystal ZnO:Al nanoparticles directly synthesized using low-pressure spray pyrolysis. *Mater. Sci. Eng. B Solid-State Mater. Adv. Technol.* **2008**, *151*, 231–237. [CrossRef]

19. Gurmen, S.; Ebin, B. Production and characterization of the nanostructured hollow iron oxide spheres and nanoparticles by aerosol route. *J. Alloys Compd..* **2010**, *492*, 585–589. [CrossRef]

20. Gurmen, S.; Guven, A.; Ebin, B.; Stopic, S.; Friedrich, B. Synthesis of nano-crystalline spherical cobalt-iron (Co-Fe) alloy particles by ultrasonic spray pyrolysis and hydrogen reduction. *J. Alloys Compd.* **2009**, *481*, 600–604. [CrossRef]

21. Lojpur, V.; Mancic, L.; Rabanal, M.E.; Dramicanin, M.D.; Tan, Z.; Hashishin, T.; Ohara, S.; Milosevic, O. Structural, morphological and luminescence properties of nanocrystalline up-converting Y1.89Yb0.1Er0.01O3 phosphor particles synthesized through aerosol route. *J. Alloys Compd.* **2013**, *580*, 584–591. [CrossRef]

22. Emil, E.; Gürmen, S.; Hall, W. Estimation of yttrium oxide microstructural parameters using the Williamson–Hall analysis. *Mater. Sci. Technol.* **2018**, *836*. [CrossRef]

23. Ebin, B.; Arig, E.; Özkal, B.; Gürmen, S. Production and characterization of ZnO nanoparticles and porous particles by ultrasonic spray pyrolysis using a zinc nitrate precursor. *Inter. J. Miner. Metall. Mater.* **2012**, *19*, 651–656. [CrossRef]

24. Yassitepe, E.; Yatmaz, H.C.; Öztürk, C.; Öztürk, K.; Duran, C. Photocatalytic efficiency of ZnO plates in degradation of azo dye solutions. *J. Photochem. Photobiol. A Chem.* **2008**, *198*, 1–6. [CrossRef]

25. Tartaj, P.; Tartaj, J. Preparation, characterization and sintering behavior of spherical iron oxide doped alumina particles. *Acta Mater.* **2002**, *50*, 5–12. [CrossRef]

26. Neamjan, N.; Sricharussin, W.; Threepopnatkul, P. Effect of various shapes of ZnO nanoparticles on cotton fabric via electrospinning for UV-blocking. *J. Nanosci. Nanotechnol.* **2012**, *12*, 525–530. [CrossRef] [PubMed]

Article

Characteristics of Ti6Al4V Powders Recycled from Turnings via the HDH Technique

Mertol Gökelma [1,*], Dilara Celik [2], Onur Tazegul [2], Huseyin Cimenoglu [2] and Bernd Friedrich [3]

[1]   Department of Materials Science and Engineering, Norwegian University of Science and Technology, Trondheim 7491, Norway

[2]   Department of Metallurgical and Materials Engineering, Istanbul Technical University, Sariyer, Istanbul 34469, Turkey; dilaracelik01@gmail.com (D.C.); tazegul@itu.edu.tr (O.T.); cimenogluh@itu.edu.tr (H.C.)

[3]   IME Process Metallurgy and Metal Recycling, RWTH Aachen University, Intzestraße 3, Aachen 52056, Germany; bfriedrich@ime-aachen.de

*   Correspondence: mertol.gokelma@ntnu.no; Tel.: +47-73-412-926

Received: 28 March 2018; Accepted: 7 May 2018; Published: 9 May 2018

Abstract: The objective of this research is for Ti6Al4V alloy turnings, generated during the machining of implants, to produce powders for the fabrication of Ti base coating via the cold spray method. In order to decrease the cost of powder production and increase the recycling rate of the turnings, the hydrogenation-dehydrogenation (HDH) process has been utilised. The HDH process consists of the following sequence: surface conditioning of the turnings, hydrogenation, ball milling (for powder production), and dehydrogenation. Afterwards, the properties of the recycled powder were analysed via phase, chemical, and morphological examinations, and size and flowability measurements. Usability of the powder in additive manufacturing applications has been evaluated via examining the characteristics of the deposit produced from this powder by the cold spray method. In short, promising results were obtained regarding the potential of the recycled powders in additive manufacturing after making minor adjustments in the HDH process.

Keywords: Ti6Al4V; HDH; powder metallurgy; powder synthesis

---

1. Introduction

Ti6Al4V alloy (Ti, 6 wt. % Al, 4 wt. % V) is $\alpha + \beta$ phase titanium alloy, which occupies about 50% of the total titanium market [1,2]. It is very commonly used in aerospace, automotive, and medical industries, due to its high strength/weight ratio, corrosion resistance, biocompatibility, and low thermal expansion characteristics. Although Ti6Al4V alloy is a high-cost material it has a loss of about 70–80% as scrap while manufacturing of engineering components [3]. Owing to the high oxygen affinity, recovery of titanium-based scraps by re-melting is a difficult and costly process. Furthermore, re-melting may cause an imbalance in alloy composition. From this point of view recovery of titanium scraps in the form of powder by following the powder metallurgy route can be attractive especially for the recently-growing additive manufacturing market (such as 3D printing and cold-spray applications, etc.). Thus, powder metallurgy of Ti6Al4V alloy can be an alternative method either to use powder as the starting material or to recycle the new scrap which comes from semi-fabrication and manufacturing operations as the powder with improved purity [4,5].

The additive manufacturing industry has been growing very quickly in recent years and reached US$5.1 billion in 2015 [6]. This increasing trend requires more input material and new production methods fulfil the requirements for Ti6Al4V powders. There are many additive manufacturing technologies in the market and the main methods are stereolithography, selective laser sintering, direct metal laser sintering, fused deposition modelling, and 3D printing [7]. 3D printing has been improved with many ongoing studies and it is expected to be used more commonly in the future [8].

It has been documented that, owing to the high oxidation tendency of atomization, a titanium melt can be made under vacuum or inert gas atmosphere at high production costs by utilising special technologies, such as electrode induction melting gas atomization, vacuum arc melting and cold hearth melting, etc. [9]. As an alternative method hydrogenation/dehydrogenation (HDH) can be an economical option for the recycling of titanium based scraps [10]. The HDH method benefits from room temperature embrittlement of the hydrate and produces powder-sized hydrate which must be dehydrated to produce the final powder product. The balance pressure and the temperature of the system are the main parameters for hydride transformation. The typical reaction can be written as $M + H_2 \leftrightarrow MH_2$ and can be reversed by the balance pressure of hydrogen gas. If the pressure is above the equilibrium pressure, hydrogen atoms enter the lattice to form metal hydrate; if it is below that level, hydrogen atoms diffuse out from the metal to from hydrogen gas [8,9]. During the hydration, some hydrogen atoms (H) enter the crystal lattice under the critical temperature and pressure. Then, while cooling under room temperature, hydrogen atoms are diffused into the crystal lattice and occupy the tetrahedral sites [11]. When titanium-based scraps are of concern, this mostly occurs in the β-Ti phase, which has more hydrogen solubility than the α-Ti phase [10–13]. Enrichment of the material by hydrogen (hydrogenation) imposes embrittlement leading to fracturing under mechanical loads (i.e., during ball milling), subsequent dehydrogenation provides some ductility as the result of outward diffusion of hydrogen. After dehydrogenation, powders must be kept under the protective atmosphere because of the oxidation affinity of the titanium. There are many studies that sintered the hydrogenated powder and performed the dehydrogenation step after the final shape was given by sintering [2,14–17].

The current work aims to present the preliminary results of Ti6Al4V powder production via the HDH method. Powder characteristics are analysed and its performance as a coating material has been evaluated. This study presents an innovation beyond the state of the art which is based on the usage of recycled Ti6Al4V scrap as a material for coating and/or additive manufacturing.

2. Experimental Methodology

In this study turnings produced during manufacturing of Ti6Al4V–ELI (0.20 wt. % of oxygen) implants were used as the scrap. Figure 1 presents flow chart for production of Ti6Al4V powder, which starts with an etching process to clean and activate the surface of the turnings. The turnings were then hydrogenated to form a brittle titanium hydrate for easy milling down to a size of less than 100 µm. Then powders of Ti6Al4V were dehydrogenated as a final step. The produced powder is characterized at the end to evaluate the usability of the product for different targets. Additionally, deposition performance of the recycled powder was analysed by low-pressure cold spray equipment.

Acid cleaning of the turning was made in H_2SO_4:H_2O solution (having a ratio of 1:6) to activate their surfaces by removing any organic or inorganic impurities. After washing with deionised water and drying at 60 °C for about 45 min turnings were placed into the chamber of the reactor for hydrogenation. Hydrogenation was performed under 2.5 bar hydrogen atmosphere and the turnings were heated at a rate of 400 °C/h up to about 700 °C. After reaching a peak temperature of 700 °C the reactor was switched off and the turnings were allowed to cool to room temperature in the reactor. Diffusion of hydrogen into the turnings (activation) started around 470 °C according to Figure 2, representing the applied hydrogenation process as a "temperature vs. time" plot. The activation can be observed by the increasing temperature with a higher rate. After switching off the reactor temperature remained at about 600 °C for a certain interval as a result of the exothermic reaction imposed by the diffusion of hydrogen atoms. Then hydrated turnings were milled at room temperature in the steel ball milling equipment (MM 301, Retsch GmbH, Haan, Germany) with the frequency of 20 s^{-1} for up to four minutes and sieved under 100 µm.

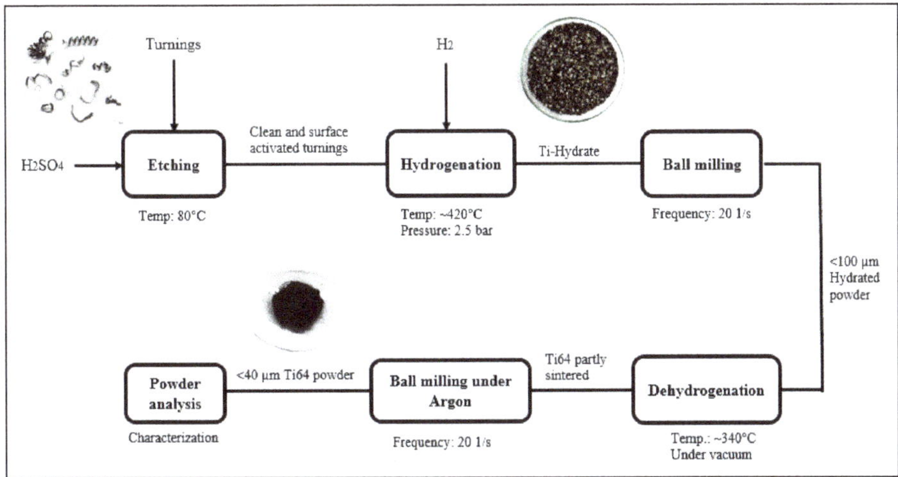

Figure 1. Experimental procedure of powder production via the HDH process.

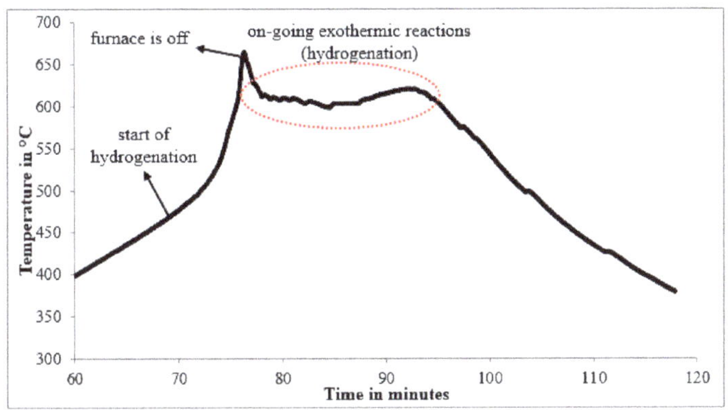

Figure 2. Temperature measurement time during hydrogenation of the Ti6Al4V turnings.

The sieved powder was placed again in the same reactor for dehydrogenation under continuous vacuum. By the time, the system was cooled down under vacuum (1 mbar) to room temperature. Figure 3 presents the variation of temperature and pressure with respect to dehydrogenation duration. While the heating rate was gradually increasing to the peak temperature of about 700 °C pressure increased after 95 min as the result of the start of dehydrogenation. Pressure gradually increased for a duration of about 110 min, above which it sharply increased to over-pressure (1000 mbar is the detection limit of the unit) due to the high amount of hydrogen release. After switching off the reactor pressure sharply reduced while temperature gradually decreased. The dehydrogenised powder was taken out from the reactor in a glovebox with the oxygen concentration of <30 ppm. It was noticed that the powder was loosely sintered during the dehydrogenation process. Therefore, the dehydrated powder was milled once again in the steel-ball milling unit under an argon atmosphere. After a final sieving, the powder under 45 μm is sent for characterization.

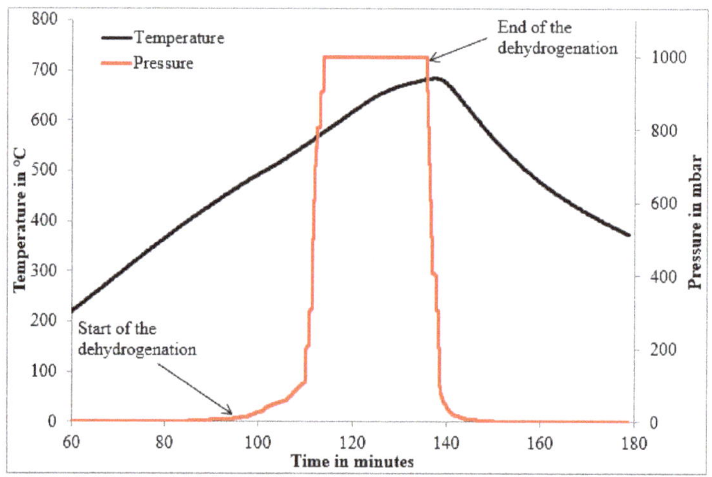

Figure 3. Temperature measurement time during dehydrogenation of the hydrate powders.

The recycled powder was characterized by the following methods: elemental analysis by atomic absorption (for Al, V, Fe) solid state infrared absorption (for oxygen) and thermal conductivity (for hydrogen), phase analysis by XRD (X-ray Diffraction, Bruker™ D8 Advance Series 35 kV 40 mA, Billerica, MA, USA), flowability measurement by AOR (angle of repose), morphology by SEM (scanning electron microscope, JEOL™ JCM-6000Plus NeoScope, Peabody, MA, USA) and particle size distribution by laser diffraction (Mastersizer 2000, Malvern Panalytical, Worcestershire, UK). Additionally, deposition efficiency of the powder on commercial purity titanium substrates by low pressure (600 kPa) cold gas dynamic spray equipment (Rusonic Model K201, Rus Sonic Technology, Arcadia, CA, USA) having a converging-diverging tubular nozzle was evaluated. Characteristics of the deposits were determined by SEM examinations and XRD analysis.

3. Results

Figures 4 and 5 present the size distribution and morphology of the powder recycled from Ti6Al4V-ELI turnings, respectively. Fifty percent of the powder is under 13 μm and 90% is under 37 μm. The morphology of the particles is irregular, as expected, with the circularity of 0.72 ± 0.04 and the form factor of 1.42 ± 0.15 (max-axis/min-axis) [18]. The flowability of the powder was measured as 38.93 ± 1.55, which is accepted as "fair-aid not needed" or "some cohesiveness" [19].

The elemental composition of the powder is given in Table 1 along with the standard composition of Grade 5 and Grade 23 Ti6Al4V alloys. In general, the concentrations of the alloying elements are in an acceptable range according to the standards of ASTM B988-13 Grade 5, except the hydrogen content. Additionally, the average oxygen content of the dehydrogenated powder is slightly higher than that prescribed by ASTM B988-13 Grade 5. Moreover, oxygen and hydrogen contents of dehydrogenated powder are not in the range of Grade 23, which is called extra low interstitial (ELI).

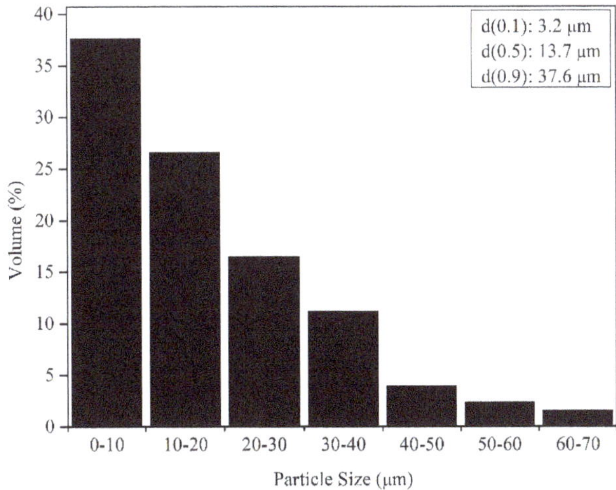

Figure 4. Size distribution of Ti6Al4V particles synthesized by the HDH method.

Figure 5. Electron microscopy picture of powder synthesized by the HDH method.

Table 1. Elemental analysis of dehydrogenated powder and relevant standards.

Sample	Al wt. %	V wt. %	Fe wt. %	O wt. %	H wt. %	Ti wt. %
Dehydrogenated powder	5.57 ± 0.6	3.97 ± 0.5	0.2 ± 0.09	0.32 ± 0.15	0.06 ± 0.03	balance
ASTM B265-15/B348-13/B381-13 Grade 23 [20–22]	5.5–6.5	3.5–4.5	0.25 max	0.13 max	0.0125 max	balance
ASTM B988-13 Grade 5 [23]	5.5–6.75	3.5–4.5	0.4 max	0.3 max	0.015 max	balance

XRD patterns of the as-received turnings, and powders (in hydrogenised and dehydrogenised states) are shown in Figure 6. The turnings consisted of α and β-Ti phases as expected. On the XRD patterns of the powders peaks of titanium hydrides (in the form of TiH_2 and $TiH_{1.5}$) appeared. As compared to the hydrogenised state, dehydrogenation caused domination of α and β-Ti phase peaks along with peaks of titanium hydrides by considering the intensity ratios of the Ti peaks with those of TiH$_{1.5}$ and TiH$_2$. However, the applied dehydrogenation process did not completely remove titanium hydrides from the powders.

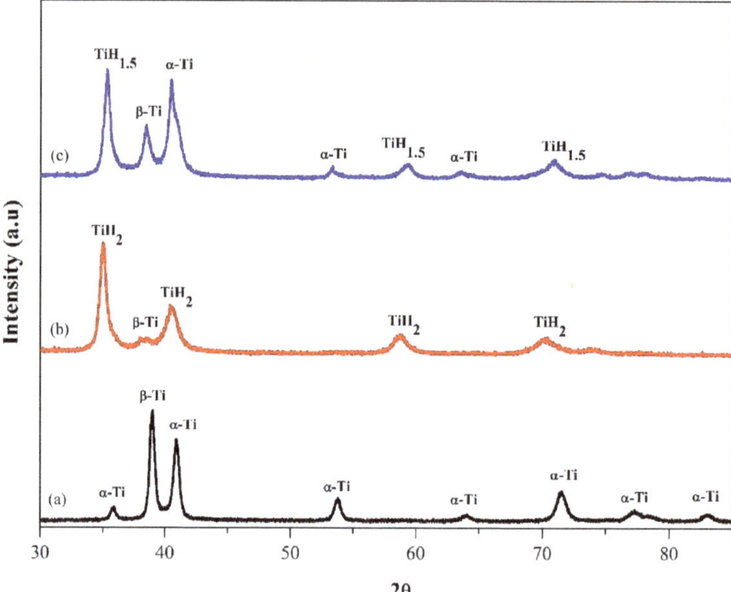

Figure 6. X-ray diffraction analysis of the raw material (**a**), after hydrogenation (**b**), and after dehydrogenation (**c**).

In order to evaluate the deposition characteristics of the recycled powder, some attempts have been made by the cold spray method which is based on the acceleration of particles in a process gas over the supersonic velocity through the convergent-divergent type nozzle [24,25]. In this system, particles of the feedstock powder are deposited as they impact on the surface of the substrate [24]. The bonding of the cold sprayed powder is a result of the mechanical interlocking between cold sprayed particles and the substrate. Increasing of the coating thickness is provided by the deposition of the following particle on the previous ones. In this process air, argon, helium, or nitrogen can be used as the process gas. In terms of acceleration capability and applicability, helium is the most favourable for high deposition efficiency. On the other hand, air appears as the best option when the cost of the coating is of concern. For those reasons, helium and air have been chosen as the process gases in this study to evaluate the deposition characteristics of the recycled powder with the parameters listed in Table 2. It should be emphasized that recycled powder was not successfully deposited with the utilized low pressure cold spray equipment over commercial purity titanium (Cp-Ti) with air as the process gas, unlike He. This observation can be associated with approximately three times higher acceleration capability of helium as compared to that of air due to its low molecular weight [24]. In order to overcome this problem 3 wt. % Al has been added into the feedstock as a binder according to our experience when using air as the process gas [26]. Additionally, the traverse speed has been reduced five times to further assist the binding of the feedstock in air.

Table 2. Low-pressure coating parameters of cold spray process.

Carrier Gas	He	Air
Feedstock	Recycled powder	97 wt. % Recycled powder + 3 wt. % Al
Gas pressures (bar)	6	6
Traverse speed (mm/s)	5	1
Number of pass	3	2
Beam distance (mm)	2	2
Standoff distance (cm)	1	1

A cross-section of the coatings deposited on the Cp-Ti substrate by cold spraying of the powder produced is shown in Figure 7. When He was used as the process gas powders were deposited on the substrate without remarkable discontinuities at the coating/substrate interface. However, a high amount of porosity was detected within the coating showing lower integration among powder particles. When air was used as the process gas successful deposition was not obtained unless addition of aluminium powder (at concentration of 3 wt. %) into the feedstock was done. Thus, Al acted as a binder for successful deposition of the powder on Cp-Ti (without noticeable porosities in the coating and discontinuities at the coating/substrate interface) when air was used as the process gas. In this respect, the dark coloured regions in the coating shown in Figure 7b were identified as Al particles by EDX (Energy-dispersive X-ray spectroscopy) analysis conducted during SEM surveys.

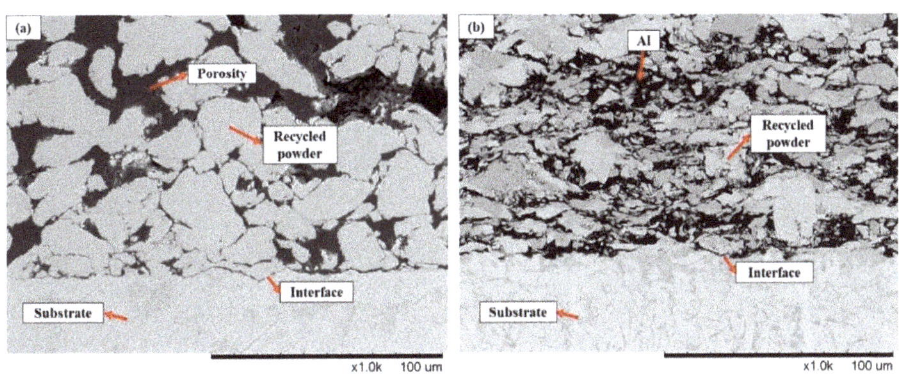

Figure 7. SEM micrographs of cold spray deposits produced by utilization of (**a**) He and (**b**) air as process gases (deposition parameters are listed in Table 3).

XRD patterns of the deposits shown in Figure 7 are depicted in Figure 8. Peaks of $TiH_{1.5}$ and Ti were detected on the XRD pattern of the deposit formed by using He as the process gas (Figure 8a). Since Al powder has been added into the feedstock, additional Al peaks appeared on the XRD pattern of the deposit produced by utilization of air as the process gas (Figure 8b). It should be noted that the peaks of $TiH_{1.5}$ which were detected in the XRD pattern of the recycled powder (Figure 6c) also remained after the cold spray process.

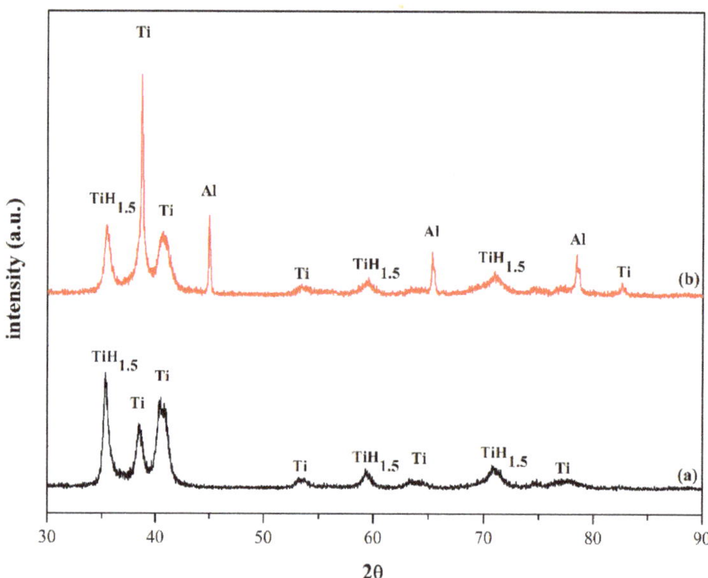

Figure 8. X-ray diffraction patterns of the coatings deposited by using (**a**) He and (**b**) air as the process gas.

4. Discussion

Ti6Al4V powder recycled from turnings via the HDH method have a mostly irregular morphology, which is also shown in this study. In spray-based deposition processes, like the cold spray method, irregular powders favour higher particle velocities than spherical ones because of their higher drag force [27–30]. In this respect, irregular powder particle morphology has some advantage on the final product of the cold-spray method [31]. In contrast, spherical particle morphology is more desirable than irregular ones for 3D printing, in terms of flowability requirements of the process [32–34]. Flowability is mostly related to the particle shape of the powder. Irregular shapes generally have lower flowability as compared to spherical ones. Although, the smaller-sized particles would have an acceptable flowability range, decreasing the particle size under the critical level could affect the productivity of the dehydrogenation process [34]. Apart from there being no special requirement of flowability for the cold spray process, the mean particle size, and the oxygen and nitrogen contents of the powder are more important factors for good coating properties and deposition efficiency [27]. When the average particle size is of concern, the HDH method provided an average particle size ($d_{0.5}$) of 13.7 µm, which is within the acceptable range for almost every coating method and additive manufacturing processes [18]. The interstitials, such as oxygen and nitrogen in the powder, deteriorate the deposition efficiency by reducing the plastic deformation capacity of the Ti powders [35].

Even after dehydrogenation $TiH_{1.5}$ still remains in the Ti6Al4V powder. This result matches with the research of Bhosle et al. [34], where they reported that the dehydrogenation occurs in a two-step process: $TiH_2 \rightarrow TiH_x \rightarrow \alpha - Ti$. The finer hydride particles contain lower hydrogen in the second hydride phase (TiH_x) and they are thermally more stable at higher temperatures than the TiH_2 phase, which may result in the remaining hydrates even after the dehydrogenation process. This is also caused by a poor vacuum setup. In Table 3 the parameters of vacuum, temperature, and duration are listed for successful dehydrogenation of titanium powder from previous study, suggesting that higher vacuum or longer duration would provide complete dehydrogenation of the powder obtained by recycling. Moreover, it should be noted that the XRD peaks shifted to lower angles as compared to that of the turnings due to the heavy deformation imposed during the ball milling process applied after hydrogenation. Additionally, the oxygen content of the recycled powder should be minimized according to the relevant standards [20–23]. The oxygen content is slightly higher than the desired range for the standard B265-15/B348-13/B381-13, but it is an acceptable range for the standard B988-13. After the HDH process, Ti6Al4V alloy powder should be within the chemistry limits of international standards and correspond to the industrial specification, which include medical, aerospace, and defence applications [12]. One of the important parameters to be considered to reduce the oxygen content is the vacuum level applied during the hydrogenation stage. Oxidation of very fine hydrate particles during milling can also be a source of oxygen due to very high surface area [36]. Therefore, it is suggested that adjusting the vacuum level and processing time of the dehydrogenation stage could also provide a solution for reducing the oxygen level of the recycled powder, which the process needs during dehydrogenation.

Table 3. Important dehydration parameters used in different researches.

Sample	Vacuum (Pa)	Temperature ($^\circ$C)	Duration (Hour)
Titanium rod [10]	6.7×10^{-3}	700	2
Titanium sponge [36]	10^{-2}	625	3
Titanium sponge [37]	8×10^{2}	700	36

Deposition characteristic of the recycled powder have been analysed by using the low-pressure cold spray method. Usage of helium as a process gas generated a deposit with relatively high porosity content. It is possible to improve the quality by using high-pressure cold spray systems generating higher velocities as compared to low-pressure cold spray systems. Moreover, reducing the hydrogen and oxygen content of the recycled powder, which imposes brittleness, would lead to the generation of denser deposits even by utilizing air as the process gas (without the addition of Al powder). Irregular morphology of the recycled powder also affected the deposition capacity. Although these powders can reach higher velocity than spherical ones [29,31], it is not possible to have a dense coating because of the irregularity. In summary, the coating trials gave promising results about the usability of the Ti6Al4V powder recycled from scrap via the HDH process. However, some adjustments are necessary to increase the success of this powder in additive manufacturing processes.

5. Conclusions

The results of the current work can be summed up in the following points:

- Morphology of the powders synthesized by the HDH method is irregular and perfect spherical shapes cannot be achieved. However, in the future the morphology of the powder can be changed by using an extra spheroidization process at the end.
- The particle size of the recycled powder product is under 45 µm. The size distribution is suitable to use in both additive manufacturing and coating methods. The average particle size and range of powder could be changed by using different milling conditions according to the further process where the powder will be used.
- The concentration of alloying elements, such as Al, V, and Fe, is in an acceptable range in terms of standards. However, the oxygen content of the final powder is near the acceptable limit of the Grade 5 standard, but this content could be reduced by using better vacuum conditions to reduce the air concentration in the reactor. Optimizing the particle size distribution and more effective dehydrogenation parameters could reduce the oxygen content. Furthermore, the hydrogen level is higher than the specified range (0.06 ± 0.03 wt. %) of the relevant standards due to the remaining hydrate phases in the powder. Therefore, the dehydrogenation procedure applied in this study must be improved (better vacuum, longer dehydrogenation time, optimized particle size range) to remove all hydrate phases.

- Low pressure cold spray deposition studies revealed the potential of the powders obtained by the HDH process for coating. In order to evaluate the usability of this powder produced from turnings scrap, the characteristics of the deposits from this powder are being investigated and the results will be presented in a follow-up publication.

Author Contributions: M.G. and D.C. planned the content of the paper and conducted the experimental work. O.T. and D.C. performed the characterisation of samples and coating. B.F. and H.C. were supervisors of the research work and participated in the writing and assessment of results.

Funding: This research received no external funding.

Acknowledgments: Dilara Celik expresses her deep thanks to European Commission for the provision of student researcher mobility at the University of RWTH Aachen under the Erasmus program.

Conflicts of Interest: The authors declare no conflict of interest.

References

1. Fang, T.Y.; Wang, W.H. Microstructural features of thermochemical processing in a Ti-6Al-4V alloy. *Mater. Chem. Phys.* **1998**, *56*, 35–47. [CrossRef]
2. Pan, L.; Wang, C.M.; Chen, Y.G.; Xiao, S.F. Recovery of Ti-6Al-4V Alloy Scrap by Hydrogenation and Hydride Sintering. *Adv. Mater. Res.* **2014**, *1061–1062*, 492–496. [CrossRef]
3. Ustinov, V.S.; Petrunko, A.N.; Olesov, Y.G.; Ognev, R.K. *A Novelty in the Field of Titanium Powder Metallurgy*; Springer: New York, NY, USA, 1982; pp. 2315–2321.
4. Goonan, T.G. Titanium Recycling in the United States in 2004. In *Overview of Flow Studies for Recycling Metal Commodities in the United States*; U.S. Geological Survey: Reston, VA, USA, 2004; pp. 1–16.
5. American Society for Metals. *Titanium and Titanium Alloys*; ASM International: Geauga County, OH, USA, 1982.
6. Prakash, P. Additive Manufacturing in Aerospace and Defence Sector: Strategy of India. *J. Def. Stud.* **2018**, *12*, 39–60.
7. Jardini, A.L.; Larosa, M.A.; Kaasi, A.; Kharmandayan, P. Additive Manufacturing in Medicine. *Ref. Modul. Mater. Sci. Mater. Eng.* **2017**, 1–21. [CrossRef]
8. Brookes, K.J.A. 3D printing materials in Maastricht. *Met. Powder Rep.* **2015**, *70*, 68–78. [CrossRef]
9. Rotmann, B.; Lochbichler, C.; Friedrich, B. Challenges in titanium recycling—Do we need a new specification for secondary alloys? In Proceedings of the European Metallurgical Conference 2011 Resources Efficiency in the Non-Ferrous Metals Industry, Optimization and Improvement, Düsseldorf, Germany, 26–29 June 2011; pp. 1–15.
10. Oh, J.M.; Roh, K.M.; Lee, B.K.; Suh, C.Y.; Kim, W.; Kwon, H.; Lim, J.W. Preparation of low oxygen content alloy powder from Ti binary alloy scrap by hydrogenation-dehydrogenation and deoxidation process. *J. Alloys Compd.* **2014**, *593*, 61–66. [CrossRef]
11. Mitkov, M.; Božić, D. Hydride-dehydride conversion of solid Ti6Al4V to powder form. *Mater. Charact.* **1996**, *37*, 53–60. [CrossRef]
12. McCracken, C.G.; Robison, J.W.; Motchenbacher, C.A. *Manufacture of HDH Low Oxygen Titanium-6aluminium-4vanadium (Ti-6-4) Powder Incorporating a Novel Powder De-Oxidation Step*; The European Powder Metallurgy Association: Shrewsbury, UK, 2009; pp. 7146–7152.
13. Gao, W.; Li, W.; Zhou, J.; Hodgson, P.D. Thermodynamics approach to the hydrogen diffusion and phase transformation in titanium particles. *J. Alloys Compd.* **2011**, *509*, 2523–2529. [CrossRef]
14. Tosetti, J.P.; Beneduce, F.; Rodrigues, D. Evaluation of Different Routes for HDH Processing of Ti-6Al-4V Alloy. *Mater. Sci. Forum* **2003**, *416–418*, 323–328. [CrossRef]
15. Azevedo, C.R.F.; Rodrigues, D.; Neto, F.B. Ti-Al-V powder metallurgy (PM) via the hydrogenation-dehydrogenation (HDH) process. *J. Alloys Compd.* **2003**, *353*, 217–227. [CrossRef]
16. Peng, Q.; Yang, B.; Liu, L.; Song, C.; Friedrich, B. Porous TiAl alloys fabricated by sintering of TiH_2 and Al powder mixtures. *J. Alloys Compd.* **2016**, *656*, 530–538. [CrossRef]
17. Peng, Q.; Yang, B.; Friedrich, B. Porous Titanium Parts Fabricated by Sintering of TiH_2 and Ti Powder Mixtures. *J. Mater. Eng. Perform.* **2018**, *27*, 228–242. [CrossRef]
18. Valdek, M.; Helmo, K.; Pritt, K.; Besterci, M. Characterization of Powder Particle. *Methods* **2001**, *7*, 22–34.

19. Geldart, D.; Abdullah, E.C.; Hassanpour, A.; Nwoke, L.C.; Wouters, I. Characterization of powder flowability using measurement of angle of repose. *China Part* **2006**, *4*, 104–107. [CrossRef]

20. Specification, S. *Standard Specification for Titanium and Titanium Alloy Forgings 1*; ASTM International: West Conshohocken, PA, USA, 2017.

21. Specification, S. *Standard Specification for Titanium and Titanium Alloy Bars and Billets 1*; ASTM International: West Conshohocken, PA, USA, 2017.

22. Specification, S. *Standard Specification for Titanium and Titanium Alloy Strip, Sheet, and Plate 1*; ASTM International: West Conshohocken, PA, USA, 2017.

23. Containing, M.; Than, L.; Percent, T.; Bars, T.A. *Standard Specification for Powder Metallurgy (PM) Titanium and Titanium Alloy*; ASTM International: West Conshohocken, PA, USA, 2017; pp. 12–15.

24. Singh, H.; Sidhu, T.S. Cold Spray Technology: Future of Coating Deposition Processes. *Frat. Integr. Strutt.* **2012**, *22*, 69–84.

25. Moridi, A.; Guagliano, M.; Dao, M. Cold spray coating: Review of material systems and future perspectives. *Surf. Eng.* **2014**, *36*, 369–395. [CrossRef]

26. Cetiner, D.; Paksoy, A.H.; Tazegul, O.; Baydogan, M.; Guleryuz, H.; Cimenoglu, H.; Atar, E. A novel fabrication method for a TiO_2 layer over CoCr alloy. *Surf. Eng.* **2018**. [CrossRef]

27. Vo, P.; Goldbaum, D.; Wong, W.; Irissou, E.; Legoux, J.; Chromik, R.R.; Yue, S. *Cold-Spray Processing of Titanium and Titanium Alloys*; Elsevier Inc.: Philadelphia, PA, USA, 2015.

28. Ajdelsztajn, L.; Jodoin, B.; Schoenung, J.M. Synthesis and mechanical properties of nanocrystalline Ni coatings produced by cold gas dynamic spraying. *Surf. Coat. Technol.* **2006**, *201*, 1166–1172. [CrossRef]

29. Xiong, Y.; Xiong, X.; Yoon, S.; Bae, G.; Lee, C. Dependence of Bonding Mechanisms of Cold Sprayed Coatings on Strain-Rate-Induced Non-Equilibrium Phase Transformation. *J. Therm. Spray Technol.* **2011**, *20*, 860–865. [CrossRef]

30. Wong, W.; Rezaeian, A.; Irissou, E.; Legoux, J.G.; Yue, S. Cold Spray Characteristics of Commercially Pure Ti and Ti-6Al-4V. *Adv. Mater. Res.* **2010**, *89–91*, 639–644. [CrossRef]

31. Wong, W.; Vo, P.; Irissou, E.; Ryabinin, A.N.; Legoux, J.G.; Yue, S. Effect of particle morphology and size distribution on cold-sprayed pure titanium coatings. *J. Therm. Spray Technol.* **2013**, *22*, 1140–1153. [CrossRef]

32. Bose, S.; Vahabzadeh, S.; Bandyopadhyay, A. Bone tissue engineering using 3D printing. *Mater. Today* **2013**, *16*, 496–504. [CrossRef]

33. Sun, P.; Fang, Z.Z.; Xia, Y.; Zhang, Y.; Zhou, C. A novel method for production of spherical Ti-6Al-4V powder for additive manufacturing. *Powder Technol.* **2016**, *301*, 331–335. [CrossRef]

34. Bhosle, V.; Baburaj, E.G.; Miranova, M.; Salama, K. Dehydrogenation of TiH_2. *Mater. Sci. Eng. A* **2003**, *356*, 190–199. [CrossRef]

35. Bhattiprolu, V.S.; Johnson, K.W.; Ozdemir, O.C.; Crawford, G.A. Influence of feedstock powder and cold spray processing parameters on microstructure and mechanical properties of Ti-6Al-4V cold spray depositions. *Surf. Coat. Technol.* **2018**, *335*, 1–12. [CrossRef]

36. Wang, C.; Zhang, Y.; Wei, Y.; Mei, L.; Xiao, S.; Chen, Y. XPS study of the deoxidization behavior of hydrogen in TiH_2 powders. *Powder Technol.* **2016**, *302*, 423–425. [CrossRef]

37. Goso, X.; Kale, A. Production of titanium metal powder by the HDH process. *J. S. Afr. Inst. Min. Metall.* **2011**, *111*, 203–210.

MDPI

St. Alban-Anlage 66

4052 Basel

Switzerland

Tel. +41 61 683 77 34

Fax +41 61 302 89 18

www.mdpi.com

Metals Editorial Office

E-mail: metals@mdpi.com

www.mdpi.com/journal/metals